CIGR Handbook
of Agricultural Engineering
Volume V

# CIGR Handbook of Agricultural Engineering

## Volume V
## Energy and Biomass Engineering

**Edited by CIGR–The International Commission of Agricultural Engineering**


**Volume Editor:**
**Osamu Kitani**
*Nihon University, Japan*

**Co-Editors:**
**Thomas Jungbluth**
*University Hohenheim, Germany*
**Robert M. Peart**
*University of Florida, Florida USA*
**Abdellah Ramdani**
*I.A.V. Hassan II, Morocco*



**Published by the American Society of Agricultural Engineers**

LCCN 98-93767 ISBN 0-929355-97-0



For Information, contact:

Manufactured in the United States of America

# Editors and Authors

**Volume Editor**
Osamu Kitani
*Department of BioEnvironmental and Agricultural Engineering, College of Bioresource Sciences, Nihon University, 1866 Kameino, Fujisawa, Kanagawa, 252-8510, Japan*

**Co-Editors**
Thomas Jungbluth
*Institut fuer Agrartechnik, Universitaet Hohenheim, Garbenstrasse 9, D-70599 Stuttgart, Germany*

Robert M. Peart
*Agricultural and Biological Engineering Department, Rogers Hall, University of Florida, Gainesville, FL 32611, USA*

Abdellah Ramdani
*I. A.V. Hassan II, Department of Agricultural Engineering, B. P. 6202, Rabat Instituts, Rabat, Morocco*

**Authors**
Phillip C. Badger
*Manager SERBEP, Tennessee Valley Authority, 104 Creekwood Circle, Florence, AL 35630, USA*

R. Nolan Clark
*Conservation and Product Research Laboratory, ARS, United States Department of Agriculture, P.O. Box 10 Bushland, Texas 79012, USA*

Albert Esper
*Institute for Agricultural Engineering in the Tropics and Subtropics, Hohenheim University, D-70599 Stuttgart, Germany*

Yasushi Hashimoto
*Agricultural Engineering Department, Ehime University, Tarumi 3-5-7, Matuyama-shi, Ehime 790-8566, Japan*

J. L. Hernanz
*E.T.S. de Intenieros Agronomos, Ciudad Universitaria s/n, 28040 Madrid, Spain*

Bryan M. Jenkins
*Biological and Agricultural Engineering Department, University of California, Davis, CA 95616, USA*

Thomas Jungbluth
*Institut fuer Agrartechnik, Universitaet Hohenheim, Garbenstrasse 9, D-70599 Stuttgart, Germany*

Isao Karaki
*Management Department, Japan Alcohol Trading Company Limited, Shinjuku NS Building, 4-1, Nishi-Shinjuku 2-Chome, Shinjuku-ku, Tokyo 163-0837, Japan*

Osamu Kitani
*Department of BioEnvironmental and Agricultural Engineering, College of Bioresource Sciences, Nihon University, 1866 Kameino, Fujisawa, Kanagawa, 252-8510, Japan*

Raymond L. Legge
*Department of Chemical Engineering, University of Waterloo, Waterloo, Ontario N2L 3G1, Canada*

Takaaki Maekawa
*Institute of Agricultural and Forest Engineering, University of Tsukuba, Tennoudai 1-1-1, Tsukuba-shi 305-8572, Japan*

Fred R. Magdoff
*Northeast Region SARE Program, Hills Building, University of Vermont, Burlington, VT 05405, USA*

Marco Michelozzi
*Istituto Miglioramento Genetico delle Piante Forestali, Consiglio Nazionale delle Ricerche, Via Atto Vannucci 13, 50134 Firenze, Italy*

Werner Muehlbauer
*Institute for Agricultural Engineering in the Tropics and Subtropics, Hohenheim University, D-70599 Stuttgart, Germany*

Hiroshige Nishina
*Agricultural Engineering Department, Ehime University, Tarumi 3-5-7, Matuyama-shi, Ehime 790-8566, Japan*

Jaime Ortiz-Canavate
*E.T.S. de Intenieros Agronomos, Ciudad Universitaria s/n, 28040 Madrid, Spain*

Robert M. Peart
*Agricultural and Biological Engineering Department, Rogers Hall, University of Florida, Gainesville, FL 32611, USA*

Kingshuk Roy
*Department of BioEnvironmental and Agricultural Engineering, College of Bioresource Sciences, Nihon University, 1866 Kameino, Fujisawa, Kanagawa, 252-8510, Japan*

Giovanni Riva
*Institute of Agricultural Engineering, University of Ancona, c/o Institute of Agricultural Engineering, University of Milan, Via Celoria 2, IT-20133 Milano, Italy*

Takashi Saiki
*R&D Department, Japan Alcohl Association Nishishinbashi 2-21-2, Dai-ichi, Nan-Oh Bld. Minato-ku, Tokyo 105-0003, Japan*

Valentin Schnitzer
*Hydro Power, Industriestrasse 100, D-69245 Bammental, Germany*

Gerhard Schumm
*Fachhochschule Jena, Fachbereich Technische Physik Tatzendpromenade 1b, 07745 Jena, Germany*

Donald Scott
*Department of Chemical Engineering, University of Waterloo, Waterloo, Ontario N2L 3G1, Canada*

F. Sissot
*Institute of Agricultural Engineering, University of Milan, Via Celoria 2, IT-20133 Milano, Italy*

Veriano Vidrich
*Dipartimento di Scienze del Suolo e Nutrizione della Pianta, Università degli Studi di Firenze, Piazzale delle Cascine 16, 50144 Firenze, Italy*

# Editorial Board

# Contents

# Foreword

This handbook has been edited and published as a contribution to world agriculture at present as well as for the coming century. More than half of the world's population is engaged in agriculture to meet total world food demand. In developed countries, the economic weight of agriculture has been decreasing. However, a global view indicates that agriculture is still the largest industry and will remain so in the coming century.

Agriculture is one of the few industries that creates resources continuously from nature in a sustainable way because it creates organic matter and its derivatives by utilizing solar energy and other material cycles in nature. Continuity or sustainability is the very basis for securing global prosperity over many generations—the common objective of humankind.

Agricultural engineering has been applying scientific principles for the optimal conversion of natural resources into agricultural land, machinery, structure, processes, and systems for the benefit of man. Machinery, for example, multiplies the tiny power (about 0.07 kW) of a farmer into the 70 kW power of a tractor which makes possible the production of food several hundred times more than what a farmen can produce manually. Processing technology reduces food loss and adds much more nutritional values to agricultural products than they originally had.

The role of agricultural engineering is increasing with the dawning of a new century. Agriculture will have to supply not only food, but also other materials such as bio-fuels, organic feedstocks for secondary industries of destruction, and even medical ingredients. Furthermore, new agricultural technology is also expected to help *reduce* environmental destruction.

This handbook is designed to cover the major fields of agricultural engineering such as soil and water, machinery and its management, farm structures and processing agricultural, as well as other emerging fields. Information on technology for rural planning and farming systems, aquaculture, environmental technology for plant and animal production, energy and biomass engineering is also incorporated in this handbook. These emerging technologies will play more and more important roles in the future as both traditional and new technologies are used to supply food for an increasing world population and to manage decreasing fossil resources. Agricultural technologies are especially important in developing regions of the world where the demand for food and feedstocks will need boosting in parallel with the population growth and the rise of living standards.

It is not easy to cover all of the important topics in agricultural engineering in a limited number of pages. We regretfully had to drop some topics during the planning and editorial processes. There will be other requests from the readers in due course. We would like to make a continuous effort to improve the contents of the handbook and, in the near future, to issue the next edition.

This handbook will be useful to many agricultural engineers and students as well as to those who are working in relevant fields. It is my sincere desire that this handbook will be used worldwide to promote agricultural production and related industrial activities.

Osamu Kitani
*Editor-in-Chief*

# Preface

Energy technology is one of the key elements of agricultural engineering. Without energy, engineering can neither do work nor produce anything. If fuels were not supplied to farm tractors, livestock housings, and processing plants, food, feed, and other organic feedstocks in agriculture cannot be effectively produced.

It is, however, true that modern technology relies too heavily on the energy from fossil resources. This dependence in turn creates problems of environmental and resource management. Running out of petroleum threatens the sustainability of human activity and even the very existence of mankind. Because modern agriculture cannot function without petroleum fuels, agricultural engineers must develop a sustainable energy system to fuel world agriculture. Natural energy derived from the sun, wind, water, and biomass energies from ethanol, biodiesel and biogas are renewable and can be used in sustainable ways. Biomass is also important as an industrial feedstock that can replace petroleum. This is why Volume V., *Energy and Biomass Engineering*, was planned and now makes up one of the five volumes of the *CIGR Handbook.*

Energy is also important from the environmental viewpoint. Most of the energy issues, such as greenhouse effect and acid rain, are associated with energy production. Ironically, the improvement of environment usually needs additional energy input. In this sense, energy and environment represent the two sides of a coin: new technologies to produce energy without pollution and new technologies to control environment with minimum energy. The various methods of energy analyses and energy-saving in terms of environmental protection are the indispensable parts of this volume.

Volume V of the handbook would not have been completed without the great endeavor of its authors and co-editors. I would like to express my sincere thanks to the co-editors, Prof. T. Jungbluth, Prof. R. M. Peart, and Prof. A. Ramdani, for their tremendous efforts to edit this volume. Deep gratitude is also expressed to the authors of this volume who contributed excellent manuscripts for this handbook.

To the members of the Editorial Board of the CIGR Handbook, I extend my deep gratitude for their valuable suggestions and guidance during the board meetings. Special thanks are expressed to Prof. J. Daelemans who reviewed the complete manuscript of the volume. Mrs. D. M. Hull, ASAE director, and Ms. S. Napela, of the ASAE books and journals department, made kind and skillful handling of the publishing process of the volume.

The editorial expenses of this volume as well as those incurred during the compiling of the other volumes of the handbook was totally covered by the donations of the following companies and foundations: Iseki & Co., Ltd., Japan Tabacco Incorporation, The Kajima Foundation, Kubota Corporation, Nihon Kaken Co., Ltd., Satake Mfg. Corporation, The Tokyo Electric Power Co., Inc., and Yanmar Agricultural Equipment Co., Ltd. Sincere gratitude is extended to their generous donations to this handbook project.

Prof. Carl W. Hall, former president of ASAE and co-editor of the *Biomass Handbook* published by the Gordon and Breach Science Publishers Inc. in 1989, gave me important advice on the editing of this volume.

Dr. Kingshuk Roy, visiting researcher at my Laboratory of Bio-Environmental System Engineering, Nihon University, steadily assisted in my editorial work of this massive handbook.

Osamu Kitani
Editor of the Vol. V

# Acknowledgments

At the World Congress in Milan, the CIGR Handbook project was formally started under the initiative of Prof. Giuseppe Pellizzi, the President of CIGR at that time. Deep gratitude is expressed for his strong initiative to promote this project.

To the members of the Editorial Board, co-editors, and to all the authors of the handbook, my sincerest thanks for the great endeavors and contributions to this handbook.

To support the CIGR Handbook project, the following organizations have made generous donations. Without their support, this handbook would not have been edited and published.

Iseki & Co., Ltd.
Japan Tabacco Incorporation
The Kajima Foundation
Kubota Corporation
Nihon Kaken Co., Ltd.
Satake Mfg. Corporation
The Tokyo Electric Power Co., Inc.
Yanmar Agricultural Equipment Co., Ltd.

Last but not least, sincere gratitude is expressed to the publisher, ASAE; especially to Mrs. Donna M. Hull, Director of Publication, and Ms. Sandy Nalepa for their great effort in publishing and distributing this handbook.

Osamu Kitani
CIGR President of 1997–98

# Acknowledgments

# 1 Natural Energy and Biomass

## 1.1 Post-Petroleum Energy and Material

*O. Kitani*

### 1.1.1 World Population and Environment

With the increase in world population and the rise of living standards, the demand for energy in the world is steadily increasing. Global environmental issues and exhaustion of fossil resources also pose serious problems for energy consumption. Environment-friendly energy technology and a shift to nonfossil energy resources such as natural energy and biomass are expected. In this section, the issues and the prospects of biomass energy technology in the world are briefly described. To cope with increasing demands for biomass energy and feedstocks, integrated systems for biomass production, conversion, and utilization of photosynthetic resources should be developed.

According to the United Nations, the world population in 2025 could reach 8.5 billion, which is almost five times that at the beginning of this century. It has doubled in the last 39 years, as indicated in Table 1.1, in contrast to the 1600 years it took after the beginning of Anno Domini to double. The primary energy consumption of the world in the same period has tripled because of the increase of both population and capita consumption (Table 1.1).

A rapid increase in world population also demanded a huge amount of food, which is another form of essential energy for mankind. Table 1.1 shows that cereal and meat production in the world increased 2.67 and 4.17 times, respectively, during the years 1955–1994. This production increase covered rapid population growth and also the rise in living standards. A change in food habits to more meat consumption requires more primary calories, on average approximately seven times compared with the direct intake from plants. A number of countries are now importing feed for livestock.

A serious problem for the present world is that the food and feed production drastically changed after the mid 1980's. The average annual increase in rates of cereal, pulses, meat, and fish production from 1955 to 1984 were 5.06, 3.11, 6.96, and 6.83, respectively. However, after 1985, they dropped to 0.68, 1.93, 3.81, and 2.99, respectively, as shown in Table 1.1. This was caused by the decrease in cropland, less input means such as irrigation facilities, and fertilizer after the mid 1980s. Note that the traditional breeding

**Table 1.1. Change of population, energy, crop land, and food**

| | Year | | Rate | Year | | Rate | Ratio |
|---|---|---|---|---|---|---|---|
| Item | 1955 | 1984 | *A*(%/yr) | 1985 | 1994 | *B*(%/yr) | *B*/*A* |
| Population (1000) | 2,691,000 | 4,764,044 | 2.65 | 4,836,789 | 5,629,804 | 1.82 | 0.67 |
| Primary energy consumption (1000 ton oil equivalent) | 2,353,520 | 6,234,290 | 5.68 | 6,399,629 | 7,880,602 | 2.57 | 0.45 |
| Energy consumption per capita (kg oil equivalent) | 874.5 | 1,308.6 | 1.71 | 323.1 | 1,395.0 | 0.60 | 0.35 |
| Crop land (1000 ha) | 1,370,000 | 1,476,761 | 0.26 | 1,476,483 | 1,450,838 | −0.19 | −0.74 |
| Irrigated area (1000 ha) | 24,077 | 219,715 | 28.01 | 220,312 | 249,549 | 1.47 | 0.05 |
| Cereal production (1000 ton) | 731,000 | 1,803,975 | 5.06 | 1,841,002 | 1,954,550 | 0.68 | 0.13 |
| Pulses production (1000 ton) | 25,200 | 47,940 | 3.11 | 49,226 | 57,782 | 1.93 | 0.62 |
| Meat production (1000 ton) | 47,650 | 143,868 | 6.96 | 148,210 | 199,111 | 3.81 | 0.54 |
| Fish production (1000 ton) | 28,120 | 83,851 | 6.83 | 86,335 | 109,585 | 2.99 | 0.43 |
| Fertilizer production (1000 ton) | 19,700 | 139,739 | 21.01 | 139,739 | 136,431 | −0.26 | −0.01 |

Compiled from *Production Yearbook*, FAO 1955–1994, and *Energy Statistics*, MITI.

and chemical applications as the main tools for the Green Revolution have not been effective anymore in the past decade and new emerging technology is now expected.

The same tendency could be detected in energy consumption. The primary energy consumption in the world increased 5.68% annually from 1955 to 1984 and then dropped to a half (2.57%) from 1985 to 1994. The annual increase in the rate of energy consumption per capita has also decreased from 1.71% before 1984 to 0.60% thereafter. Energy demand increases with the world population and an improved qualify of life. But the oil crisis and the environmental issues restricted the expansion of energy consumption. Improved energy conversion and a utilization system for effective use of energy with less environmental load is now needed.

Improved quality of life also demands more living necessaries and utensils. Their production demands energy and industrial feedstocks. Both currently come mainly from fossil resources. Metal, plastics, and other materials are considered to be the secondary energy or embodied energy in that sense. Recycling them or alternating them with renewable resources is another important measure to be taken.

### 1.1.2 Energy and Environmental Issues

The greenhouse effect and acid rain, for example, are mainly associated with the use of fossil energy. The carbon cycle in nature is basically balanced, but the artificial emission of $CO_2$ by the combustion or disintegration of fossil resources and other organic matters is the cause of the increase in $CO_2$ in the air [1]. Other gases like $NH_3$ and $N_2O$ also can be the cause of the greenhouse effect, but their weight is smaller compared to $CO_2$. Nuclear energy poses the problems of radioactive pollution and diffusion of nuclear weapons. Energy and environment currently are two sides of one coin. To separate one from another, the world needs more renewable energy in the future. Natural energy—like solar, wind, hydraulic, and geothermal energy—can be free from environmental problems. Biomass energy is considered to be $CO_2$ neutral insofar as its production and consumption are balanced. Biomass is also noted for less S content and, thus, less likely to cause acid rain.

An increase in food, feed, and industrial feedstock production in the future requires more energy. The reduction and the effective use of fossil energy are essential in every sector of economic activities. Technology for utilization and conversion of natural energy and biomass should be developed. Fixation of $CO_2$ by use of plants and algae needs to be promoted. Recycling or cascade use of photosynthetic resources before their final combustion is also important to reduce the environmental load.

## 1.2 Natural Energy

*R. M. Peart*

### 1.2.1 Main Sources of Natural Energy

Natural energy is classified here as energy directly utilized from the sun, wind, and natural hydraulic sources. Of course, man-made devices—solar collectors, windmills, and dams—are required for capturing these natural forms of energy. Geothermal, tidal, and wave action are not covered here because they are available in only a few locations and are of little significance in a worldwide energy handbook.

Solar energy is by far the largest energy source, and all life on earth depends upon it. The amount of solar energy absorbed by the earth is so large that it is likely to confuse any discussion of its utilization because of the tiny proportion of the global surface that can practically be used to capture solar energy. In any given location, solar energy is available only during the daytime, and during that time, clouds can cut it drastically. On the other hand, it is available at every location on the globe, except a few locations on the shady side of steep mountains. Solar energy is difficult to store. A fluid such as water may be heated to store solar thermal energy, but heat losses are a problem. Electricity generated with solar photovoltaic cells may be stored in batteries, but these are heavy and expensive.

Wind and hydraulic energy vary from solar energy in such characteristics, so the three major characteristics of the forms of natural energy are discussed: (1) density, (2) storability, and (3) dynamics. The principles of utilization of these three forms of natural energy are then discussed.

### 1.2.2 Characteristics of Natural Energy

When energy is studied, it is important to recognize various characteristics or properties of energy sources that can vary greatly from one source to another. Also, these properties may vary in their importance from one application to another.

***Density***

Density is a concept that is difficult to define as used here, but it is important to recognize. It includes the properties of availability and portability. If one wishes to design a facility with power requirements of, say, 1000 kW at a given site that is 1 ha in size (10,000 $m^2$), the geographic requirements for natural energy capture come into question. The maximum solar energy that can be received on the earth's surface is in the range of 1 $kW/m^2$. Then the designer must reduce this power to account for the inefficiency of the particular solar-collecting device and for the reduction in solar energy received in the early and late hours of the daytime. Further reductions must be made to account for cloudy periods. These calculations will show that a rather large part of this 1-ha site must be devoted to solar collectors because of the rather low geographic density of solar energy. On the other hand, the geographic availability of solar energy is excellent, as the collector may be located anywhere on the property in this example, even atop buildings so as to require little extra space.

Compare this, for example, with the geographic density of wind energy, which in general is less than that of solar energy, although these are difficult to compare on a strict land-area basis. Hydraulic power from a major dam at a deep reservoir will have a much more dense form if the space for the entire reservoir or lake is not taken into account.

All these sources can be converted to electricity, which is relatively easy to transport over transmission lines, but which is not yet practically portable (batteries) for large power units, i.e., a 100-kW tractor.

Availability is widely different for the three forms of natural energy. Solar energy is limited by the diurnal cycle and by the probabilities of cloudy weather. Wind energy may be available throughout the diurnal cycle, but weather and shifting land and sea

**Table 1.2. U.S.A. Electricity generation capacity, 1984 by fuel type, Ref. Hansen[2]**

| Fuel type | Total capacity GW($10^9$ W) | Percent of total |
|---|---|---|
| Coal | 298.1 | 44.1 |
| Oil and Gas | 219.3 | 32.4 |
| Nuclear | 73.7 | 10.9 |
| Conventional hydraulic | 65.3 | 9.7 |
| Pumped Storage | 16.0 | 2.4 |
| Exotic, wind, geothermal, etc. | 3.4 | 0.5 |
| Solar | 0.0(minuscule) | 0.0(minuscule) |
| Total capacity | 675.8 | 100.0 |

breezes affect its availability greatly. The major advantage of hydraulic energy is its ready availability within the limits of the capacity of the reservoir.

***Storability***

Storability is one of the most important properties of natural energy, as most energy requirements are for a continuous source over time. This is a problems with solar energy and wind energy, as they are usually converted into electric energy, which is difficult and expensive to store. By contrast, hydraulic energy has a huge advantage in storability. The magnitude of the capacity of hydraulic energy is shown in Table 1.2 from US energy data by Hansen [2]. Note that pumped storage is listed separately, so the total generation capacity for all hydraulic power is the total of these two, or 81.3 GW, 12.1% of the total. Pumped storage is an outstanding example of the storability of hydraulic energy. Table 1.2 shows that 2.4% of the total US electrical generating capacity came from pumped storage in 1984. These systems include a hydraulic reservoir that is drained and then refilled on a daily basis. Water turbine-powered generators provide power when the entire system is under peak-load requirements, draining the storage; then the same turbine generators are converted to electric motor pumping systems to refill the reservoir when excess low-cost electric-generating capacity is available from nuclear-generating systems.

***Dynamics***

Dynamics, or the temporal fluctuations in the availability of the natural energy source, either diurnal or seasonal, are a very important property, yet are often overlooked when evaluations are made of various forms of natural energy. The dynamics of the availability of the source and, equally important, the dynamics of the demand for the energy are directly related to storability. If the cost of storing the energy is low, then the dynamics of the availability of the natural energy can be handled economically. Also, if the dynamics have a relatively short cycle, as in the diurnal cycle of solar energy availability, storage costs are lower than for an annual cycle of availability. For example, solar panels to power small electric pumps or electric fence applications in fields away from a central electric service, with batteries to store enough electric energy to cover the nighttime and cloudy periods, are practical today.

### 1.2.3 Utilization Systems

Because of the properties of natural energy mentioned above, and also because of the properties of the application, there are several principles we can define to guide the efficient application of natural energy to a need.

***Load Matching***

Match source dynamics to the dynamics of the load and use storability of the natural source as needed. The most widespread example is the water reservoir for an electric-power-generating dam. Huge amounts of potential energy are stored in the reservoir, and the power generator is automatically controlled to match the load requirements throughout the day and the night. Another example that is not yet widely used is solar-heated air for crop drying. The dynamics of solar availability and needs of the crop to be dried are important here. In temperate climates, solar energy can be utilized more effectively for drying crops that are harvested in the warmer and longer days of the summer (such as hay) than for those such as maize that are harvested in the shorter days of the fall season.

***Design Factors***

Account for density of the natural energy source along with storability and distance to load. For example, a solar panel for an electric fence or a small water pump is now practical because the density of the solar power allows a relatively small panel at the load to serve small loads that are distant from the central electric power system. However, solar panels for a 20-kW irrigation pump may not be practical because the density of the solar energy would require a relatively large panel for these larger loads and central electric service or diesel energy may be competitive.

***Combined System***

Consider several natural energy sources in combination with several storage methods. Should the design call for pumping water directly with a windmill and storing water in an elevated tank or pumping water as needed with a battery-powered electric pump and storing wind-generated electric energy in the batteries?

In conclusion, the many important characteristics of the three major forms of natural energy require creativity and attention to the principles mentioned here for the development of our natural, renewable sources of energy. Further, more detailed information on how to design specific systems is found throughout this handbook.

## 1.3 Biomass Resources

*O. Kitani*

### 1.3.1 Principles of Biomass Utilization

Biomass is a renewable resource so far as its production is continued in a sustainable way. It is in huge amount: 1800 billion tons of C as stock resource on the Earth and 170 billion tons of C as flow per year. It forms, in principle, a closed carbon cycle and therefore is $CO_2$ neutral. It is diversified in species, properties, and ways of utilization. It distributes more evenly on the Earth than do other natural resources.

Present biomass production needs a certain amount of energy, which comes mainly from fossil resources. In general, it is true that some amounts of both direct and indirect energy input to agriculture are necessary to get better yield. Actually the reduced annual farm production rate after the mid 1980's was caused by the decrease in input means to agriculture. However, if we rely too much on the expiring fossil energy and pollute our environment, sustainability of production itself cannot be secured. As a part of the energy production system utilizing the photosynthetic ability of plants, minimum input should be made to a biomass production process so that the system efficiency is kept as high as possible and at the same time achieves minimal production cost.

Biomass conversion is a process to convert photosynthetic material into a more useful form. Energy and material inputs also are needed in this process. Moreover, the heat value of biomass itself usually decreases. Hence the system efficiency of the process is always less than 1 and is expected to be made as high as possible.

A broad spectrum of biomas may be called 7f utilization, because biomass has been used in many ways as food, feed, fertilizer, feedstocks, fuels, fibers, and for fine chemicals. Since animal protein and fat will be limited in the coming century by a vast population, plant proteins in the form of leaf protein, steam-exploded plant tissues, and single cell protein, for example, are expected. They make it possible to get necessary protein with much less energy input.

### 1.3.2 Biomass Energy

Fuels from biomass can take various forms such as liquid, gas, and solid. They are used for electric or mechanical power generation and heat according to their properties and economy. A sustainable biomass system forms a closed cycle of carbon; fuel from biomass is therefore important from the viewpoint of global and regional environments. Utilization of biomass wastes as fuel is important in a practical sense, because they must be disposed of anyway to avoid pollution and are mostly more advantageous in economy. Energy crop cultivation or plantation is now becoming important because world agriculture is changing after the General Agreement on Tariffs and Trade for free trade and UNCED for global environment protection.

***Liquid Fuel from Biomass***

*Ethanol*

Ethanol has a smaller heat value but a higher octane rate than gasoline, which enables higher of engine efficiency with a larger compression ratio. At present it is most advantageous in terms of energy ratio and cost to get it from sugar crops, especially sugar cane in tropical countries. New sugar crops are also being searched for. In the temperate regions, ethanol is usually obtained from the fermentation of starch crops like corn and potato. Reductions in the cost of the fermentation process, including the pretreatment, are ongoing. New technologies are being developed to get it from cellulosic biomass with a simultaneous saccharification and fermentation method, which will enable us to utilize most parts, except lignin, of the grassy plants.

Ethanol is used for spark-ignition engines either in the form of a 20%–23% mixture with gasoline or in its pure form. The latter requires a newly designed engine with a higher compression ratio and hence can achieve higher efficiency.

Methanol can be synthesized from biomass pyrolysis gas and can be used as an alternative fuel to gasoline. It is, however, more easily processed from natural gas.

*Biodiesel*

Vegetable oils from rapeseed, soybean, sunflower, and others can be used for diesel engines. Raw vegetable oils are usually so viscous and their cetane numbers are so low for high-speed diesels that transesterification with methanol is performed. The development of a special engine to take refined raw vegetable oil is still ongoing. Emission from biodiesel is characterized with low a $SO_x$ content. It is reported that in some cases $NO_x$ increases, but with the adjustment of valve timing, $NO_x$ could be kept on the same level as that of conventional diesel fuel.

### *Gas Fuel from Biomass*

*$CH_4$*

$CH_4$ production is advantageous to get fuel from biomass with a high moisture content. Steady production with a simple fermentation tank is, however, not so easy. A large-scale reactor with more sophisticated control is suited for steady and efficient operation. A two-tank reactor is better in principle but needs a certain level of control. $CH_4$ can be used for heat and stationary power without an emission problem.

*Pyrolysis Gas*

Pyrolysis gas from a biomass gasifier is usually obtained through the reduction process of $CO_2$ and consists of CO and $H_2$. Filtered gas could be used for an internal combustion engine. Pure cracking is also possible with a sophisticated reactor that could produce not only gas but also liquid fuel or biocrude oil as a basic feedstock from biomass.

### *Solid Fuel from Biomass*

Fuel wood and charcoal are common biomass solid fuels in many countries, but heat efficiency of a wood furnace is generally low. Energy loss in charcoal production is also considerably large. An improved furnace and kiln were developed, but promotion of their use needs to be accelerated.

Agricultural wastes such as straw or husk can be used for heat or electric generation. Rice husks, for example, could generate more electricity and heat than a country elevator consumes. Forestry wastes are often used for power generation by the Rankine cycle with turbine in wood-processing factories. Solid municipal wastes from a city with a large population can run a power plant. Fluctuation of the heat value of the waste must be adjusted somehow.

Energy plantation is becoming important. Short-rotation intensive culture of fast-growing trees or grasses can run a small power plant. Since the transportation cost of biomass is high, the cultivation area around the power plant should be limited.

The cost of biomass fuel is always a problem. However, economic feasibility should be investigated by taking into consideration various factors, including job opportunities, environmental effect, and national or rural economy.

Table 1.3. **Present energy consumption (commercial and biomass) (in 1988)**

| Regions | Commercial energy (PJ) | Biomass energy *B* (PJ) | Total energy *T* (PJ) | Weight of biomass *B*/*T* (%) | Per capita energy (GJ) |
|---|---|---|---|---|---|
| Developing | 70,430 | 42,631 | 113,061 | 38 | 30 |
| Africa | 7363 | 9160 | 16,524 | 55 | 28 |
| Central America | 5488 | 8.68 | 6356 | 14 | 45 |
| South America | 8328 | 2858 | 11,186 | 26 | 40 |
| Asia | 49,146 | 29,689 | 78,835 | 38 | 28 |
| Oceania | 105 | 56 | 161 | 35 | 28 |
| Industrialized | 235,456 | 7312 | 242,768 | 3 | 203 |
| North America | 91,636 | 4027 | 95,662 | 4 | 355 |
| Europe | 64,832 | 1527 | 66,359 | 2 | 134 |
| USSR | 57,690 | 1720 | 59,411 | 3 | 210 |
| Asia (industrial) | 17,106 | 0.06 | 17,112 | 0 | 135 |
| Oceania (industrial) | 4192 | 0.31 | 4223 | 1 | 216 |
| World | 305,885 | 49,943 | 355,829 | 14 | 71 |

Source: J. Woods and D. O. Hall. 1994. *Bioenergy for Development.* Rome: FAO (Table 1).

***Energy Demand***

In Table 1.3, J. Woods and D. O. Hall describe the biomass energy consumption in the world during 1988 [3]. The weight of biomass energy in developing countries was 38% and only 3% in industrialized countries. In Africa it reached 55%. The average weight in the world was 14%. Some industrialized countries, like Sweden, now use as much as 16% biomass energy.

***Potential of Biomass Energy***

It is often pointed out that the total amount of biomass in the world is quite large, but the actual amount for sustainable usage is rather limited. Matsuda *et al.* [4] pointed out in 1982 that the potential contribution of biomass energy to the total energy consumption in the world was estimated to be 16.6% and, in some regions, a larger amount was being consumed than should be.

The stagnation of food production in the late 1980's in contrast to the population increase creates serious conflict between food and feed production and between energy and feedstock production in those regions in which land for biomass production is limited. Development of new crop varieties to expand biomass cultivation area or to avoid the conflict with conventional food crop cultivation is really needed.

### 1.3.3 Biomass Material

Biomass has been used as fibers and construction material. Cotton and pulp are the major fibers from plants. Recycling of fiber is one of the important movements to make the best use of bioresources and to reduce the load on ecosystems. Wood is still widely used for building houses and making furniture because of its amenity to mankind.

Vegetable oil, resin, and wax have been used as industrial material for many purposes; solvents, inks, paints, rubber, and soaps are some examples. Petroleum has been used for industrial material for inks, plastics, and many other products, but it will be gradually replaced by the feedstocks from biomass. Printing inks, for example, are now being replaced by vegetable oils because of safety and environmental concerns. Plastics from biomass are now being developed for their biodegradability.

Medical ingredients and other chemicals, such as essential oils, vitamins, disinfectants, and others from biomass, are extracted and utilized. Biochemicals for pesticides and insecticides as well as organic fertilizers will be more important for a lower input of chemicals and less load on the environment in agriculture and forestry in the future.

### 1.3.4 Environmental Considerations

Biomass production must be sustainable; otherwise it will disturb the ecological balance and cause environmental pollution. Biomass cultivation must be carried out in a sustainable way with a smaller load on the environment. The application of chemicals and too much tillage resulting in erosion should be avoided. Sustainable cultivation also will contribute to reduce the production cost and to achieve higher efficiency as an energy production system.

Biomass conversion and utilization should be made in such a way to reduce the environmental effect. The rise of conversion efficiency and energy efficiency is a continuous challenge for agricultural engineers. Hence development of the following technologies is expected and is underway:

1. New crops or crop varieties resistive to drought, diseases, salinization, and nutrient deficiency with higher content of required ingredients. New crops will enable us to expand the cultivation of land and turn marginal lands like deserts into arable land.
2. Cultivation practices with appropriate tillage, fertilizing, pest, and insect control, as well as efficient harvesting methods.
3. Improvement of transportation and storage of bulky biomass.
4. New conversion technology to make better use of biomass resources and to raise the conversion efficiency.
5. Utilization system of biomass with higher efficiency and lower cost for various kinds of biomass for diversified usage.

### 1.3.5 Biomass Systems

There are more than 300,000 species of plants. The conditions for their production, conversion, and utilization are quite different from one area to another. The combinations of factors ruling biomass systems are astronomical in terms of numbers and complexity. Hence it is worthwhile to carry out vast research and development on biomass energy with international cooperation and to exchange or transfer the results. Since biomass technology must be adjusted to each local condition, newly developed basic technologies must be upgraded to practical ones by trained scientists and engineers. For this purpose, career building programs must be incorporated in the international cooperative research and development on biomass systems.

Biomass resources for energy and feedstocks require a solid infrastructure. Good statistics on biomass and energy are essential to make production plants and control the resources. A basic survey of soil and water may be necessary in some areas where no precise information exists on suitable cropping. Local research stations and extension services will be necessary for cases in which innovative technologies are to be introduced.

Social and economic systems to enable new biomass technology are essential. Legal measures to protect and control resources may be necessary. Economic incentives to initiate new biomass energy production will be needed. An appropriate price setting for electricity, for example, may accelerate power generation from biomass waste. Credits and taxes are sometimes effective in introducing new technologies for biomass production, conversion, and utilization.

## References

1. Houghton, R. A. 1989. Global circulation of carbon. *Biomass Handbook*, eds. Kitani, O. and C. W. Hall, pp. 56–61. New York: Gordon & Breach.
2. Hansen, H. J. 1989. Types of fuels for electricity generation. *Electrical Energy in Agriculture*, ed. McFate, K. L., (B. A. Stout, Editor-in-Chief). Energy in World Agriculture, Vol. 3, pp. 13–19. Elsevier Science Publ., New York.
3. Woods, J. and D. O. Hall. 1994. *Bioenergy for Development*, Rome: FAO.
4. Matsuda, S., H. Kubota, and H. Iwaki. 1982. Biomass energy: Its possibility and limitation. *Science* 52: 735–741.

# 2 Energy for Biological Systems

## 2.1 Energy Analysis and Saving

### 2.1.1 Energy Analysis

*J. Ortiz-Cañavate and J. L. Hernanz*

***Definition and Scope of Energy Analysis***

Energy analysis, along with economic and environmental analyses, is an important tool to define the behavior of agricultural systems. Economics, Energy, and Environment are the three E's that necessarily have to be considered in all agricultural projects.

One of the most important goals of mankind throughout history has been to handle and control energy in all its different forms. In agriculture, as in other economic activities, in the last century the amount of energy dedicated to crop production has increased substantially, more in developed countries than in developing countries.

The reasons for this increase of energy consumption in agriculture at a world level are: (1) the continuous growth of world population, (2) the migration of the labor force from rural to urban areas, and (3) the development of new production techniques.

Energy analysis started as a relevant subject in agricultural production in the 1970's [1] as a result of the dramatic increase of oil-derivative prices. The consequences were the rationalization of energy consumption, the use of new energy sources, and the aim for more efficient working methods.

The establishment of methodologies to identify and evaluate the different energy flows that take part in agricultural production is the basis of an energy analysis.

The field of application is as wide as defined or needed, but the goal is clear: to reduce energy inputs or to look for other renewable energy sources in agricultural processes. This goal is combined, if possible, with the reduction of production costs and environmentally friendlier production methods as part of a better management system.

The use of advanced computer programs and worldwide information networks nowadays allow the analysis of energy problems with powerful tools. Consequently, farmers can take more appropriate and accurate decisions in relation to energy and economic resources. In this respect, generalized use of the global position system techniques in agriculture is foreseeable in the future. Modern management through precision farming will allow the saving of energy by application of the right quantities of seed, fertilizer, and pesticides according to the local variations of production in each field.

The influence of fuel (used in mechanized operations) and fertilizer (especially nitrogen) in energy balances in agriculture represents—as we see in this section and the next—more than 70% of the total input energy, and this has to be taken into consideration to reduce these inputs.

***Methodologies and Approaches***

The methodologies applied to energy analysis are quite different in their bases and approaches. This occurs with the energy analysis school established by Odum [2] and the energy evaluation system suggested in 1974 by the International Federation of Institutes for Advanced Study, Energy Analysis Workshop [1].

The first approach considers all types of energies, renewable and nonrenewable, stating criteria in relation to their quality; this school proposes interconnections between the energy systems supplied by man and by natural ecosystems; it considers manpower as a high-quality energy source and evaluates energy flows in the establishment of the net energy analysis.

In the second school, the analysis procedure relies on assigning nonrenewable energy amounts to each production factor. Total energy in a process is established by adding partial energies associated with each step without assigning a quality factor. A second workshop of the International Federation of Institutes for Advanced Study, in 1975, compared energy analysis and economic analysis and attempted to unify them.

In an energy analysis of production systems it is necessary to consider the following steps [3]:

- Set a limit in the process or system to be analyzed in such a way that all inputs and outputs, which pass that limit in a certain time interval, are evaluated. For example, in crop production, it is necessary to quantify the energy requirements of selected inputs like fuel tractors, farm machinery, pesticides, fertilizer, labor, transportation, etc.
- Assign energy requirements to all inputs.
- Multiply the amount of inputs by their corresponding assigned energy value and add these values to obtain the total sequestered energy in the process.
- Identify and quantify all outputs, establishing criteria for energy embodied in the main products and that corresponding to by-products.
- Relate output energy to total sequestered energy to obtain the energy ratio (relationship between output and input energies) and the energy productivity (units of product obtained per unit of input energy).
- Apply energy analysis results. An evident application is the production of biofuels, but it can be of interest for any agricultural product to compare alternatives of production, complementing economic and environmental impact analysis.

One of the problems of this methodology of energy analysis is unifying criteria for assigning amounts of energy to each input. The lack of reliable data for each country or region forces us, in many cases, to take values from other countries for which circumstances are different. That is the case of fertilizer production: the amount of energy needed per fertilizer unit (e.g., nitrogen) depends to a great degree on the technical level of the manufacturing industry and also on the distance of transportation, which is variable but can be taken as an average value for a region.

Another problem is considering different qualities of energy: whether it contaminates more or less, whether it is renewable, how much it costs, etc. For example, electricity from a hydroelectric plant (renewable, more efficient in producing mechanical power) is of higher quality than coal (nonrenewable and less efficient).

There are also problems with the energy assigned in cases of multiple outputs. Fluck [1] recommended that the energy apportioned to agricultural systems that have multiple outputs should be proportional to the relative values of the products. Let us take the example of cereal production: In a cereal crop it is impossible to separate the energy needed to form grain from the one needed to form straw. It has to be assigned according to the use given to the straw and to its corresponding value.

When an energy analysis is made, it is important to specify the procedure used for establishing the amount of energy assigned to each item.

***Energy Ratio, Net Energy Gain, and Energy Productivity***

The energy ratio (ER) is defined as the ratio between the caloric heat of the output products and the total sequestered energy in the production factors. This index allows us to know the influence of the inputs expressed in energy units in obtaining consumer goods, normally related to the food production, but that can be applied appropriately to the energy balance of biomass or biofuel production.

Fluck [1] analyzes this concept, stating that, in a strict sense, "the energy ratio can be applied to the use of energy in isolated societies, in which it is important that the output energy is greater than the input one in order to assure their subsistence." In industrialized agriculture, farming for energy must be energetically sound, but there is no reason why farming for other products should cost less energy than is produced. In short, in developed countries at present, energy to produce human food is not a limiting factor.

Net energy gain (NEG) or net energy production is the difference between the gross energy output produced and the total energy required for obtaining it. In agricultural processes this energy is normally related to the unit of production (e.g., 1 ha).

Energy productivity (EP) is the measure of the amount of a product obtained per unit of input energy. Its relationship to the ER is direct: Their ratio is the calorific value of the product.

EP is specific for each agricultural product, location, and time. It can serve as an evaluator of how efficiently energy is utilized in different production systems that yield a particular product.

To improve EP in a process, it is possible either to reduce the energy sequestered in the inputs or to increase the yield of product—that is, to reduce losses. All these items are analyzed in more detail in Section 2.1.2: "Energy saving in crop production."

***Energy Inputs***

Energy consumption in agricultural systems is associated with all inputs that take part in the production processes. These inputs have to be defined and quantified according to their energetic intensities.

The total energy per production unit (e.g., hectare) is established by the addition of the partial energies of each input referred to the unit of production. After the yield produced is obtained, it is possible to calculate the ER, the NEG, and the EP.

The energy associated with these inputs comes from different sources: renewable and nonrenewable. The energy analysis makes no distinction between them.

In energy analysis, it is necessary to distinguish between inputs that are completely consumed in the period in which they are used (fuels, fertilizers, chemicals, seeds, etc.) and inputs that participate in different processes during a longer period (tractors, farm machinery, irrigation equipment, etc.). In the latter case, the energy of materials, manufacturing and maintenance has to be divided into fractions throughout its useful life. If the duration of the intervention of each factor is known, it is possible to establish the associated energy for a process.

Energy inputs can be classified in two main groups: direct-use energy and indirect-use energy.

*Direct-use Energy*

Direct-use energy is taken from a fossil fuel or a renewable energy source like a biofuel for direct application in a process.

At present, most of the energy used directly in agriculture of developed countries comes from a fossil origin such as diesel fuel, gasoline, liquid petroleum gas, coal, and from electricity, which can be produced from a fossil fuel or other sources (such as hydropower plants, nuclear plants, etc.).

Tractors and self-propelled farm machinery are powered generally with diesel engines because diesel engines are sturdier, have a higher efficiency and a longer life, and are considered less polluting than gasoline engines. Diesel fuel is the most widely used of all direct energies in agriculture (60%–80% of the total); liquid petroleum gas fuel is used mainly for heating and drying and electricity for irrigation plants.

Developing countries also use these fuels, although in a smaller amount, using other kinds of direct energy sources such as animal traction, for which the real fuel is the caloric value of the feed the animals need to work. Their needs are covered with crops produced in the same farm or from the surrounding area.

To establish the values of the energy sequestered in these inputs, it is necessary to consider their heating value (enthalpy), adding the energy needed to make their energy available directly to the farmer. For example, a liter of diesel fuel contains 38.7 MJ. However, to mine, refine, and transport a liter of diesel fuel to the farmer, an additional 9.1 MJ are needed. Thus the energy cost to consume the liter of fuel totals 47.8 MJ (see Table 2.1).

In Table 2.2 diesel fuel consumption for different farm equipment is given.

*Indirect-use Energy*

Although approximately one-third of the energy consumed on the farm is for direct use, nearly two-thirds of the energy is consumed indirectly. Indirect use refers to the energy used to produce equipment and other materials that are used on the farm. The major indirect use for energy is for fertilizers and primarily for nitrogen fertilizers. Other important items are farm machinery and biocides. In irrigated areas, irrigation can also be relevant.

***a. Farm Machinery.*** The production and the repair of farm machinery are important issues in the total energy balance. We consider several steps in calculating this energy,

**Table 2.1. Energy values for various fuels**

| Energy Source | Energy Content | Production | Total Energy Costs |
|---|---|---|---|
| Gasoline | 38.2 MJ/L | 8.1 MJ/L | 46.3 MJ/L |
| Diesel | 38.7 MJ/L | 9.1 MJ/L | 47.8 MJ/L |
| Fuel oil | 38.7 MJ/L | 9.1 MJ/L | 47.8 MJ/L |
| LPG | 26.1 MJ/L | 6.2 MJ/L | 32.3 MJ/L |
| Natural gas | 41.4 MJ/m$^3$ | 8.1 MJ/m$^3$ | 49.5 MJ/m3 |
| Coal | 30.2 MJ/kg | 2.4 MJ/kg | 32.6 MJ/kg |
| Electricity | 3.6 MJ per kWh | 8.4 MJ per kWh | 12.0 MJ per kWh |

Adapted from [4].

**Table 2.2. Diesel fuel consumption estimated for different farm equipment**

| Equipment | Consumption (L/ha) | Equipment | Consumption (L/ha) |
|---|---|---|---|
| Moldboard plow | $25 \pm 7$ | Centrifugal fertilizer | $2 \pm 0.5$ |
| Disc plow | $22 \pm 5$ | Manure spreader | $7 \pm 2$ |
| Chisel (straight arm) | $13 \pm 2$ | Mounted sprayer | $1.5 \pm 0.5$ |
| Chisel (curved arm) | $10 \pm 2$ | Trailed sprayer | $3 \pm 0.5$ |
| Heavy disc harrow | $9 \pm 3$ | Grain drill | $5 \pm 0.5$ |
| Medium disc harrow | $7 \pm 2$ | Row planter | $5.5 \pm 1$ |
| Heavy cultivator | $10 \pm 2$ | Ridge planter | $7 \pm 1.5$ |
| Light cultivator | $8 \pm 1$ | Combine | $18 \pm 2$ |
| Vibrocultivator | $6 \pm 1$ | Cutterbar mower | $4 \pm 0.5$ |
| Rotary hoe | $4 \pm 1$ | Rotary mower | $5.5 \pm 0.5$ |
| Roller | $4 \pm 1$ | Swather | $3 \pm 1$ |
| Rotary cultivator | $20 \pm 4$ | Baler | $5 \pm 1$ |
| Hoeing toolbar | $2 \pm 0.5$ | Forage harvester | $25 \pm 5$ |
| Combined equipment for plowing and seed bed preparation | $24 \pm 6$ | Sugarbeet harvester | $60 \pm 10$ |

taking into consideration the procedure established by Bowers [5]: first, the energy used in producing the raw materials (like steel, 22–60 MJ/kg); second, the quantity of energy required in the manufacturing process (mean value of these two first steps, 87 MJ/kg); third, the transportation of the machine to the consumer (estimated, 8.8 MJ/kg); and fourth, the energy sequestered in repairs. The total energy sequestered in different types of farm machinery is given in Table 2.3.

Pellizzi [6] establishes a mean value per kilogram and year for a group of tractors and agricultural machines (Table 2.4).

To evaluate the energy input for machinery and equipment per hectare, it is necessary to know the weight of the machinery used in the farm, its working life span, and the average surface on which it is used annually. When we assign the energy value that we consider more appropriate for the machinery used, it is possible to establish the energy input for the farm machinery per hectare, having calculated previously the average

**Table 2.3. Example of energy sequestered in different types of farm machinery**

| Equipment | Energy (MJ/kg) |
|---|---|
| Tractor | 138 |
| Plow | 180 |
| Disc harrow | 149 |
| Planter | 133 |
| Fertilizer | 129 |
| Rotary hoe | 148 |
| Combine | 116 |

Adapted from [5].

**Table 2.4. Primary energy incorporated in farm machinery refered to the unit of mass (kg) and per year (a)**

| Equipment | Energy (MJ/kg. a) |
|---|---|
| Tractors and self-propelled machines | 9–10 |
| Stationary equipment | 8–10 |
| Agricultural machinery and implements | 6–8 |

Adapted from [6].

weight of machinery utilized per hectare and per year. For example, for mechanized corn production, ~55 kg of equipment is used per hectare and year.

***b. Fertilizers.*** Fertilizers are those chemical elements that, being incorporated into the soil or directly into the plants, are necessary for the nutrition and normal growth of the plants.

Crop plants take up nutrients at different rates, so nutrients have to be incorporated into the soil to maintain its productive potential. The process involved in this task is known as fertilization.

There are 15 major elements that constitute plants. Oxygen, hydrogen, and carbon form 90%–92% of the plants' weight. The other elements can be classified into three groups:

- Primary nutrients (nitrogen, phosphorus, and potassium).
- Secondary nutrients (calcium, magnesium, and sulfur).
- Micronutrients or oligonutrients (copper, iron, manganese, molybdenum, boron, and zinc).

According to the source, fertilizers are divided into three categories:

- Chemical: These are manufactured from the air to obtain nitrogen fertilizers and also from geological material to obtain phosphate and potash fertilizers.
- Organic: These are obtained from crop residues and animal wastes.
- Biological: These are related to nitrogen fixation by micro-organisms placed in the cells of the roots of legume crops.

**Table 2.5. Energy embodied in main mineral fertilizers**

| Fertilizer | Energy (MJ/kg) | | |
|---|---|---|---|
| | Production | PTA[a] | Total |
| N | 69.5 | 8.6 | 78.1 |
| $P_2O_5$ | 7.6 | 9.8 | 17.4 |
| $K_2O$ | 6.4 | 7.3 | 13.7 |

[a] PTA: packaging, transportation, and application.
Adapted from [7].

Chemical fertilizers are widely used in agricultural crops and require high rates of direct energy for their production, mainly in the nitrogen fertilizer industry. Ammonia is the basic source for commercial nitrogen fertilizers; it is produced by the synthesis of hydrogen with atmospheric nitrogen under high-pressure conditions. The chemical processes need natural gas both to heat steam and to react with water to obtain hydrogen at temperatures ranging between 400 and 1200°C during certain stages of production [7].

Rock phosphate is the source of all phosphate fertilizers. This raw material can be applied directly after a previous process of beneficiation, drying, grinding, etc. Most commercial phosphate fertilizers come from phosphoric acid, which is the result of the treatment of phosphate rock with sulfuric acid.

The basic material for potash fertilizer products is potassium salt. The extraction can be performed by mining or solar evaporation, when the raw material comes from lake beds.

Average values for energy intensities of nitrogen, phosphorus, and potassium are shown in Table 2.5. Overall energy includes production, packaging, transportation, and application. For mixed fertilizers, it is necessary to add 1.14 MJ/kg for fluid application.

Organic residues contain different degrees of nitrogen, phosphorus, and potassium, depending on the origin of the material and the method of processing and storage. Nevertheless, an approach can be made by consideration of the composition in a percentage on a dry weight basis and the energy intensities for mineral nitrogen, phosphorus, and potassium production—70, 8, and 6 MJ/kg, respectively.

*c. **Chemical Biocides.*** During the past 50 years, chemical biocide consumption in the world has increased substantially. Most farmers need them every year to control weeds, pests, and diseases, and the quickest way to attain that compared with the use of non-chemical control systems is by spraying chemical products on plants or soil.

Nowadays, biotechnology provides new products that, applied in low amounts, allow control of weeds, pests, and diseases more efficiently. The energy embodied in active ingredient production has to include production, formulation, and packaging (Table 2.6). To establish the overall energy, it is necessary to add transportation and application [8].

An active ingredient requires both direct and indirect energy for its formulation. Direct energy includes fuel, electricity, and steam to synthesize the organic compounds as the basis of the final product. Indirect energy comprises the amount of crude oil or natural

**Table 2.6. Energy inputs for various biocides**

| Product | Energy (MJ/kg[a]) | Product | Energy (MJ/kg[a]) |
|---|---|---|---|
| Herbicides | | Fungicides | |
| MCPA | 130 | Ferbam | 61 |
| 2,4-D | 85 | Maneb | 99 |
| 2,4,5-T | 135 | Captan | 115 |
| Dicamba | 295 | Benomyl | 397 |
| Fluazilop b. | 518 | Insecticides | |
| Alachlor | 278 | Methyl parathion | 160 |
| Chlorsulfuron | 365 | Carbofuran | 454 |
| Atrazine | 190 | Lindane | 58 |
| Paraquat | 460 | Cypermethrin | 580 |
| Glyphosate | 454 | Malathion | 229 |
| Linuron | 290 | | |

[a] Active ingredient.
Adapted from [8].

gas to produce ethylene, propylene, or methane that will be transformed into the organic products mentioned above.

The active ingredient must be formulated in emulsive oil, wettable powders, and granules. Energy for formulation is 20, 30, and 20 MJ/kg, respectively.

Packing and distributing formulated biocide require additional energy, ranging from 3 to 8 MJ/kg, depending mainly on the distance of distribution.

Application includes both direct energy for tractor diesel fuel and indirect energy embodied in the equipment (sprayer) and the tractor during the time of application.

***d. Crop Propagation.*** Agricultural crops can be propagated by seeds, bulbs, tubers, etc., so that in the energy analysis, the energy required for their production must be included. Unfortunately, not much information about this is available. On the other hand, for the same input, the associated energy depends on processes to be obtained later. For example, grain seed requires different energy rates, depending on whether it is produced on the farmer's own farm or purchased from a seed producer company.

Heichel [9] stated different methods for its assessment. He considers that all steps, preharvest and postharvest, must be taken into account for different inputs and processes in each case. Values for energy costs for different seeds used in crop production are shown in Table 2.7.

***e. Irrigation.*** Water is an essential element for life, but unfortunately there are many areas where this element is limited. Increasing the need to produce food to nourish the world population forces people to redistribute hydraulic resources where crop production is not possible or is scarce. This objective requires infrastructures for storing and transporting large amounts of water to agricultural farms. In other cases, it is necessary to drill the ground soil to make use of well water from acquifers. After that, it is needed to lift or pressurize the water to irrigate the soil where the crop is planted.

Energy assessment in irrigation systems depends on both the direct use (DE) and the indirect use (IE). The former includes the energy consumption to lift or pressurize ($H$)

Table 2.7. Energy cost for seed and tuber production

| Crop | Energy (MJ/kg) | Crop | Energy (MJ/kg) |
|---|---|---|---|
| Alfalfa | 230 | Rice | 17 |
| Clover | 135 | Sugarbeet | 54 |
| Corn hybrid | 100 | Forage hay | 88 |
| Wheat | 13 | Oil seed rape[a] | 200 |
| Barley | 14 | Sunflower[a] | 20 |
| Oats | 18 | Potato | 93 |
| Soybean | 34 | Cotton | 44 |

[a] Estimated.
Adapted from [9].

the overall rate of water required by crop considered per hectare. Direct energy can be expressed by the following equation:

$$\mathrm{DE} = (\delta g H Q)/(\eta_1 \eta_0) \tag{2.1}$$

where DE is direct-use energy (in joules per hectare), $\delta$ is the water density (1000 kg/m$^3$), $g$ is gravity (9.8 m/s$^2$), $H$ is the total dynamic head, including friction losses (in meters), $Q$ is the overall rate of water, including losses by evaporation, drainage run-off, etc. (in m$^3 \cdot$ ha$^{-1}$), $\eta_1$ is the pump efficiency, and $\eta_0$ is the overall efficiency of the power device, electric or diesel.

Pump efficiency is a function of vertical height to lift, speed, and flow water pumped. It ranges between 70% and 90%.

Overall efficiency is considered for both electric- and fuel-powered devices; it ranges between 18% and 22%. In this factor, for the electric motor, the generating plant, the transmission line, and the motor efficiency are included. Diesel efficiency is approximately 25%–30%, but the energy to produce and transport fuel must also be considered [10].

Indirect energy includes raw materials, manufacturing, and transportation of the different elements that constitute an irrigation system with the same treatment as other infrastructures in their expected total life. It is difficult to establish this value so that a percentage of direct-use energy can be considered for the irrigation systems, ranging from 18% for the traveling sprinkler to 375% for the surface with a run-off recovery system [11].

*f.* ***Transportation.*** Agriculture is basically a transportation industry. Energy is required for moving inputs to the farm from their point of origin, for moving labor, machinery, and products to and on the farm, and for moving farm products to market.

Energy requirements in transport are normally expressed as energy intensity, the energy needed per unit of weight and per unit of distance traveled (in MJ $\cdot$ t$^{-1} \cdot$ km$^{-1}$).

In agriculture, transportation is done mainly by truck or by internal means (tractors and trailers) and little is done by rail, air, or water. Fluck [11] establishes the value of energy intensity transportation by truck of 1.6–4.5 MJ t$^{-1}$ km$^{-1}$ (see Table 2.13 in Section 2.1.2).

***g. Labor.*** Energy in labor is expended by agricultural workers when processes are being carried out. There are several methods to assess this input, which are based on different criteria. For primitive agricultural systems, the most accurate method is directly related to partial or total metabolic energy of the food consumed. In the mechanized agriculture of developed countries, the caloric energy of food is insufficient to be considered only. Indirect energy involved in all the processes in which the food industry, transportation, and distribution take part should be added. Both direct and indirect energy in this last case are higher than metabolic energy.

Fluck [11] summarizes nine methods. The first five methods include muscular energy and energy sequestered in food, ranging from 1 to 93.2 MJ/day, and the last four methods consider indirect energy based on the above-mentioned considerations and the style of life. Values range between 510 and 1450 MJ/day, although the latter seems too high to be taken into account.

***Energy Balance***

Let us consider as an example the sugar beet crop, which produces bioethanol (Table 2.8). In Fig. 2.1 the mass and the energy balance per hectare of sugar beet are given [12].

Considering ethanol, beet cuts as fodder for animals, and crowns and leaves for organic fertilizer as useful outputs results in a NEG of $(115 + 40 + 13) - (32.9 + 15.2 + 80.7) = 168 - 128.8 = 39.2$ GJ/ha and an ER of $168/128.8 = 1.3$.

**Table 2.8. Energy consumption in sugar beet production**

| Input | Quantity | Energy Intensity | Energy (MJ/ha) | Energy (%) |
|---|---|---|---|---|
| Direct-use energy | | | | |
| Diesel | 120 (L/ha) | 47.8 (MJ/L) | 5740 | 17.5 |
| Electricity | 600 (kW h/ha)[a] | 12 (MJ/kW h) | 7200 | 21.9 |
| Indirect-use energy | | | | |
| Machinery | 180 (kg/ha) | 9 (MJ/kg a) | 1620 | 4.9 |
| Fertilizers | | | | |
| N | 130 (kg/ha) | 78.1 (MJ/kg) | 10,100 | 30.9 |
| $P_2O_5$ | 50 (kg/ha) | 17.4 (MJ/kg) | 870 | 2.7 |
| $K_2O$ | 150 (kg/ha) | 13.7 (MJ/kg) | 2055 | 6.3 |
| Biocides | 10 (kg/ha)[b] | 250 (MJ/kg) | 2500 | 7.6 |
| Seed | 1.9 kg/ha | 50 (MJ/kg) | 95 | <0.1 |
| Irrigation | (20% of consumed electricity[a]) | — | 1440 | 4.4 |
| Labor | 45 (h/ha)[c] | 26.6 (MJ/ha) | 1200 | 3.7 |
| | | | 32,900 | 100.0 |
| Crop production | | | 60,000 kg | |
| Energy productivity in crop | | | 1.82 kg/MJ | |

[a] Electricity for pumping in irrigation.
[b] Active ingredient.
[c] Estimated, 640 MJ/day working 8 h/day.

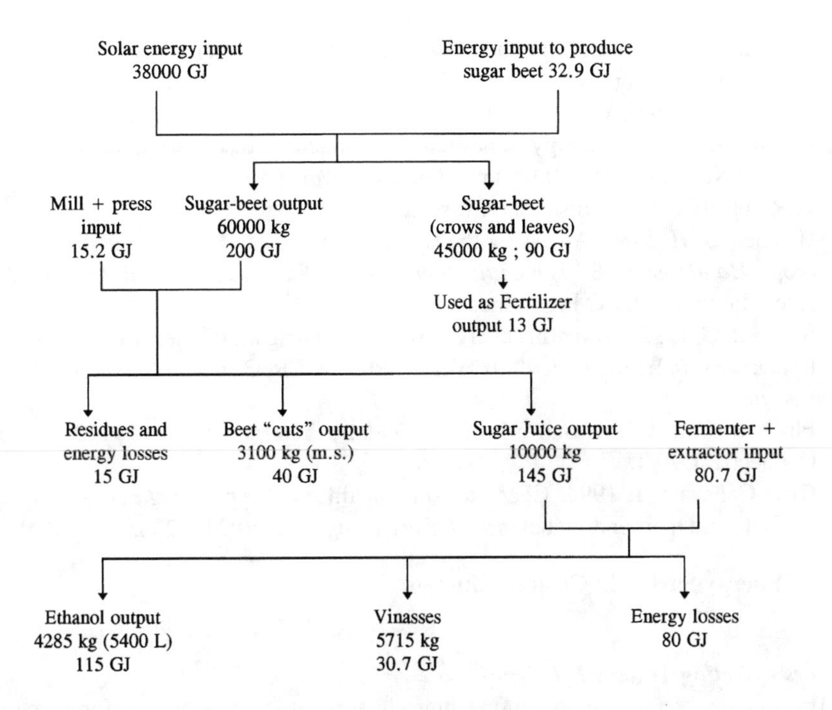


**Figure 2.1. Energy and mass balance for 1 ha of sugar beets.**

If we consider ethanol as the main product, then EP = (5400 L/ha)/(128.8 GJ/ha) = 41.9 (L/GJ).

Therefore bioethanol derived from sugar beet can be considered as a possible biofuel source, taking into account that the energy balance of this crop is positive.

## References


1. Fluck, R. C. and C. D. Baird. 1982. *Agricultural Energetics*, pp. 41–71. Westport, Connecticut: AVI.
2. Slesser, M. 1977. Letter to editor. *Science* 196:259–261.
3. Fluck, R. C. 1992. Energy analysis for agricultural systems. Energy in Farm Production. *Energy in World Agriculture*, Vol. 6, ed. Fluck, R. C., pp. 45–53. Amsterdam: Elsevier.
4. Cervinka, V. 1980. Fuel and energy efficiency. *Handbook of Energy Utilization in Agriculture*, ed. Pimentel, D. pp. 15–21. Boca Raton, FL: CRC Press.
5. Bowers, W. 1992. Agricultural field equipment. Energy in farm production. *Energy in World Agriculture*, Vol. 6, ed. Fluck R. C., pp. 117–129. Amsterdam: Elsevier.
6. Pellizzi, G. 1992. Use of energy and labour in italian agriculture. *J. Agric. Eng. Res.* 52:231–238.

7. Mudahar M. S. and T. P. Hignett. 1987. Energy requirements, technology and resources in fertilizer sector. Energy in Plant Nutrition and Pest Control. *Energy in World Agriculture*, Vol. 2, ed. Helsel, Z. R., pp. 25–61. Amsterdam: Elsevier.
8. Green, M. B. 1987. Energy in pesticide manufacture, distribution and use. Energy in Plant Nutrition and Pest Control. *Energy in World Agriculture*, Vol. 2, ed. Helsel, Z. R., pp. 165–195. Amsterdam: Elsevier.
9. Heichel, G. H. 1980. Assessing the fossil energy costs of propagating agricultural crops. *Handbook of Energy Utilization in Agriculture*, ed. Pimentel, D., pp. 27–34. Boca Raton, FL: CRC Press. Inc.
10. Sloggett, G. 1992. Estimating energy use in world irrigation. Energy in Farm Production. *Energy in World Agriculture*, Vol. 6., ed. Fluck, R. C., pp. 203–218. Amsterdam: Elsevier.
11. Fluck, R. C. and C. D. Baird. 1982. *Agricultural Energetics*, pp. 87–120. Westport, Connecticut: AVI.
12. Ortiz-Cañavate, J. 1994. Characteristics of different types of gaseous and liquid biofuels and their energy balance. *J. Agric. Eng. Res.* 59:231–238.

### 2.1.2 Energy Saving in Crop Production

*J. L. Hernanz and J. Ortiz-Cañavate*

***Factors Affecting Tractor Efficiency***

Tractors are the most important machines in agriculture, and their performance is of the greatest importance in achieving high efficiency in crop production.

*Engines*

The latest developments in the diesel engines of tractors are related to:

- higher pressure of the injection pumps (~1000 bar),
- integrated pump-nozzle units,
- electronic injection control,
- a broad introduction of turbocharging and intercooling.

Unfortunately, the turbocharger heats the air going into the engine. The heat reduces the density of the air. Intercooling reduces the temperature of the air inlet in the combustion chamber, which thereby increases the air density, which in turn increases the power output. In some tractor turbocharged diesel engines, an intercooler with a water-jacket coolant is used. Such an installation limits the aftercooling to ~25°C above the water-jacket temperature. Using air to cool the air from the turbocharger is more effective in lowering the temperature; however, the size of an air-cooled intercooler usually discourages its application for mobile equipment.

The efficiency of internal-combustion engines can be described satisfactorily by engine performance maps. These maps show lines of equal specific fuel consumption and consequently constant engine efficiency, within the diagram of torque/engine speed. They can be used to control engines by use of microcomputers to achieve optimal conditions of operation (power–torque–fuel consumption). For most agricultural tractors, performance maps are available from OECD Tests Reports.

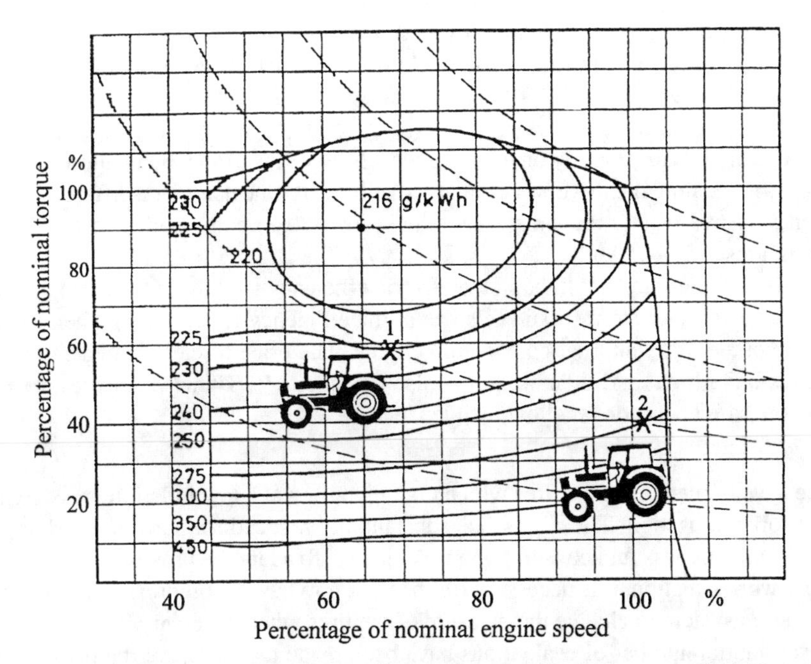


**Figure 2.2. Tractor running at the same power with different amounts of specific fuel consumption.**

As an example, in Fig. 2.2 a tractor is considered: It runs at the same speed with 40% of maximum power in two different gears. At point 1 fuel consumption is 225 g/kW h and at point 2, it is 275 g/kW h. If the tractor has a nominal power of 80 kW, the fuel saved working at point 1 in relation to point 2 is

$$F = 0.4 \times 80 \times (275 - 225) = 1600\,\text{g/h}.$$

If we assume for diesel fuel a density of 850 g/L, we find that

$$F = 1600/850 = 1.88\,\text{L/h}.$$

Therefore there is a substantial saving in fuel when the proper gear is used to obtain the optimum position in the engine performance map. In the case of figure 2.2, when the engine is run at a high speed (point 2), fuel consumption is greater than that at point 1 and also the life of the engine is reduced.

*Transmissions*

In modern tractor design there is a change from synchronized to power-shift transmissions.

Power shifting means changing from one gear ratio to another without the use of a clutch or any interruption of power. Power-shift transmissions save energy, allowing us to change gears without having to disengage the power coming from the engine. In general, planetary gears are used in power-shift transmissions. Recently, reversers for all speeds have been also introduced in medium-sized tractors of 60–90 kW [1].

Hydrostatic transmissions are used in special agricultural equipment, but not so much in standard tractors. In some designs, the hydraulic motor is separated from the pump, and motors are located directly on the drive wheels. Such an arrangement allows us more flexibility in locating the drive wheels and also eliminates the need for a differential. The efficiency of hydrostatic transmissions is low (50%–70%) compared with mechanical transmissions (85%–95%).

Electric transmissions: Although the electric efficiency is high (95%–98%), the production of electricity from fossil fuels has a low efficiency (30%–35%) when used as a power source of an internal-combustion engine. The price of electric generators and motors is quite high, and their weight is also considerable. Their advantages are easy regulation, safe operation, and low maintenance.

*Tires*

The development of radial tires with a high ply rating at a low inflation pressure may reduce soil compaction. The slip is also substantially reduced for a given drawbar pull, and consequently less fuel consumption is achieved. To adapt the pressure of radial tires to every work situation, it is necessary for the tractor to have a compressor and a central tire inflation system to change the pressure of the tires when necessary.

A substantial number of evaluations have been made that compare the performance of bias and radial agricultural tires on both two-wheel- and four-wheel-drive tractors. Significant performance improvements (wheel slip reduction of 6%–7%, drawbar power increase of 10%–12%, and fuel savings of up to 20%) have been achieved in both types of tractors when radial tires at low inflation pressures are used instead of bias tires [2].

*General Considerations in Tractor Operations*

Several recommendations to achieve greater efficiency in tractor applications are considered [3]:

a. ***For Heavy Traction Works.***
   - The axles of the tractor should be ballasted with different weights and the tires should also filled with water in such a way that the total weight limit is not surpassed and the soil is not compacted.
   - The tractors should be in the four-wheel traction mode and the differential connected.
   - The depth of plowing should be established in such a way that the slip

$$s = 1 - v_a/v_t \tag{2.2}$$

   would be in the 10%–15% range, where $s$ is the wheel slip, $v_a$ is the actual travel speed, and $v_t$ is the theoretical wheel speed.
   - The gear should be chosen in such a way that the real work speed will be as high as possible.

b. ***For Light Traction Works in the Field.***
   - Reduce the weight of the tractor to avoid formation of tracks in the ground.
   - The wheels should be big and wide or use twin wheels.

- Regulate the right (low) pressure in the tires.
- The width of the implement and the weight of the tractor should be adapted in such a way that the slip does not surpass 10%.

***c. For Jobs Done with Mounted Machines (Like Fertilizer Spreaders, Sowing Machines, or Sprayers).***

- The weight of the machine and that of the tractor should be maintained as low as possible to reduce rolling resistance.
- If narrow wheels are used to work on lines, the wheels should have as big a diameter as possible and low pressure.
- In this last case, always try to drive in the same track to affect only a small percentage of the soil surface.

***d. For Transport Work.***

- Set up high pressure on the tires to reduce rolling resistance (it is advantageous to have a compressor on the tractor).
- Disconnect the four-wheel drive. Only for special heavy traction forces (as in a big slope) should the four-wheel drive be connected during the time needed.

In general, it is important to attune the combination of tractor and machine. When the machine is too small for a tractor, much fuel is unnecessarily wasted to drive a large tractor. The engine load will be relatively low and the fuel consumption rather high. For this reason it is important to adapt the capacity of the implements to the real power of the tractor.

***Energy Saving in Tillage Systems***

Tillage may be defined as a modification of soil structure due to the mechanical work of implements. This work involves large amounts of energy necessary to cut, break down, invert soil layers, reduce clod size, rearrange aggregates, etc.

The main objective of tillage is to obtain the best soil conditions for crop growing and development. These conditions differ from site to site, depending on soil and weather conditions.

Implement actions on soil require direct use of energy, mainly fuel, transformed to mechanical energy by means of tractor engines. The energy is conducted from the engine to the implement through several components in which important losses can be expected.

Overall efficiency ($\eta_{\text{OE}}$) in the use of fuel energy is the result of multiplying partial efficiencies at different steps (engine, transmissions, wheels, and implement):

$$\eta_{\text{OE}} = \eta_E \eta_T \eta_{\text{TW}} \eta_I \tag{2.3}$$

where $\eta_E$ is the engine efficiency (0.3–0.4), $\eta_T$ is the transmission efficiency (0.8–0.9), $\eta_{\text{TW}}$ is the tractive wheel efficiency (0.5–0.7), and $\eta_I$ is the implement efficiency (0.4–0.8).

These values allow establishing the overall efficiency ranges from 2.5% to 10%, depending on field operation. In many cases, tillage for seed-bed preparation needs at least two or three steps so that $\eta_{\text{OE}}$ is close to 1%, according to Chancellor [4], who estimates 0.25%. Overall efficiency in all operations may be stated as a relationship between the energy directly applied on the soil to obtain a desired particle size and the energy of the fuel required for obtaining the same effect. The direct measurements to assess direct energy in breaking down soil samples include the drop shatter, the crushing test, etc. [5].

*Tillage and Soil Compaction*

An important objective for tillage is to reduce soil compaction resulting from natural factors (rain, swelling–shrinkage processes, etc.) and anthropic actions (traffic, some tillage practices, etc.).

Severe soil compaction reduces both air and water permeability, root penetration, nutrient uptake, and crop yield; on the other hand, it increases serious environmental problems such as soil degradation, nutrient loss by run-off, and gasification, and it decreases biological soil activity.

Soil compaction increases energy consumption in tillage operations for two main reasons:

- The specific draft force increases in the same proportion as the bulk density at the same soil moisture content, and higher tractor slippage can be expected. Both variables increase fuel consumption.
- If the soil is cloddy after primary tillage, a greater number of passes on the field will be necessary to achieve the appropriate size and distribution of the soil aggregates in secondary tillage operations.

There are several ways to reduce these constraints [6]:

- Practice controlled traffic.
- Carry out tillage operations at the optimum soil moisture content for each implement.
- Reduce the tire pressure.
- Use four-wheel-drive tractors.
- Increase the capacity of field operations.
- Improve the organization and planning of field transport.

*Conventional Tillage*

Conventional tillage may be defined as traditional practices carried out in seed-bed preparation. In most cases, it is associated with the use of a moldboard plow or a disk plow for primary tillage. Both implements invert soil layers to bury crop residues, weeds, fertilizers, and loose soil to increase its porosity.

Chisels and heavy cultivators also can be considered implements for primary tillage and even heavy disk harrows. Deep tillage varies according to the implements, soil type, weather conditions, and crop requirements. With moldboard and disk plows, it ranges from 20 to 35 cm and even more; with chiseling and cultivating, deep tillage penetrates ~20 cm, and with disk harrowing up to 15 cm.

Secondary tillage is required for creating optimum soil conditions for sowing and for advancing germination and plant emergency. Implements commonly used must be classified as nonpowered, for example, light cultivators, vibrocultivators, light harrows, rollers, etc., and powered, such as rotary tillers, oscillating harrows, rotary harrows, etc. In any case, a seed-bed layer can reach to 10 or 12 cm on a average.

Two main factors affect energy consumption in tillage practices [7]:

***a. Soil Type and Tillage Conditions.*** The soil moisture content has a big influence on the mechanical properties of cohesion, friction, and adhesion. Clayey soils are cohesive, and this variable is strongly conditioned by moisture content. Under dry conditions cohesion can reach values up to ten times higher than that under wet conditions, so that the specific draft force increases in the same proportion. On the other hand, when the tilth is cloddy, more energy is required for breaking up clods into fine aggregates. Under wet conditions soil is sticky and the adhesion forces are predominant, making the tilth difficult.

Sandy soils have low cohesion and high friction, so the moisture content does not have a significant influence on the specific draft force. The optimum moisture content of most soils is close to its plasticity limit.

***b. Implements.*** Energy consumption is related to the working action of tools. For primary tillage, moldboard plowing requires more energy than chiseling. Both systems mean two different ways to produce soil breakup. By plowing, soil is cut, lifted, turned, and projected onto a side furrow. The specific draft force depends on the moldboard type, the travel speed, the bulk density, and the soil moisture content. By chiseling, coulters push soil forward and laterally to produce its breakup. The affected soil section is similar to an inverse triangle in optimum soil moisture. This implement requires ~30%–40% of the energy required by the moldboard plow under the same tillage conditions.

For secondary tillage, nonpowered implements require less energy than powered ones but require more field operations for obtaining the same size of aggregates. Powered implements need extra energy in power take-off to propel their own tools. A great advantage of these powered implements is their possibilities with associated implements to reduce field operations. Table 2.9 shows specific works for the main tillage implements.

*Minimum Tillage*

Minimum tillage means either the reduction of the depth in tillage or the number of field operations. For primary tillage, chisels, cultivators, and disk harrows are implements commonly used. Minimum tillage would mean reducing energy consumption by ~50%, if we take into account that deep tillage reaches down to 15 or 20 cm. In some cases of minimum tillage, weeds are not accurately controlled, so a herbicide might be necessary.

**Table 2.9. Specific work (kJ/m$^3$) for the main tillage implements for several soil types**

| | Soil Type | | | | |
|---|---|---|---|---|---|
| Implement | Sand | Sandy Loam | Silt | Clay Loam | Heavy Clay |
| Moldboard plow | 30 ± 5 | 40 ± 5 | 60 ± 8 | 80 ± 10 | 120 ± 20 |
| Chisel plow | 20 ± 4 | 27 ± 4 | 40 ± 6 | 55 ± 7 | 80 ± 10 |
| Cultivator | 18 ± 3 | 24 ± 4 | 36 ± 5 | 48 ± 6 | 65 ± 8 |
| Disk harrow | 18 ± 3 | 22 ± 4 | 30 ± 4 | 40 ± 5 | 55 ± 7 |
| Rotary tiller (spike rotor) | 150 ± 15 | 175 ± 20 | 210 ± 25 | 250 ± 30 | 320 ± 40 |
| Oscillating harrow | 45 ± 8 | 60 ± 10 | 88 ± 12 | 118 ± 15 | 175 ± 20 |
| Powered harrow (vertical rotor) | 120 ± 10 | 135 ± 15 | 165 ± 20 | 190 ± 25 | 250 ± 30 |

Adapted from Perdok. U. D. and G. van de Werken [8].

Secondary tillage is carried out with the same implements. Seed-bed preparation may be achieved in one or two passes in which a combination of chisel, powered implement, roller and drill, or planter is frequently used, mainly in humid regions when time for sowing is short. Minimum tillage can be included in the conservation tillage systems on the condition that a minimum of 30% of soil surface should stand covered with crop residues after seeding.

*Mulch Tillage*

For areas where erosion is a problem, mulch tillage is an accurate practice for its prevention; therefore this technique is included in conservation tillage systems [8]. Soil is loosened down to 10 or 13 cm in most cases by chisels, cultivators, harrows, and V-sweep implements. The latter provides good weed control. Other implements also used are rod weeders and stubble treaders.

If soil presents compaction problems, a deep loosener (paraplow) is used to increase macroporosity and water infiltration. With mulch tillage, a minimum of 33% of the previous crop residues is left on the soil surface.

*Strip Tillage*

Strip tillage is performed in bands where seeds are planted. Crop residues cover the undisturbed soil surface, which is ~50% of the whole field surface. Planting may be carried out in flat or ridged rows associated with implements such as rotary, sweep, and disk strip tillage. Strip units are in front of each row planter.

*No Tillage*

No tillage is also called direct drilling or direct planting. The soil is disturbed in only a narrow band, ~3–8 cm wide, by the openers of seeders, mainly coulters and disks. In place of tools, herbicides, which are applied before seeding, control the weeds. This system requires the least amount of energy compared with the other systems mentioned, but it is not possible to apply it in all cases. Good natural soil drainage is the first condition required for producing accurate crop growing. On the other hand, weeds, pests, and disease must be controlled as soon as possible; otherwise they will seriously limit the success of this system.

Large amounts of crop residues on the soil surface may be a limiting factor to seeding quality, mainly with drills, if they have not been previously accurately treated (chopping, spreading, etc.). Both drills and planters in direct drilling or planting need additional weight because the soil has not been previously loosened by tillage implements. In humid areas, where high rates of crop residues can be expected, soil disk openers are frequently used. In dry regions, with hard soils and low crop residue content, hoe (knife) openers allow a good seeding and are lighter than disk openers. After seeding, ~70%–90% of the soil surface is covered with previous crop residues. Table 2.10 shows overall energy consumption per hectare for different tillage systems.

***Fertilizer Management***

As stated in Section 2.1.1 in the subsection on energy inputs, both the fuel and the mineral fertilizers require the highest amounts of energy compared with the other inputs in crop production.

**Table 2.10. Estimated values of fuel consumption for different tillage systems**

| Tillage Systems | Overall Energy[a] (MJ/ha) |
|---|---|
| Conventional tillage | |
| Moldboard plow + disk harrow or cultivator (two passes) + drill | 2200 ± 350 |
| Moldboard plow + powered implement/roller + drill | 1900 ± 300 |
| Chisel plow (two passes) + disk harrow or cultivator + drill | 1400 ± 250 |
| Chisel plow + disk harrow or cultivator + drill | 1100 ± 200 |
| Chisel plow + powered implement/roller + drill | 1300 ± 250 |
| Minimum tillage | |
| Chisel plow (15-cm depth) + disk harrow or cultivator + drill | 1000 ± 200 |
| Cultivator (two passes) + drill | 800 ± 150 |
| Disk harrow or cultivator + drill | 620 ± 100 |
| Powered implement/roller/drill | 720 ± 100 |
| Mulch tillage | |
| Paraplow + V-sweep plow + cultivator + drill | 1450 ± 250 |
| V-sweep plow + seed conditioner + drill | 860 ± 150 |
| Cultivator/rod weeder + drill | 630 ± 100 |
| Strip tillage | |
| Rotary strip tillage/flat planting | 400 ± 75 |
| Sweep strip tillage/flat planting | 350 ± 50 |
| Disk strip tillage/ridges planting | 400 ± 75 |
| No tillage | |
| Disk planter | 300 ± 50 |
| Disk drill | 400 ± 75 |
| Coulter drill | 450 ± 75 |

[a] Fuel equivalent is considered to be 47.8 MJ/L.
*Sources*: [9] and [10].

Accurate fertilizer management can improve energy use, mainly if we refer to nitrogen. This nutrient has two important characteristics: the high energy content incorporated during manufacture and its ability to be lost quickly. Therefore it becomes necessary to increase its profitability with the crops and reduce losses by improving management practices.

*Fertilizing Rates*

In most cases, a continual assessment of soil nutrient levels allows us to know what the optimum rates to be applied are. Experiments with different fertilizer rates lead us to make certain recomendations to the farmers to attain the greatest efficiency. From the energetic point of view, the procedure is similar to the economic evaluation. Crop yield can be characterized as a function $y = f(x)$, quadratic or exponential, of the nutrient supply $x$.

The most profitable application rate of a fertilizer is the optimum energetic rate, which, by definition, occurs when the energy involved in the last increment applied equals the value of the last additional yield that resulted from an increment of fertilizer [11]. To obtain this rate it is necessary to solve

$$d[(E_c/E_f)y]/dx = 1, \tag{2.4}$$

where $y$ is the crop yield (in $kg \cdot ha^{-1}$), $x$ is the fertilizer rate supply (in $MJ \cdot ha^{-1}$), $E_c$ is the gross energy intensity of the crop (in $MJ \cdot kg^{-1}$), and $E_f$ is the energy intensity of the fertilizer (in $MJ \cdot kg^{-1}$). This procedure allows us to establish comparisons among different management practices for any crop with the same soil and weather conditions when a specific nutrient is considered as a limiting factor.

*Loss Reduction*

Fertilizer loss means, in many cases, not only crop yield reduction, but economic and environmental problems such as contamination of ground waters, aquifers, lakes and rivers. As mentioned above, nitrogen is the most delicate nutrient to be managed accurately. Besides the problems already mentioned, losses by gasification ($NO_2$, $N_2$, $NH_3$) can be expected.

The main processes involved in nitrogen losses are erosion, leaching, denitrification, and ammonification. Dusbury et al., cited in [12], showed that for every 100 kg/ha of nitrogen applied in the USA, 50% is harvested in the crop, 25% is leached, 5% is lost by run-off, and 20% is lost in primary denitrification. Table 2.11 shows the influence of some factors in the different processes in which nitrogen is lost.

Reducing losses can be achieved by accurate fertilizer management if the following aspects are considered [13]:

***a. Fertilizer Location.*** In many cases, fertilizers are distributed by broadcast applications. This method is adapted to high rates of a specific nutrient when it is a limiting factor. Equipment for granular application allows high field capacities because rotary spreaders broadcast fertilizers more than 24 m wide. In this case the control of the overlap is important in getting a uniform distribution on the field. Drop-type and pneumatic equipment can be used for both broadcast and banded applications. They work with smaller widths than rotary spreaders but with greater uniformity. In conservation tillage systems, important amounts of nitrogen are immobilized by crop residues on soil surface. Sideband applications in which nutrients are located close to, at the side of, or below the seed increase the chances that the roots of the plants absorb the fertilizer. Efficiency is increased because losses are reduced. This method is succesfully used in no-tillage systems. In most cases, seeding and fertilizing are carried out in one operation with combined equipment.

***b. Time of Application.*** Much loss can be avoided if the fertilizer is applied in several sessions, especially in humid areas where leaching is a serious problem. Fertilizer is more efficient when it is supplied just before the time of greatest plant consumption.

***c. Crop Rotation with Legumes.*** Legumes present an important source of increasing the nitrogen level of soil by symbiotic fixation. This means a reduction in mineral rates and consequently a lower energy in crop production. For alfalfa, fixed nitrogen can be higher than 200 kg/ha; for clover, it ranges from 115 to 200 kg/ha; for vetch, 80 to 100 kg/ha; for peas, 70 to 80 kg/ha; for beans, 60 to 90 kg/ha; for soybean, 50 to 100 kg/ha, etc.

**Table 2.11. Influence of soil and management practices on main nitrogen losses**

| | Nitrogen Losses[a] | | | |
|---|---|---|---|---|
| Factor | Erosion | Leaching | Denitrification | Ammonification |
| Soil type | | | | |
| Light | + | +++ | + | + |
| Medium | ++ | ++ | ++ | + |
| Heavy | +++ | + | +++ | ++ |
| Field slope | | | | |
| Low | + | ++ | ++ | ++ |
| High | +++ | + | + | + |
| Tillage system | | | | |
| Conventional | +++ | +++ | ++ | + |
| Minimum | + | ++ | ++ | + |
| Zero | + | + | + | ++ |
| Soil compaction | | | | |
| Low | + | ++ | + | + |
| High | +++ | + | +++ | ++ |
| Soil moisture | | | | |
| Low | + | + | + | ++ |
| High | +++ | +++ | +++ | +++ |

[a] Legend: + + +, high influence; ++, moderate; + Low; – no influence.
Adapted from [11] and [12].

***Biocide Management***

Nowadays in crop production it would be difficult to attain such high yields if weeds, pests, and disease were not controlled. The overall energy required in crop protection is lower than other inputs, although it is of great significance because yields can be seriously affected if control is not accurately performed.

Energy reduction is not the main factor to be taken into account in planning crop protection, because environmental and economic considerations are more important. To reduce the negative effects of both weeds and parasites on crops, different and complementary strategies, such as preventive, physical, cultural, biological, and chemical ones, must be considered [14].

*Weed Control Strategies*

Weeds compete with the crop plant for light, water, and nutrients so that their control in earlier stages of growth, or even before their emergence, is of prime importance to save money and energy and to reduce the environmental effects. For weed control, several methods must be considered.

***a. Preventive Weed Control.*** These practices permit farmers to avoid further problems of weed infestation. At the time of sowing, weed seeds must be separated from crop seeds by accurate cleaners. Although certified seed is more expensive than seed recovered from a previous crop or from another farm, the risk of infestation is greatly reduced with it. Another important aspect is keeping field borders free of weeds.

***b. Physical Weed Control.*** Mechanical weed control is the oldest system used by growers. In primary cultivation, the moldboard plow is the more efficient implement for weed control when the weeds are completely buried within the soil, mainly in humid areas, where a helicoidal bottom is commonly used. Nevertheless, a high amount of energy is required. On the other hand, soil layer inversion improves the conditions of weed seed germination when these seeds are located near the top of the soil.

Chiseling and cultivating are less efficient in weed control than the moldboard plow because the weeds are partially buried. To improve their control, V-sweep tools are mounted on the shanks and in many cases even a rod weeder is attached behind the implement. An interrow cultivator is efficient and easy to manage. Its energy consumption is low and allows for weed control for most of the growing season, depending of the crop type.

Perennial weeds present serious problems when controlled by tillage, especially rhizomes, which can be increased by the sectioning of some mechanical practices such as disking.

Another physical manner of weed control is burning them by propane combustion close to the crop plants. This system is expensive, requires a high amount of energy (150–200 kg/ha of propane), has a travel speed of ~1 to 2 km/h, and weed control is low. Today this method does not seem appropriate for most of the crops.

***c. Cultural Weed Control.*** This refers to the different techniques used in crop management. Cultural practices can be classified into the following categories [15]:

- Competitive or allelopathic crops. Each crop differs in its competition with weeds according to its growth rate and the spacing and the life cycle of seeds. In this latter case, for winter crops the activity of weeds is reduced comparatively with that of spring and summer crops. Even for the same species, the cultivars have significant influence on the competitiveness.
- Allelopathic crop residues. Many mulch crops such as rye, wheat, oats, and barley contain allelochemicals, which deter weed germination.
- Intercropping. This management is often performed in the tropics. Two crops, such as soybean and corn, are overlapped at some point during their growing.

***d. Biological Weed Control.*** A biological agent, normally a fungal, is released on a field to control annual weeds before they cause losses in crop yield. These agents are called mycoherbicides. Insects are also used as biological agents for weed control.

***e. Chemical Weed Control.*** Since the beginning of their use 50 years ago, chemicals are the most widely used products for weed control today. Continuous research is being made, both to improve their effectiveness with new active ingredients and to develop new application technologies. Nevertheless, environmental problems are arising as a consequence of inappropriate utilization.

To reduce the quantity of active ingredients, it is necessary to act in earlier stages of weed growth. If application is delayed, the rate of herbicide could be increased 100% more in relation to the amount needed at the optimum time. To attain efficient weed control, the following aspects must be considered:

- The best concentrations in water of a market product range from 1% to 2% for conventional applications with sprayers.
- A flat nozzle is more accurate than a turbulence nozzle for herbicide application.
- The operating liquid pressure is generally 100 to 250 kPa.
- Travel speed must be maintained constant throughout the treatment.
- Nozzles, when operating at the same pressure, should be rejected when the flow rate per unit of time is 15% higher than the values specified by the manufacturer.
- Screen protectors against drift allow an increase in the effectiveness of control.

*Pest and Disease Control*

As stated for weed control, pest and disease control practices follow similar trends for improvement. The term integrated pest management is defined as "A strategy of pest containment that seeks to maximize the effectiveness of biological and cultural control factors, utilizing chemical controls only as needed and with a minimum of environmental disturbance." This means that different strategies are involved with the objective of chemical reduction [16].

***a. Cultural Control.*** Cultural control involves management systems in which the pest's life cycle is the most vulnerable. By conventional tillage, many insects are buried into the soil, breaking down their natural cycle. This can also be achieved by straw burning, but in this case, ecological disadvantages predominate over benefits. Crop rotation also allows some control of pests and diseases but its efficiency is limited for mobile insects and pathogenic spores; it is more accurate on rootworms and wireworms.

***b. Biological Control.*** Many species exert a natural control on insects, pathogens, and weeds. This creates trophic linkages that maintain the crop injury level beneath the economic threshold. Chemicals, in many cases, have a negative influence because these ecosystems are broken down. For biological control, knowledge of the different species and ecology is important.

***c. Biotechnological Control.*** In recent years, biotechnology has promoted the development of new resistance mechanisms in crops by genetic engineering, but we are still in an early stage of knowledge. Maybe in the near future this will be a powerful tool for pest and disease control.

***Irrigation***

Energy saving in irrigation systems is influenced by the following factors [17].

*Power Unit Efficiency*

In diesel engines, both torque and speed have to be regulated to obtain the lowest specific fuel consumption. In most engines the speed is nearly 60%–75% of its maximum value and the torque is ~75%–80% of its maximum. When the unit pump engine is selected, the optimum speed of the pump must be known in order to achieve the best transmission between pump and engine.

For electric-powered units, efficiency depends on the generating plant, for which thermal energy presents an average efficiency of ~20%, nuclear 25%, and hydroelectric

50%. Losses in electric transmission lines are ~15% and efficiency in converting electric in mechanical power can reach 90%.

*Lifting Equipment*

Depending on the pump type and its performance, there are different efficiencies. In low- to medium-capacity centrifugal pumps, efficiency ranges from 50% to 80%. These are used by individual farmers. In high-volume propeller, mixed-flow, and turbine pumps, for which the discharge rate is higher than 500 $m^3$/h, efficiency can be up to 90%. These kinds of pumps are normally operated to supply water to several farms.

*Conveyance and Distribution*

Large amounts of water can be lost if infrastructures are not maintained in good condition. These losses are seepage, evaporation, and spillage. Efficiency ranges from 50% to 98%. The highest values are for drip irrigation.

*Irrigation Systems*

The irrigation gravity system presents the lowest efficiency, mainly because of wild flood; losses can reach up to 80%. For sprinklers, depending on this system, efficiency ranges from 50% to 80% (the lowest value refers to big gun equipment). The most efficient systems are drip and trickle irrigation, for which an efficiency of up to 90% can be obtained. These last systems allow farmers to incorporate nutrients into the water so its effectiveness is also improved.

***Harvesting***

The efficient use of harvesting equipment is associated with the application of global position system (GPS) techniques and the timeliness of the operation.

The global position system, or more accurately, the differential global position system (DGPS), uses satellite navigation with a fixed reference point located near where the machine operates. It can locate the machine in the field with a precision of a few centimeters.

For cereal harvesting, the grain collected by the combine is registered by a continuous weight sensor, and consequently a yield map can be obtained. This map gives information for optimizing the inputs that have to be set into the crop: seed, fertilizer, and pesticide. In this way, much energy can be saved in crop production. This method can be applied to any crop.

According to the ASAE [18], timeliness can be defined as the ability to perform an activity at such a time when quality and quantity of product are optimized. Energy productivity may be substantially increased by an improvement in timeliness. Yields are increased with better timeliness for different crop operations, including harvesting.

Losses of timeliness can be calculated by

$$L(\text{kg/ha}) = KDY/Z = lY(\text{kg/ha}). \tag{2.5}$$

$K$ is defined as the timeliness coefficient (per day), $D$ is the number of days within the time span in which the operation should be accomplished, $Y$ is the yield per area (in $\text{kg} \cdot \text{ha}^{-1}$), $l$ is the coefficient of losses related to unity (adimensional), and $Z$ is a factor

**Table 2.12. Examples of values for the timeliness factor**

| Crop and Location | $K(d^{-1})$ |
|---|---|
| Shelled corn, Iowa | 0.003 |
| Soybeans, Illinois | 0.006–0.010 |
| Wheat, Ohio | 0.005 |
| Rice, California | 0.009 |
| Cotton, Alabama | 0.002 |
| Sugar cane, Queensland, Australia | 0.002–0.003 |
| Hay making, Michigan (June) | 0.018 |

*Source*: [18].

(adimensional) that is equal to 4 if the operation can be balanced evenly at approximately the optimum time and equal to 2 if the operation either commences or terminates at the optimum time.

Values of $K$ have to be determined for each crop in a specific location. Table 2.12 gives information about values of $K$ in different places derived from crop research reports.

For example, if rice is harvested in California during $D = 25$ days starting at the optimum day (e.g., 20 September; therefore $Z = 2$), losses produced for the operation duration are

$$l = KD/Z = 0.009 \times 25/2 = 0.113 = 11.3\%.$$

To adapt the size of machines to the amount of surface cultivated, early, medium, and late varieties of crops should be programmed in such a way that they can be planted and harvested, one after the other, at their optimum or the most near-optimum date to reduce timeliness losses.

***Transportation***

In developed countries, approximately one half of the energy consumed in the food system goes to transportation, marketing, and household preparation. In general, transportation of agricultural commodities from their open-air production to the consumption point (even if the distance is several thousand kilometers) is more economical from the point of view of energy and real costs than producing them in greenhouses [19].

Table 2.13 lists the energy intensities for different transport modes that can be used to transport agricultural products.

Some measures to save energy in agricultural transportation could be

- The choice and the purchase of the most economical vehicles for the load to be carried and with the minimum fuel consumption.
- Proper maintenance of trucks, tractors, and trailers.
- Good driving habits.
- Good planning to reduce trips.

Vehicle loading is an important factor for efficient energy use. Loading the vehicle up to its maximum capacity reduces energy intensiveness. This is also important in relation to energy saving in transport: the use of concentrated fertilizers and pesticides,

**Table 2.13. Energy intensities for different transport systems**

| Transport System | Energy Intensity ($MJ\ t^{-1}\ km^{-1}$) |
|---|---|
| Boats | 0.3–0.8 |
| Railroads | 0.4–0.9 |
| Trucks | 1.6–4.5 |
| Airplanes | 1–30 |

*Source*: Adapted from [19].

the removal of moisture from commodities, and the processing of commodities on the farm to eliminate residues that do not have to be transported.

## References

1. Renius, K. T. 1992. Tendencies in Tractor Design: *Proc. Int. Conference on Agricultural Engineering AgEng'92*: pp. 5–6. Uppsala (SW).
2. Stout, B. A. and M. McKiernan. 1992. New technology—energy implications. *Energy in Farm Production*. Energy in World Agriculture, Vol. 6, ed. Fluck R. C., pp. 131–170. Amsterdam: Elsevier.
3. Kutzbach, H. D. 1989. *Allgemeine Grundlagen Ackerschlepper Fördertechnik*, pp. 168–169. Hamburg: Paul Parey.
4. Chancellor, W. 1982. Energy efficiency of tillage systems. *Soil Water Conservation* 37:105–108.
5. Dexter, A. R. 1988. Advances in characterization of soil structure. *Soil Tillage Res.* 11:199–238.
6. Hakansson, I. 1995. Prediction of soils from mechanical overloading by establishing limits for stresses caused by heavy vehicles. *Soil Tillage Res.* 35:85–99.
7. Srivastava, A., C. Goering, and R. Rohrbach. 1993. *Engineering Principles of Agricultural Machines*, pp. 149–220. St. Joseph: ASAE.
8. Perdok, U. D. and V. de Werken. 1983. Power and labour requirements in soil tillage—a theorical approach. *Soil Tillage Res.* 3:3–25.
9. FAO, 1989. *Energy Comsumption and Input-Output Relations of Field Operations.*, eds. Pick, O. and V. Nielsen. REUR Technical Series 10:9–87.
10. Frye, W. W. 1995. Energy use in conservation tillage. Farming for a Better Environment: pp. 31–33. Ankeny, Iowa: Soil and Water Conservation Society.
11. Fox, R. H. and V. A. Bandel. 1986. Nitrogen utilization with no tillage. *No-Tillage and Surface-Tillage Agriculture. The Tillage Revolution*, eds. Sprague, M. A. and G.B. Triplett, pp. 117–149. New York: Wiley.
12. Soane, B. D. and C. van Ouwerkerk. 1995. Implications on soil compaction in crop production for the quality of the environment. *Soil Tillage Res.* 35:5–22.
13. Buchholz, D. D. and L. S. Murphy. 1987. Conservation of nutrients. *Energy in Plant Nutrition and Pest Control*. Energy in World Agriculture, Vol. 2, ed. Helsel, Z. R., pp. 101–132. Amsterdam: Elsevier.

14. Barrett, M. W. and W. Witt. 1987. Maximizing pesticide use efficiency. *Energy in Plant Nutrition and Pest Control*. Energy in World Agriculture, Vol. 2, ed. Helsel, Z. R., pp. 235–267. Amsterdam: Elsevier.
15. Regnier, E. E. and R. R. Janke. 1990. Evolving strategies for managing weeds. *Sustainable Agriculture Systems*. eds. Edwards, C. A., R. Lal, P. Madden, R. H. Miller, and G. House, pp. 174–203. Ankeny, Iowa: Soil and Water Conservation Society.
16. Luna, J. M. and G. House. 1990. Pest management in sustainable agriculture systems. *Sustainable Agriculture Systems*. eds. Edwards, C. A., R. Lal, P. Madden, R. H. Miller, and G. House, pp. 157–174. Ankeny, Iowa: Soil and Water Conservation Society.
17. Slogget, G. 1992. Estimating energy use in world irrigation. *Energy in Farm Production*. Energy in World Agriculture, Vol. 6, ed. Fluck, R. C., pp. 203–218. Amsterdam: Elsevier.
18. *ASAE. Standards*. 1996. Engineering Practices and Data. St Joseph, Mich.: ASAE.
19. Fluck, R. C. and C. D. Baird. 1982. *Agricultural Energetics*. Connecticut: AVI.

### 2.1.3 Energy Saving in Animal Housing

*T. Jungbluth*

In a constant warm climate under confinement, the most sensitive animals are pigs and poultry (see 2.1.2). The primary heat source for animal houses is the animals' metabolism. Most of the time, the heat production of animals allows an optimal temperature control by ventilation. From the energy balance from animal houses, heat losses occur through walls and by air exchange:

$$Q_A + Q_H - Q_{\mathrm{Tr}} - Q_V = 0,$$

where $Q_A$ is the heat production from animals, $Q_H$ is the heat production from the heating system, $Q_{\mathrm{Tr}}$ is the heat loss by transmission, and $Q_V$ is the heat loss by ventilation.

Houses for pigs and poultry are equipped mainly with forced-ventilation systems with a constant use of electricity (see 2.1.2, 4.2). With a low outside temperature, however, the animals' heat production will not offset the heat loss from building construction and ventilation, so additional heating will be needed. Possibilities for energy saving are

- reducing transmission heat loss,
- reducing heat loss through ventilation,
- heat recovery systems (heat exchangers, pumps)
- reducing energy demand of forced ventilation (low air velocity, correct design).

Energy saving in product storage (milk) and the cleaning of milking machines are described in Volume 2 of this handbook.

#### *Reducing Transmission Heat Loss*

In normally insulated animal houses ($k \sim 0.6$ W/m$^2$ K), only ~20%–30% of the heat losses are transmission losses. So the reduction of those losses by improving the insulation of existing buildings might be expensive. Table 2.14 gives examples for thermal properties of different insulation materials.

**Table 2.14. Thermal properties of different building materials**

| Building Material | Specific Weight ($kg/m^3$) | Heat Conductivity (W/m K) | Heat Capacity ($Wh/m^2$ K) |
|---|---|---|---|
| 5-cm rigid foam, polystyrene | 30 | 0.035 | — |
| 5-cm mineral wool | 110 | 0.035 | 0.2 |
| 20-cm spruce | 600 | 0.13 | 70 |
| 24-cm gas concrete block | 600 | 0.16 | 43 |
| $36^5$-cm light brick | 800 | 0.27 | 84 |
| 74-cm vertical core brick | 1200 | 0.51 | 219 |
| 300-cm concrete | 2400 | 2.1 | 1834 |

**Table 2.15. Comparison between an earth heat exchanger and a ground-water cooling system [1]**

| Air Flow ($m^3/h$) | Earth Heat Exchanger System[a] (kW) | Ground-Water Cooling System Test Construction[b] (kW) |
|---|---|---|
| 9600 | $24 \times 0.67 = 16.08$ | 27.10 |
| 7200 | $24 \times 0.62 = 14.88$ | 24.75 |
| 4800 | $24 \times 0.48 = 11.52$ | 19.99 |

[a] Tube length, 10 m; number of tubes, 24; outside temperature, 26°C; ground temperature, 14°C.

[b] Outside temperature, 26°C; ground-water temperature, 10°C; ground-water flow, 2 $m^3/h$.

***Reducing Heat Loss Through Ventilation***

Preheating of ventilation air with earth-tube heat exchangers in winter may lead to a reduction in the heating energy required. Different design examples for pig houses are available. Table 2.15 gives examples for the performance and the cooling capacity compared with a ground-water cooling system tested on a laboratory scale. One should keep in mind that earth-tube heat exchangers could also increase the ventilation rate in winter but reduce it in the summer and so affect the energy demand on the ventilation systems.

To reduce the energy demand of forced ventilation in summer in cooling the warm incoming air, ground-water–air-pipe heat exchangers might be profitably used for pig-fattening houses in subtropical regions.

***Heat Recovery Systems***

Approximately 70%–80% of the heat losses are used to warm the incoming air. Heat exchangers or heat pumps help limit the necessary heat capacity. Heat recovery systems take energy from the exhaust air; air–air heat exchangers transfer it directly to the incoming air at the same temperature level, whereas heat pumps use a compressor and electric energy to gain a higher temperature level. Energy might then be used for purposes other than heating the incoming air. Different types of air–air heat exchangers can be used in animal houses [2]:

- folded heat exchangers made of aluminium with plastic covers,
- foil heat exchangers made of plastic sheets,
- plate heat exchangers made of glass or aluminium,
- pipe heat exchangers made of glass or aluminium.

The efficiency of an air–air heat exchanger without condensation is [2]

$$\eta_d = q_{vi}(t_2 - t_1)/q_{vo}(t_3 - t_1),$$

where $\eta_d$ is the dry efficiency, $q_{vi}$ is the mass flow of incoming air (in kilograms air per second), $q_{vo}$ is the mass flow of outgoing air (in kilograms air per second), $t_1$ is the temperature of incoming air before the exchanger (in degrees Celsius), $t_2$ is the temperature of incoming air after the exchanger (in degrees Celsius), and $t_3$ is the temperature of outgoing air before the exchanger (in degrees Celsius).

The energy efficiency of the exchangers remains almost at a 30%–35% level. Figure 2.3 gives examples of typical efficiency coefficients.

The exhaust air of the animal house is often used as a heat source for heat pump applications. A profitable amount of energy can be recovered by cooling this air. Three types of heat pumps can be designed [2]:

- air–water heat pumps in a compact design,
- air–water heat pumps with a remote condenser,
- water–water heat pumps with an absorber in the animal house.

The heat demand varies with the season. In most cases, recovered energy is sufficient for the hot-water supply of the farmer's house. An addition to the heating system might improve the efficiency of the system.

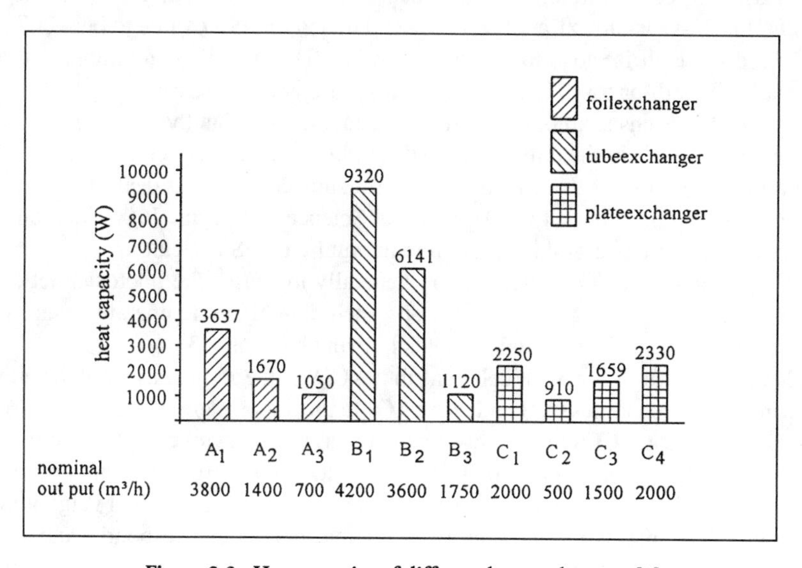


**Figure 2.3. Heat capacity of different heat exchangers [3].**

***Energy Demand of Forced Ventilation***

The energy demand of forced-ventilation systems depends mainly on the air volume and the design of the air duct system, which determines the pressure loss of the system. Typical pressure losses range from 5 to 80 Pa in pig-fattening houses. The average energy demand is between 10 and 30 kWh per pig. A reduction of pressure loss is possible by use of the correct design of inlet and exhaust air guidance system (see Section on forced ventilation).

## References

1. Huang. Y. 1992. Design of a simple cooling system for pig-fattening houses, pp. 125–126. Ph.D. Thesis, Agricultural Engineering. Hohenheim: Max-Eyth-Gesellschaft.
2. Rist, M. *et al.* 1989. Energy recuperation in animal houses: *2nd Report of CIGR Working Groups on Climatization of Animal Houses*: 44–55.
3. Krüger, J. 1985. Investigations on heat exchangers under practical conditions, p. 41, AEL Report 9. Essen: ARBETTSGEMEINSCHAFT FÜR ELEKTRIZITÄTSANWENDUNG IN DER LANDWIRTSCHAFT.

## 2.2 Energy and the Environment

*R. M. Peart*

### 2.2.1 Energy and Carbon Dioxide

The buildup $CO_2$ in the Earth's atmosphere is well documented and predictable. In their widely read book, *Limits To Growth*, Meadows *et al.* in 1972 [1] showed a plot of atmospheric $CO_2$ concentrations based on data from 1958 to 1970 and then extrapolated to 2000. This 25-year-old extrapolated curve predicted the 1997 value to be $\sim$367 parts in $10^6$ (ppm), which is close to the current value. The world's continuing, and even increasing, demand for fossil energy is the cause of this $CO_2$ increase.

The current consensus among scientists and engineers is that the long-term effect of increasing $CO_2$ levels in the atmosphere will produce a significant climate change. The amount and the timing of these changes are still subject to a great deal of uncertainty. In recent study by the Council for Agricultural Science and Technology, Waggoner [2] points out the importance of this problem to the future of US agriculture.

Although increased $CO_2$ concentration generally increases the photosynthetic rate, the predicted climate-change effects on crops, including higher temperatures and shifts in rainfall patterns, are detrimental to yields in many locations [3].

Although the burning of fossil fuels contributes $CO_2$ to the atmosphere, photosynthesis in growing plants consumes $CO_2$ and returns $O_2$ to the atmosphere. If biomass is used as fuel, it also contributes $CO_2$ to the atmosphere, but as a photosynthetic plant it consumes almost exactly the same amount of $CO_2$ during its growth, thus maintaining a zero balance of $CO_2$ (if energy use during processing of the biomass into fuel is ignored).

To study these effects in detail, crop simulation models that respond accurately to climatic variables were used. While these models do not account for diseases and insect problems, they are excellent tools for making comparisons of crop performance under different environmental conditions.

Climate-change effects were included by use of results from general circulation models (GCM's), which have been developed to study world weather dynamics. Crop simulations showed climate-change effects on both rainfed and irrigated yields and also on the amount of water required for irrigation. With these results, the effects on energy inputs into crop production and on the output yields expressed in energy units have been calculated.

This work has been supported by the US Environmental Protection Agency Division of Policy and Planning, by the Southeast Regional Climate Center, Columbia, SC, and by the Florida Agricultural Experiment Station.

***Methodology***

The analysis used two sets of simulation models: first, the results of three simulations of global climate change, and secondly, biological simulations of three different crops. GCM parameters were applied to 40 years of weather data at each location. A monthly average adjustment was applied to each precipitation event and to the daily temperatures and solar radiation. The GCM results used in this study were from the Goddard Institute of Space Studies, by Hansen *et al.* in 1988 [4]; the Geophysical Fluid Dynamics Lab, by Manabe and Wetherald in 1987 [5]; and the United Kingdom Meteorological Office, by Wilson and Mitchell in 1987 [6]. These results are based on a doubled-greenhouse-gas assumption for approximately the year 2030. $CO_2$ concentration was assumed to be 555 ppm versus ~360 ppm today, with increases in the other greenhouse gases, including $CH_4$, adding to the effect. Four 40-year weather data sets were created for each location, the historical baseline, and three climate-change cases.

These weather data sets were used as input to the soybean, the peanut, and the maize crop models: SOYGRO by Jones *et al.* in 1988 [7], PNUTGRO by Boote *et al.* in 1989 [8], and CERES/Maize by Jones and Kiniry in 1986 [9]. DSSAT, a decision support system developed to assist in agricultural decision making, by Jones *et al.* in 1990 [10], and IBSNAT in 1989 [11] were used to link the crop models and the weather data.

The crop models are physiologically based crop simulators that include processes for photosynthesis, respiration, partitioning, phenological development, soil-water balance, etc. The models include changes in the photosynthetic rate and evapotranspiration caused by changes in $CO_2$ concentration. They use a daily time step, with hourly integrations for some processes such as photosynthesis. A $CO_2$ concentration of 555 $\mu$mol mol$^{-1}$ resulted in ~30% higher average photosynthetic rates for $C_3$ crops, soybean and peanut, and ~10% higher for maize, which has a $C_4$ photosynthetic pathway.

The crop models have been tested in many different environments worldwide and have been validated against many plot trials in the US and other parts of the world. For example, SOYGRO has been extensively validated under ambient conditions by Brisson *et al.* in 1989 [12], Jones and Ritchie in 1991 [13], Egli and Bruening in 1992 [14], Colson *et al.* in 1993 [15], and Nagarajan *et al.* in 1993 [16]. Furthermore, the model predictions are in general agreement with studies conducted by Allen in 1986 [17] and Jones *et al.* in 1985 [18].

The models specify variety, or cultivar, planting date, location (for day length), soil type (with specific parameters for moisture capacity in various layers), atmospheric $CO_2$ concentration, and daily environmental inputs such as maximum and minimum temperatures, solar radiation, and rainfall. Outputs include evapotranspiration, seed yield, total biomass yield, irrigation required (if applied), and dates of phenological events such as flowering and seed maturity.

Assumptions used regarding the climate-change cases and simulations include the following: (a) Precipitation changes from the GCM's affected the amount of each rainfall event proportionately and the number of events remained the same, i.e., precipitation frequency and interannual variability remained the same. (b) The $CO_2$ level for equivalent doubling of atmospheric $CO_2$ was assumed to be 555 $\mu$mol mol$^{-1}$ compared with the 1951–1990 assumed average level of 330; the increases in the other greenhouse gases (e.g., $CH_4$, $N_2O$, and chlorofluorocarbons) were assumed to account for the additional effect on climate of doubled $CO_2$ alone. (c) Cultivars were chosen that are currently grown in the area around each location and were not changed with changed climate. (d) Soil property data were specified for generic soils similar to current local soils. (e) Current planting dates were specified as current practice around each location, and were not changed with climate-change cases. (f) For irrigation cases, water supply for irrigation was nonlimiting, and enough was applied to fill the upper 30 cm of soil to field capacity whenever the crop came under moisture stress. Soil water in the total root zone was extracted as a function of evapotranspiration demand and root density in each soil layer. The irrigation water was assumed to be applied with 100% efficiency.

Simulation runs were made for each year of the weather database for each case at each location for both rainfed and irrigated cases. Soil moisture at the beginning of each season was assumed to be at field capacity. Weather data for 27 locations in the Southeast for 40 years (1951–1990) were provided by the National Oceanic and Atmospheric Administration/National Climate Data Center (NCDC) Ashville, NC, the National Center for Atmospheric Research, Boulder, CO, and the Southeast Regional Climate Center, Columbia, SC. In this paper, two of these locations are reported: Milton, near Pensacola, FL, and Sumter, near Columbia, SC.

Climate-change cases used in the study were developed with 40-year standard historic daily weather databases for each location plus the GCM parameters [4–6] for a doubled-greenhouse-gas environment supplied by the US Environmental Protection Agency. These parameters included monthly temperature changes in degrees and monthly change ratios for precipitation for each location. The historic weather included daily minimum–maximum temperature and precipitation. A synthetic weather-generating program, WGEN, developed by Richardson in 1985 [19] was used to generate solar radiation based on temperature and precipitation.

An energy conversion spreadsheet was developed with data and conversion factors from the midwest by Doering and Peart in 1977 [20] for factors that would be typical for the southeast, and from a recent study on energy inputs in Florida agriculture by Fluck *et al.* in 1992 [21]. This spreadsheet transformed units such as kilograms of herbicide to megajoules of energy for the raw materials and the processing energy to

Table 2.16. Energy equivalents for crop production inputs and outputs

| Parameter | English Units | SI Units |
|---|---|---|
| Nitrogen fertilizer per unit of N | 25,000 Btu/lb | 58.09 MJ/kg |
| Phosphate fertilizer per unit of $P_2O_4$ | 3000 Btu/lb | 6.97 MJ/kg |
| Potassium fertilizer per unit of $KO_2$ | 2000 Btu/lb | 4.65 MJ/kg |
| Diesel fuel | 135,250 Btu/gal | 35.91 MJ/L |
| Liquid Petroleum Gas | 93,500 Btu/gal | 24.82 MJ/L |
| Electricity, input/unit of output | 10,239 Btu/kWh | 3.0 kWh/kWh |
| Pesticides per unit of active ingredients | 120,000 Btu/lb | 278.85 MJ/kg |
| Soybean seed yield | 7239 Btu/lb | 16.80 MJ/kg |
| Peanut seed yield | 9000 Btu/lb | 20.91 MJ/kg |
| Maize seed yield | 6270 Btu/lb | 14.57 MJ/kg |

produce it. Energy equivalents of the crop inputs (including fuel, fertilizers, and pesticides) and outputs (seed yields) are shown in Table 2.16. Current typical data were used for conventional tillage, fertilization, and pesticide application practices from the Midwest, which would be typical for the Southeast, except that peanut pesticide application was doubled over that used for soybean. A variation in fuel for harvesting proportional to yield and a similar factor calculating the fertilizer based on the simulated yield are included. The following expressions specify the functions used in the spreadsheet.

***Energy Inputs for Grain Production***

Fuel inputs for plowing, disking, and planting were 1478, 349, and 349 MJ/ha, respectively, for soybean and peanut, a total of 2176 MJ/ha. Plowing for maize was 1337 MJ/ha, with disking and planting the same as for soybean and peanut, totaling 2035 for all three operations. Since maize and soybean are usually alternated, plowing for maize is on harvested soybean land, which has less crop residue and roots and requires slightly less energy than does harvested maize land.

For harvesting, 441, 561, and 1121 MJ/ha were assumed for soybean, maize, and peanut, respectively, for yields of 2.35, 9.42, and 2.5 t/ha, then adjusted according to actual yield, as follows. Energy for harvesting peanut is much greater because it involves two separate operations, digging and spreading into a windrow for drying, then combine harvesting of the vines from the windrow:

$$E \text{ (harvest soy) (MJ/ha)} = 441 \times \text{yield/(2.35 t/ha)},$$

$$E \text{ (harvest maize) (MJ/ha)} = 561 \times \text{yield/(9.42 t/ha)},$$

$$E \text{ (harvest peanut) (MJ/ha)} = 1121 \times \text{yield/(2.50 t/ha)}.$$

Fertilizer for soybean and peanut, both legumes, was estimated as 352 MJ/ha plus 352 MJ/ha adjusted in proportion to the actual yield divided by 2.35 t/ha for soybean or 2.5 t/ha for peanut. This reflects an estimate of 704 MJ/ha of fertilizer for the standard yields of 2.35 t/ha for soybean and 2.5 t/ha for peanut.

These energy values account for the raw materials and the processing and the transportation energy for producing the fertilizer. Maize requires much more fertilizer than the legume crops, especially nitrogen, and energy for fertilizer was calculated based on

13,240 MJ/ha for a standard yield of 9.42 t/ha, with a yield adjustment of

$$E \text{ (fert soy) (MJ/ha)} = 352 + 352 \times \text{yield/(2.35 t/ha)},$$
$$E \text{ (fert maize) (MJ/ha)} = 7821 + 4419 \times \text{yield/(9.42 t/ha)},$$
$$E \text{ (fert peanut) (MJ/ha)} = 352 + 352 \times \text{yield/(2.50 t/ha)}.$$

Pesticide energy was estimated at 1251 MJ/ha for soybean, 1681 MJ/ha for maize, and 3363 MJ/ha for peanut.

Drying of maize and peanut was calculated on the basis of 5044 MJ/ha for a normal maize crop of 9.42 t/ha, and the actual energy was computed as proportional to the ratio of the actual yield to 9.42 t/ha. Soybeans are not normally dried after harvest:

$$E \text{ (dry maize or peanut) (MJ/ha)} = 5044 \times \text{yield/(9.42 t/ha)}.$$

Irrigation energy was calcualted based on the amount of water used as calculated by each crop simulation program. A pumping depth of 30.48 m (100 ft.), a head pressure of 2758 kPa (40 psi), and an overall engine pump efficiency of 20% were assumed. Electric pumping was assumed to have the same efficiency as internal-combustion engine pumping, including the energy losses in generating the electricity. The equation used was

$$E \text{ (irrigation) (MJ/ha)} = 287.32 \times \text{irrigated water used (cm)}.$$

The ratio of energy output to input was calculated in each case. It is always greater than 1.0 because of the solar energy used by the photosynthesis process, which is not counted as an input with the nonrenewable inputs. Also, implicit energy inputs, such as the energy used for manufacturing the equipment, are not counted.

### 2.2.2 Production Systems for Lower Emissions

When projecting possible future practices, we should note that it is feasible to improve on the energy efficiencies calculated here by using minimum tillage practices, lower-pressure irrigation, more efficient pumps and power units, and by using lower-energy drying systems for maize and peanut.

All the results shown are for production systems for lower emissions, as they show that biomass energy sources of soybean, peanut, and maize may be produced with far less input fossil fuel than is represented in the grain output. These figures show that from 6 to 12 units of biomass energy in the form of grain can be produced with 1 unit of fossil fuel input. Further processing is required for turning the grain into liquid fuel, oil, or ethanol, so more fossil fuel inputs might be required, although the biomass fuel itself, or the crop residues, could be used to provide the processing energy.

***Energy Input and Output***

Results are shown for simulations for two locations in the Southeast: Sumter, SC, near Columbia, on the Coastal Plains; and Milton, FL, near Pensacola and the Gulf of Mexico, and representative of a large area in southern Georgia, Alabama, and Mississippi, in addition to the Florida panhandle. Sumter is more representative of a wide climate area in the southeast US and is labeled "typical climate" in the tables, whereas Milton has a higher annual rainfall and is labeled "humid climate" in the tables. These results in Tables 2.17–2.25 are for simulations averaged over 40 years of historical weather data for each location. For each of the three higher-$CO_2$ climate-change cases these weather data were adjusted by the appropriate parameters and the simulations were run again. The crop simulation results for climate change are also the averages of the simulations of the three different climate-change cases.

Tables 2.17–2.19 show yields and the equivalent energy represented by these yields, based on the factors given in Table 2.16 for the three crops, two locations, and historic weather. In addition, each input factor is shown in energy value, and the amount of irrigation water used is listed. The values in Tables 2.20–2.25 in the Climate-Change columns (averages of the three cases) reflect the effects of the higher concentration of $CO_2$ and the increased temperature, changes in solar radiation, and changes in monthly rainfall totals, according to the location.

**Table 2.17. Energy output/input for soybean, 40-year average, historic weather**

| | Sumter, Typical Climate | | Milton, Humid Climate | |
|---|---|---|---|---|
| Factor | Rainfed | Irrigated | Rainfed | Irrigated |
| Input | | | | |
| Field fuel (MJ/ha) | 2761 | 2952 | 2849 | 2965 |
| Fertilizer (MJ/ha) | 818 | 971 | 889 | 981 |
| Pesticide (MJ/ha) | 1251 | 1251 | 1251 | 1251 |
| Irrigation water (mm) | 0 | 166.4 | 0 | 115.4 |
| Irrigation fuel (MJ/ha) | 0 | 4781 | 0 | 3316 |
| Energy in (MJ/ha) | 4830 | 9955 | 4988 | 8513 |
| Output | | | | |
| Seed Yield (t/ha) | 3.12 | 4.14 | 3.59 | 4.21 |
| Energy out (MJ/ha) | 52,535 | 69,709 | 60,448 | 70,888 |
| Output/input | 10.88 | 7.00 | 12.12 | 8.33 |

**Table 2.18. Energy output/input for peanut, 40-year average, historic weather**

| | Sumter, Typical Climate | | Milton, Humid Climate | |
|---|---|---|---|---|
| Factor | Rainfed | Irrigated | Rainfed | Irrigated |
| Input | | | | |
| Field fuel (MJ/ha) | 3623 | 4256 | 3700 | 4054 |
| Fertilizer (MJ/ha) | 850 | 1049 | 874 | 985 |
| Pesticide (MJ/ha) | 3363 | 3363 | 3363 | 3363 |
| Irrigation water (mm) | 0 | 134.0 | 0 | 91.0 |
| Irrigation and drying fuel (MJ/ha) | 1900 | 6506 | 1991 | 5029 |
| Energy in (MJ/ha) | 9736 | 15,173 | 9928 | 13,431 |
| Output | | | | |
| Seed yield (t/ha) | 3.54 | 4.95 | 3.71 | 4.50 |
| Energy out (MJ/ha) | 74,106 | 103,623 | 77,665 | 94,203 |
| Output/input | 7.61 | 6.83 | 7.82 | 7.01 |

**Table 2.19. Energy output/input for maize, 40-year average, historic weather**

| Factor | Sumter, Typical Climate | | Milton, Humid Climate | |
|---|---|---|---|---|
| | Rainfed | Irrigated | Rainfed | Irrigated |
| Input | | | | |
| Field fuel (MJ/ha) | 2543 | 2704 | 2725 | 2864 |
| Fertilizer (MJ/ha) | 11,819 | 13,805 | 13,250 | 14,352 |
| Pesticide (MJ/ha) | 1681 | 1681 | 1681 | 1681 |
| Irrigation water (mm) | 0 | 107.0 | 0 | 102.0 |
| Irrigation and drying fuel (MJ/ha) | 4564 | 9084 | 6198 | 10,387 |
| Energy in (MJ/ha) | 20,607 | 26,555 | 23,853 | 29,285 |
| Output | | | | |
| Seed yield (t/ha) | 8.52 | 11.22 | 11.57 | 13.92 |
| Energy out (MJ/ha) | 143,460 | 188,922 | 194,816 | 234,385 |
| Output/input | 6.96 | 7.11 | 8.17 | 8.00 |

**Table 2.20. Yields and marginal yield for irrigation of soybean, 40-year average**

| Yields | Sumter, Typical Climate | | Milton, Humid Climate | |
|---|---|---|---|---|
| | Historic | Climate Change | Historic | Climate Change |
| Irrigated (t/ha) | 4.14 | 4.72 | 4.21 | 4.47 |
| Rainfed (t/ha) | 3.12 | 3.02 | 3.59 | 3.08 |
| Marginal yield (t/ha) | 1.02 | 1.70 | 0.62 | 1.39 |
| Marginal yield (%) | 32 | 56 | 17 | 45 |

**Table 2.21. Yields and marginal yield for irrigation of peanut, 40-year average**

| Yields | Sumter, Typical Climate | | Milton, Humid Climate | |
|---|---|---|---|---|
| | Historic | Climate Change | Historic | Climate Change |
| Irrigated (t/ha) | 4.95 | 4.76 | 4.50 | 3.85 |
| Rainfed (t/ha) | 3.54 | 3.07 | 3.71 | 2.48 |
| Marginal yield (t/ha) | 1.41 | 1.69 | 0.79 | 1.37 |
| Marginal yield (%) | 40 | 55 | 21 | 55 |

**Table 2.22. Yields and marginal yield for irrigation of maize, 40-year average**

| Yields | Sumter, Typical Climate | | Milton, Humid Climate | |
|---|---|---|---|---|
| | Historic | Climate Change | Historic | Climate Change |
| Irrigated (t/ha) | 11.22 | 10.21 | 13.92 | 11.54 |
| Rainfed (t/ha) | 8.52 | 9.15 | 11.57 | 10.27 |
| Marginal yield (t/ha) | 2.70 | 1.06 | 2.35 | 1.27 |
| Marginal yield (%) | 32 | 12 | 20 | 12 |

**Table 2.23. Energy efficiency and output/input ratio of soybean, 40-year average**

| | Sumter, Typical Climate | | Milton, Humid Climate | |
|---|---|---|---|---|
| Output/Input | Historic | Climate Change | Historic | Climate Change |
| Irrigated | 7.00 | 6.42 | 8.33 | 6.11 |
| Rainfed | 10.88 | 10.60 | 12.12 | 10.77 |

**Table 2.24. Energy efficiency and output/input ratio of peanut, 40-year average**

| | Sumter, Typical Climate | | Milton, Humid Climate | |
|---|---|---|---|---|
| Output/Input | Historic | Climate Change | Historic | Climate Change |
| Irrigated | 6.83 | 6.25 | 7.01 | 5.00 |
| Rainfed | 7.61 | 6.98 | 7.82 | 6.08 |

**Table 2.25. Energy efficiency and output/input ratio of maize, 40-year average**

| | Sumter, Typical Climate | | Milton, Humid Climate | |
|---|---|---|---|---|
| Output/Input | Historic | Climate Change | Historic | Climate Change |
| Irrigated | 7.11 | 6.95 | 8.00 | 7.31 |
| Rainfed | 6.96 | 7.24 | 8.17 | 7.70 |

Table 2.17 shows that rainfed soybean is an efficient producer of biomass energy because of its low requirements for nitrogen fertilizer (it is a legume) and its relatively low need for pesticides, the main one being herbicides for weed control. The field fuels for tillage, planting, and harvesting are the major components of energy input, and fuel for drying is assumed to be zero, as soybean seeds normally need no artificial drying for safe storage, unlike maize. Irrigated soybean requires much more energy, doubling the total input in Sumter and raising it by ~60% in the more humid Milton.

Peanut (Table 2.18) is not as energy efficient as soybean, mainly because of its need for artificial drying and for more pesticides (mainly fungicides) than soybean. Energy efficiency was not reduced as much by irrigation, since the irrigation requirements were not as large as for soybean (peanut is planted later than soybean in these areas).

Table 2.19 shows the higher numbers for maize, both inputs and outputs. Maize requires high amounts of nitrogen fertilizer and large amounts of fuel for drying the high-yield crop. Maize gave a high-yield response to irrigation at both locations, so energy efficiency was only slightly affected by irrigation.

Tables 2.20–2.22 show results for climate change as well as historic weather, and they show the important marginal yield due to irrigation and the increase (in tons per hectare and in percentage) in yield due to irrigation. This figure is important because it shows the amount of incentive a grower has, or will have, under climate change, to install an irrigation system. For soybean and peanut (Tables 2.20 and 2.21), climate-change results show a much larger marginal yield for irrigation, increases of from 45% to 56%. From this, we assume that, under climate change, the demand for irrigation water will be higher and more land for these crops will be irrigated. For example, for soybean in Sumter, an increase of 1.7 t/ha converts to ~28 bu/a, which at the recent price of $7/bu means a return of $196/acre, a strong incentive.

Maize, as shown in Table 2.22, would have a decreased marginal yield for irrigation under climate change than it has in the past. Being a $C_4$ plant, maize does not respond to increased $CO_2$ as much as soybean or peanut, which are $C_3$ plants. Climate change reduces the irrigated yields, while rainfed yields are not reduced as much; in Sumter, rainfed yield under climate change is increased slightly, and thus there is a lower marginal yield for irrigation.

Tables 2.23–2.25 show the output/input ratios for the three crops and give an idea of their efficiency for use as biomass energy crops currently and under climate change. For soybean and peanut, with historic weather, the energy output/input for irrigated crops was less than that for rainfed crops in the Southeast. For maize, there was little difference; there was even a slight increase in efficiency for irrigation at Sumter. For climate change, the energy efficiencies for soybean and peanut decrease somewhat, but not drastically. A more pessimistic view results if comparisons are made between rainfed historic weather and irrigated climate change, as more land will probably be irrigated in the Southeast under climate change. However, even this comparison shows a relatively small difference for maize, which produces the highest energy output per hectare.

For soybean, requirements for irrigation water were more than doubled in Florida and increased by ~50% in South Carolina; however, this resulted in relatively small decreases in output/input due to climate change (Table 2.23).

Rainfed yields of soybean (Table 2.20) were reduced ~15% in Florida, but very little in South Carolina under climate change, while peanut yields were reduced by one-third in Florida and only 15% in South Carolina (Table 2.21). The ratios of energy output to input (Table 2.23) were reduced accordingly. Irrigated yields were higher in both locations because the increased $CO_2$ dominated the negative effects of higher temperature. The climate change effect on rainfall was, of course, nullified by irrigation.

***Effects of Climate Change***

Some writers in the 1970's claimed that energy from biomass would require more energy inputs than would be produced. In this discussion, the energy cost of converting the biomass energy into a useful fuel source must be included. Early calculations on ethanol from maize grain showed that a fairly high proportion of the energy in the grain would be required for the production and processing of the grain into liquid ethanol. The cooking, distillation, and drying of the ethanol required more energy than did the production of the grain. However, more efficient methods, especially in drying of the ethanol, are now being used, and biomass ethanol is a net positive energy source. In addition, co-products of oil and protein livestock feed come from the maize grain. Ethanol production from maize is currently stable in the US and is a substantial liquid fuel energy source. Production of ethanol from sugar cane continues at a high level in Brazil and most automobiles there run on neat (not totally dry) ethanol.

There are three separate effects of climate change on crops in the southeastern US. First, increasing the $CO_2$ concentration increases the photosynthetic rate, all other inputs remaining the same. Second, changes in climate as simulated by global change models (mainly increasing temperature and changing rainfall patterns) decreased yields and increased the energy inputs needed to obtain a given amount of maize, soybean, or peanut in the southeastern US. Third, with increased $CO_2$ concentration, canopy resistance to transpiration from leaves is increased. This in itself would reduce water loss from the crop, but the increased temperature overpowers this effect by increasing the vapor pressure difference between the leaf and the air and between the soil surface and the air, causing higher evapotranspiration.

Differences between historic weather patterns and soil characteristics at different locations in the southeastern US cause different responses to climate change, but in general the changes are negative. The three crops are different, as maize has a $C_4$ pathway for photosynthesis, while soybean and peanut are $C_3$ crops. In general, the negative effects of climate change on maize are less extensive than for soybean or peanut in the southeastern US.

As in the two locations documented here, crop yields over all the 27 locations studied were generally lower under climate change, especially rainfed soybean and peanut. In all cases, the energy input per unit of equivalent energy output in the form of seed yield was greater for climate change. Thus, for a given amount of yield, more energy inputs will be required under climate change.

Additionally, and most importantly, water demand for irrigation and the needed increased energy inputs will likely be increased for soybean and peanut under climate change for two reasons. First, irrigation under climate change generally showed a greater increase in water demand and a lower yield or a lesser increase in yield, indicating that more water would be required under climate change for a given amount of soybean or peanut production. Second, the increase in yield due to irrigation is greater under climate change than under historical weather. This means the incentive will be greater to put southeastern croplands under irrigation because of the greater increase in yields over rainfed culture. More acreage under irrigation, with each acre's requiring more irrigation water for peanut and soybean production, would add more competition for water with growing urban and industrial water demands. Already the southeastern US has areas of shortages of water for new growth. This increased demand for irrigation water and for the energy to pump it may be the most important agricultural effect of climate change in the southeastern US.

## References

1. Meadows, D. *et al.* 1972. *Limits to Growth.* The New American Library, Inc., New York.
2. Ed., Waggoner, P. E. 1992. Preparing US agriculture for global climate change. CAST Task Force Report No. 119. Ames, IA: Council for Agricultural Science and Technology.
3. Curry, R. B., R. M. Peart, J. W. Jones, K. J. Boote, and L. H. Allen, Jr. 1990. Response of crop yield to predicted changes in climate and atmospheric $CO_2$ using simulation. *Trans. ASAE* 33:1383–1390.
4. Hansen, J., I. Fung, A. Lacis, S. Lebedeff, D. Rind, R. Ruedy, G. Russell, and P. Stone. 1988. Global climate changes as forecast by the GISS 3-D model. *J. Geophys. Res.* 98:9341–9364.

5. Manabe, S. and R. T. Wetherald. 1987. Large-scale changes of soil wetness induced by an increase in atmospheric carbon dioxide. *J. Atmos. Sci.* 44:1211–1235.
6. Wilson, C. A. and J. F. B. Mitchell. 1987. A doubled $CO_2$ climate sensitivity experiment with a global model including a simple ocean. *J. Geophys. Res.* 92:13315–13343.
7. Jones, J. W., K. J. Boote, S. S. Jagtap, G. Hoogenboom, and G. G. Wilkerson. 1988. SOYGRO V5.41: Soybean crop growth simulation model. User's Guide. Florida Agr. Exp. Sta. Journal Series No. 8304, IFAS. Gainesville, FL: University of Florida.
8. Boote, K. J., J. W. Jones, G. Hoogenboom, G. G. Wilkerson, and S. S. Jagtap. 1989. PNUTGRO V1.02, IBSNAT Version, User's Guide. Gainesville, FL: University of Florida.
9. Jones, C. A. and J. R. Kiniry, eds. 1986. *CERES-Maize: A Simulation Model of Maize Growth and Development*. College Station, TX: Texas A & M U. Press.
10. Jones, J. W., S. S. Jagtap, G. Hoogenboom, and G. Y. Tsuji. 1990. The structure and function of DSSAT. *Proc., IBSNAT Symposium: Decision Support System for Agrotechnology Transfer*. Part I: Symp. Proc. Dept. of Agronomy and Soil Science: pp. 1–14. Honolulu, HI: University of Hawaii.
11. IBSNAT. 1989. Decision Support System for Agrotechnology Transfer (DSSAT v2.1), User's Guide. IBSNAT Project, Dept. of Agronomy and Soil Science. Honolulu, HI: University of Hawaii.
12. Brisson, N., S. Bona, and A. Bouniols. 1989. Adaptation of a soybean crop simulation model (SOYGRO) to Southern European conditions: *Proc. World Soybean Conference. IV*, Vol. 1, ed. Pascale, A. J., pp. 279–284.
13. Jones, J. W. and Ritchie. J. T. 1991. Crop growth models. *Management of Farm Irrigations Systems*: pp. 69–98. St. Joseph, MI: ASAE.
14. Egli, D. B. and W. Bruening. 1992. Planting date and soybean yield: evaluation of environmental effects with a crop simulation model, SOYGRO. *Agric. Meteorol.* 62:19–29.
15. Colson, J., A. Bouniols, and J. W. Jones. 1995. Soybean reproductive development: Adapting a model for European cultivars. *Agron. J.* 87:1129–1139.
16. Nagarajan, K., R. J. O'Neil, C. R. Edwards, and J. Lowenberg-DeBoer. 1993. Indiana soybean system model (ISSM): I. Crop model evaluation. *Agric. Syst.* 43:357–79.
17. Allen, L. H. Jr. 1986. Plant responses to rising $CO_2$. In *Proceeding of the 79th Annual Meeting of the Air Pollution Control Association*: pp. 86–89.
18. Jones, P., L. H. Allen, Jr., and J. W. Jones. 1985. Soybean canopy growth, photosynthesis and transpiration responses to whole-season carbon dioxide enrichment. *Agron. J.* 76:633–637.
19. Richardson, W. W. 1985. Weather simulation for crop management models. *Trans. ASAE* 28:1602–1607.
20. Doering, O. C. and R. M. Peart. 1977. Evaluating alternative energy technologies in agriculture. NSF/RA-770124, Agr. Exp. Station W. Lafayette, IN: Purdue University.
21. Fluck, R. C., B. S. Panesar, and C. D. Baird. 1992. Florida Agricultural Energy Consumption Model, Final Report, Agr. Eng. Dept. Gainesville, FL: University of Florida.

# 2.3 Solar Energy

## 2.3.1 Present Situation

*W. Mühlbauer and A. Esper*

An analysis of the present situation of world agriculture shows a completely contrasting situation between industrialized and developing countries. This disparity has greatly influenced the possibilities of utilizing solar energy in agriculture.

In industrialized countries, the intensive mechanization of agricultural production resulted in high labor productivity. The use of high-quality seeds, pesticides, fertilizers, and the mechanization of almost all farm operations have significantly increased crop yields.

At present the energy supply in industrialized countries is still sufficient. Electricity and fossil fuels are available at relatively low prices. Almost all farms are connected to the public grid. As a result, solar technologies that can be used independently of the weather conditions have to compete with the highly efficient and reliable conventional technologies.

Utilization of low-density and fluctuating energy sources such as solar energy requires highly sophisticated technologies. Under these conditions, solar technologies could be competitive only if the production costs are considerably reduced without lowering their efficiencies and reliabilities. Energy saving alone is not an incentive for farmers to invest in solar technologies. In some cases, however, there are niches for which solar energy applications are feasible. Solar dryers for selected agricultural products, solar heating of greenhouses, aeration of fish ponds, and electric fences are some examples.

In developing countries, it is expected that, in the near future, food production will no longer supply the needs of the rapidly growing population. The lack of appropriate preservation and storage systems causes considerable losses, thus significantly reducing the food supply. In addition, the tremendous reduction of the prices for tropical products has worsened the economic situation of the rural population.

The increase in crop yields, the improvement of product quality, and the reduction of losses are directly related to the availability of energy. Fossil fuels that have to be imported, whenever they are available, by most developing countries, are expensive in rural areas. Therefore providing energy from alternative sources is an urgent necessity. The migration of rural people to the urban areas because of adverse living conditions has caused social problems. Providing electricity for lighting, telecommunication purposes, and the operation of small machinery like mills, water pumps, and fans should then be given high priority.

In developing countries, the possibilities of utilizing solar energy are economically feasible compared with its use in industrialized countries. High solar insolation, decentralized use, and the low energy demand favor its use.

After the energy crisis in 1973, great attempts were made to utilize solar energy for drying agricultural products, heating animal houses and greenhouses, pumping water, and providing electricity for farm households. Numerous investigations were carried out, and demonstration units were installed and tested. Despite these great efforts, only a limited number of solar dryers, solar pumps, and solar home systems for providing electricity for farm houses were commercially produced and disseminated. Investigations and tests have shown that most solar technologies are technically feasible, but the economic feasibility is, in most cases, still questionable.

### 2.3.2 Principles of Solar Energy Applications

***Solar Radiation***

The application of solar energy is completely dependent on solar radiation, a low-grade and fluctuating energy source. The Sun radiates energy, which is generated by a hydrogen-to-helium thermonuclear reaction in the form of corpuscular rays and electromagnetic radiation. The Sun's radiation corresponds to that of an equivalent blackbody temperature of 5762 K. The average amount of solar radiation in near-Earth space is called the solar constant. Its measured value amounts to 1353 $W/m^2$ [1]. Because of the eccentricity of the Earth's orbit, the solar constant varies by $\pm 3\%$ during a year.

Attenuation of the solar radiation passing through the Earth's atmosphere occurs because of absorption by gas molecules, scattering by air molecules, water vapor, and dust, as well as reflection by gas molecules.

The spectral distribution of the extraterrestrial and the terrestrial radiation normal to the Earth's atmosphere is shown in Fig. 2.4.

Global radiation is the sum of direct (or beam) and diffuse radiation incident upon a surface. Beam radiation strikes the surface without having been scattered by atmosphere, but weakened. The diffuse radiation, scattered by the Earth's atmosphere, does not form any shade; neither can it be concentrated nor attached to any direction.

Beam radiation depends on the season, time of the day, latitude, and the inclination of the irradiated surface against the horizontal. When the laws of the spherical trigonometry are applied, the incidence angle of beam radiation on any inclined surface can be calculated [2–6].

A solar radiation map of the world is given in Fig. 2.5. According to the latitude, a wide variation in availability of solar radiation can be found. Depending on the weather conditions and pollution, the transmissivity of the Earth's atmosphere is further impaired. In Central Europe the annual mean of total radiation on a horizontal surface is in the

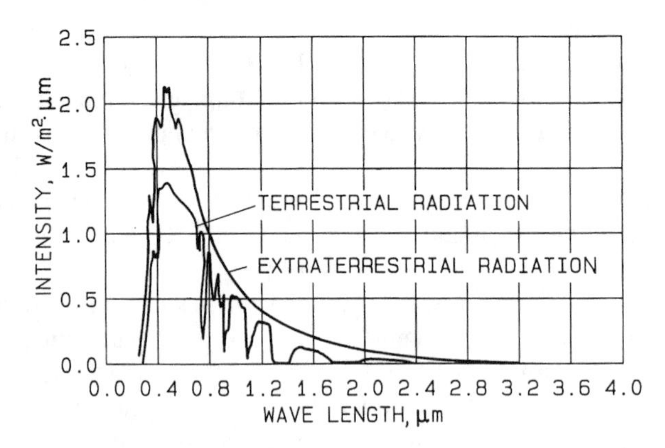


**Figure 2.4. Spectral distribution of the extraterrestrial and terrestrial radiation normal to the Earth's atmosphere [2].**

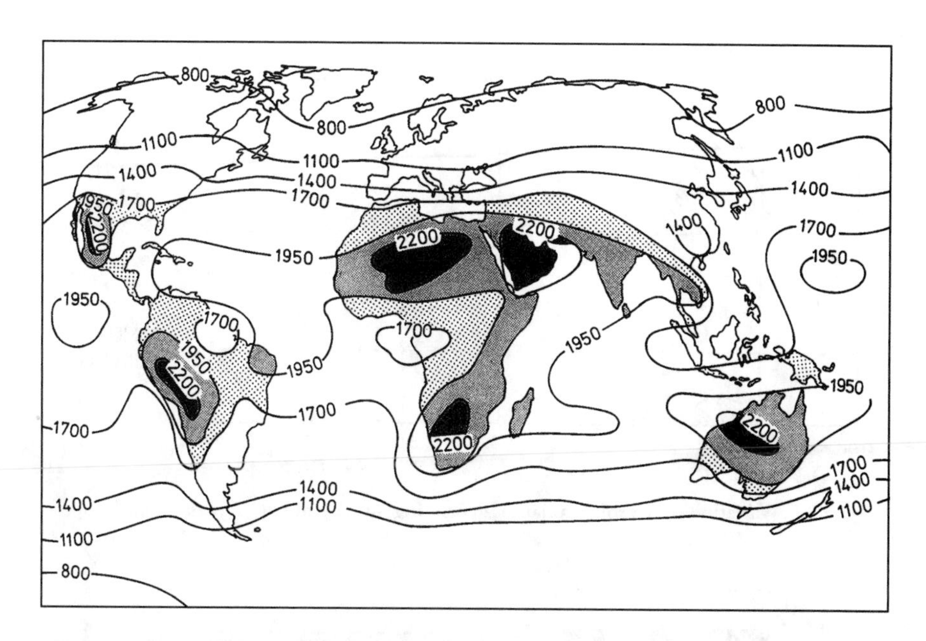


**Figure 2.5. Annual means of the total radiation on a horizontal surface (in kilowatt hours per square meter)** [7].

range between 950 and 1150 kWh/m$^2$. In southern California or in the Sahara, 2 to 2.5 times the availability in Central Europe can be expected.

An intrinsic difficulty in using solar energy is given by the wide variation in the solar radiation intensity. The availability of solar radiation depends not only on the location, but also on the season. Extreme differences are experienced between summer and winter, and from day to day. The duration of sunshine as well as the total radiation varies from day to day (Fig. 2.6). Therefore, for performance prediction or calculation of the availability at a particular site within a certain period, e.g., during harvesting season, measured mean values of the monthly or daily radiation over at least one decade are required.

***Flat-Plate Collectors***

In general, solar water and solar air heaters are flat-plate collectors, consisting of an absorber, a transparent cover, and backward insulation [9, 10] (Fig. 2.7).

Despite the similarity in designs, the different modes of operation and the different properties of the heat transfer medium greatly affect the thermal performance and the electric energy consumption for forcing the heat transfer medium through the collector.

Solar water and solar air heaters can be distinguished by the mode of operation. Solar water heaters are operated as a closed-loop system whereas, in most cases, solar air heaters are operated in the open-loop mode. This can cause problems since the moist and dust-contaminated air is continuously passing between absorber and collector.

Furthermore, the two collector types differ in the properties of the heat transfer medium. Compared with air, water has a higher density and specific-heat capacity. To

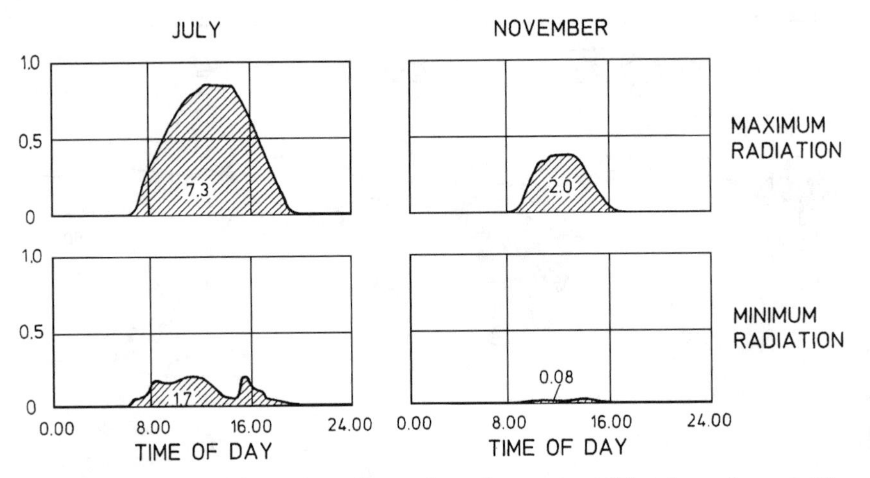


**Figure 2.6. Variation of the solar radiation depending on time of day, day, and month [8].**

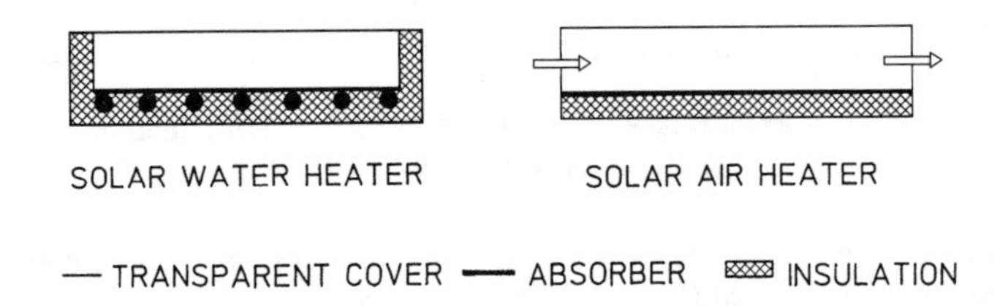


**Figure 2.7. Flat-plate collectors for water and air heating puroses [9].**

transfer the same amount of heat flow, the volume of air passing through the collector should be ~3200 times higher than the volume of water. To minimize the pressure drop inside the air heaters and to reduce the electrical power requirement of the fan, the cross section of the air ducts has to be made much larger than the cross section of the water tubes. In addition, to reach the same temperature rise, air heaters have to be made longer because of the lower heat transfer between the absorber and the air compared with that of water heating systems.

Because of the enlarged dimensions, solar air heaters are fixed without any tracking device. The optimum orientation of the collector can be computed based on geographic location and the time of the year in which the solar air heater is in operation. In practice, optimum orientation cannot be followed because air heaters are normally incorporated into either the roofs or the walls of the buildings. However, deviation of the optimum orientation by ±20° results in a reduction of the thermal efficiency of only ~10% and is therefore tolerable.

Depending on the application, the importance of the different factors affecting the performance varies. For heating purposes, useful energy gain is the most important factor, whereas for drying agricultural products, temperature rise is of major importance. In any

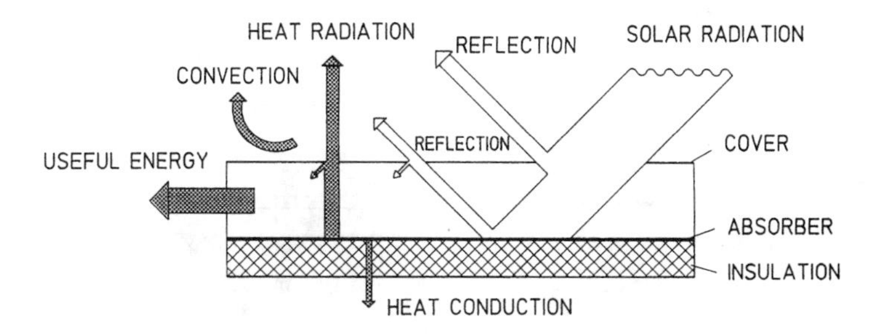


**Figure 2.8. Heat flow inside a flat-plate collector [11].**

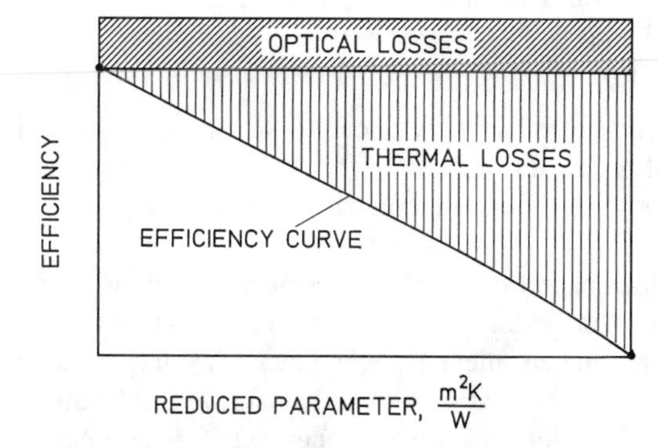


**Figure 2.9. Efficiency curve of a solar air heater [10].**

case, the pressure drop inside the collector should be minimized to reduce the energy requirement of the fan. This is to ensure that more energy is gained by the solar air heater than that used for driving the fan.

The performance of solar air heaters is mainly influenced by meteorological parameters (direct and diffuse radiation, ambient temperature, wind speed), design parameters (type of collector, collector materials), and flow parameters (air flow rate, mode of flow).

The thermal performance of a solar air heater operating under steady-state conditions can be determined from the energy balances of the different collector components and the energy and mass balance of the heat transfer medium (Fig. 2.8). The efficiency curve characterizes the performance of the solar air heater in terms of optical efficiency and the overall heat loss coefficient (Fig. 2.9). The optical efficiency depends strongly on the optical properties of the absorber and the transparent cover, while the overall heat loss coefficient is influenced mainly by the heat losses at the top cover.

For air heating purposes, several types of collectors were developed (Fig.2.10). Solar air heaters can be classified based on the mode of air circulation. In the bare plate collector, which is the most simple solar air heater, the air passes through the collector underneath the absorber. This kind of solar air heater is suitable for a temperature rise of only between 3 and 5 K because of the high convection and radiation losses at the surface.

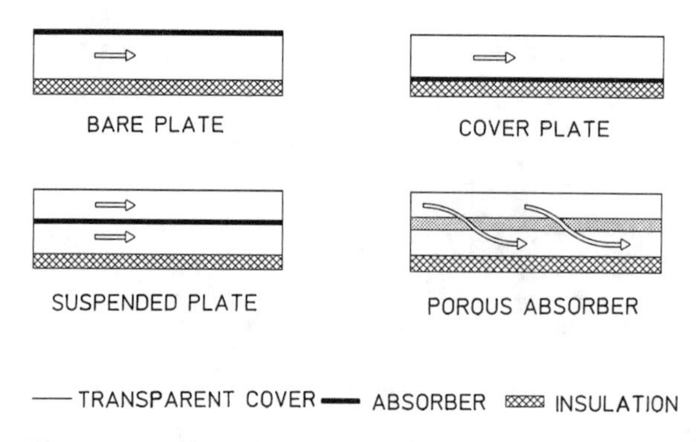


**Figure 2.10. Solar air heaters classified based on the mode of air circulation [10].**

The top losses can be reduced significantly when the absorber is covered with a transparent material of low transmissivity for infrared radiation. The air flow occurs in this kind of solar air heater either underneath the absorber or between the absorber and transparent cover.

Because of the transparent cover, the incident radiation on the absorber is reduced slightly, but because of the reduction of the convective heat losses, a temperature rise of between 20 and 50 K can be achieved, depending on insulation and air flow rate.

A further reduction of the heat losses can be achieved if the air is forced above and underneath the absorber, since this doubles the heat transfer area. The heat losses due to radiation will be reduced by this process because of a lower absorber temperature. However, there is simultaneous reduction in the absorptivity of the absorber due to dust deposit if the air flow is above or on both sides of the absorber.

The best heat transfer can be achieved with a porous material as absorber. The hot air passes underneath the absorber; thus heat losses at the top are reduced. Compared with other modes of air circulation, the porous absorber shows a higher air resistance and tends to clog if dusty air passes through.

The efficiency of solar air heaters is influenced both by the design and air circulation and by the properties of the material used for cover, absorber, and insulation. Considering its short useful life, the material and the production costs are the main selection criteria. The influence of air circulation on the efficiency is relatively low for a low temperature rise, while solar air heaters with air flow underneath or above the absorber show significantly higher values of efficiency in a higher temperature range.

A higher efficiency is obtained at a higher temperature if the solar air heater is double covered. However, this is economical only for a plastic-film solar air heater because of the high labor requirement for construction and the short useful life.

Glass shows significant advantages with respect to the optical properties and deterioration compared with those of all other materials. However, high cost and weight, high construction labor requirement necessary for a watertight and airtight fixing, and compensation for thermal stretches limit the use of glass as a transparent cover for solar air heaters.

In spite of the lower transmittance, reinforced Fiberglas can be used economically for incorporating the solar air heater into roofs or walls of buildings. The best plastic-film material for a solar air heater is polyvinyl chloride because of its low transmittance of infrared radiation and deterioration. Air-bubble foil is especially preferred for a greenhouse-type dryer because of its high mechanical strength and simple fixing method.

Solar air heaters consisting completely of air-inflated transparent and black plastic foils were developed to reduce the collector cost. These air-inflated collectors can be easily rolled up and stored when not in use, which increases the life span considerably. However, it should be emphasized that deflated collectors can be destroyed if the stagnation temperature is higher than the heat resistance of the plastic materials.

Selective coated absorbers show great advantages compared with all other black materials, particularly if high temperature is desired. High costs as well as high aging of the selective coated materials available in the market make them uneconomical to use. Black fabric, black-painted aluminum, or steel are the most common absorber materials at the moment.

Increasing the absorbing area either by corrugations or fins leads to better heat transfer. But because of the high labor requirement for construction and the resulting higher air resistance, the use of such an absorber surface is not economical.

Glass or mineral wool plates and polyurethane foam plates are well suited for heat insulation purposes. Styrofoam plates cannot be used because of the low temperature resistance and the high susceptibility for rodents. It is expected that polyurethane foam plates will be replaced by Polypropylen (PP) foam plates in the near future.

Compared with those of solar water heaters, the production costs for solar air heaters are much lower. Commercially produced solar air heaters made from durable materials are available at prices between 200 and 300 US $\$/m^2$. Additional cost is required for system integration. Incorporating the solar air heaters into new buildings reduces the investment cost considerably because the roof or the wall structure can be used as the base for the solar air heaters. Furthermore, the costs for the roof tile can also be saved. In some industrialized countries, do-it-yourself collectors for drying agricultural products were developed. Utilizing prefabricated components lowers the collector cost to less than 60 US $\$m^2$. A further reduction of the collector cost to $\sim$30 US $\$m^2$ can be achieved with UV-resistant, cheap plastic foils as cover and absorber material.

### 2.3.3 Solar Drying

***Industrialized Countries***

In industrialized countries, combine harvesters are used for harvesting cereals, oil seeds, and grain legumes. Green forages are harvested with self-loading wagons. Successful use of these highly efficient machines requires adequate dryers for preserving the crop immediately after harvest. Consequently, sun drying was almost completely replaced by mechanical dryers that use fossil fuels as energy source to heat the drying air and electricity to force the drying air through the crop. Only crops such as hay and alfalfa are still dried in the field. Studies have shown that the successful application of solar energy to high-temperature drying systems is neither technically nor economically feasible without lowering the capacity and the reliability [12]. In addition, the use of solar air heaters for preheating the drying air is not economically feasible, even for small-scale batch dryers, under the prevailing low fuel oil prices because of the high initial investment for the solar air heaters.

High-temperature drying requires significant quantities of fuel oil or gas. As a consequence of the energy crisis in 1973, low-temperature in-storage drying of low-moisture crops such as cereals, grain legumes, and hay became extensively used [13, 14]. Low-temperature drying uses ambient air or slightly preheated ambient air as the energy source and can be characterized as a passive solar system. Its advantages include low energy consumption, minimal investment, and uniform drying of the crop. Because of the required temperature rise of only 3–5 K, the application of solar energy is technically and economically more feasible compared with that of high-temperature dryers. In addition, low-temperature drying allows intermittent or variable heat input. However, to prevent deterioration in the upper layer, supplementary heat is necessary during the night or during cloudy or rainy periods.

Investigations have shown that heating the drying air with the solar collector could increase the drying rate and lower the drying time significantly. However, it could also lead to overdrying of the bottom layer of the crop. For crops intended for marketing, partial overdrying is not advisable since it reduces the income of the farmer. Therefore solar-assisted low-temperature in-storage drying is economically feasible for only those drying crops intended for on-farm feeding purposes for which the effect of overdrying is insignificant [15]. In regions with occasionally extended adverse weather conditions, an auxiliary heating system during the harvesting period is required for preventing deterioration of the crop. The additional investment for this system, however, remains a barrier in justifying the economical advantage of using solar energy in a low-temperature drying system (Fig. 2.11).

Another promising alternative to utilize solar energy for drying agricultural products is the use of cheap plastic tunnel greenhouses [17]. A greenhouse can be used either as solar collector for heating the drying air or for drying the crop inside the greenhouse. However, the low capacity of the system in relation to the costs of its construction prevents its wide acceptance. Currently only a few small-scale farms are using plastic greenhouses for drying tobacco.

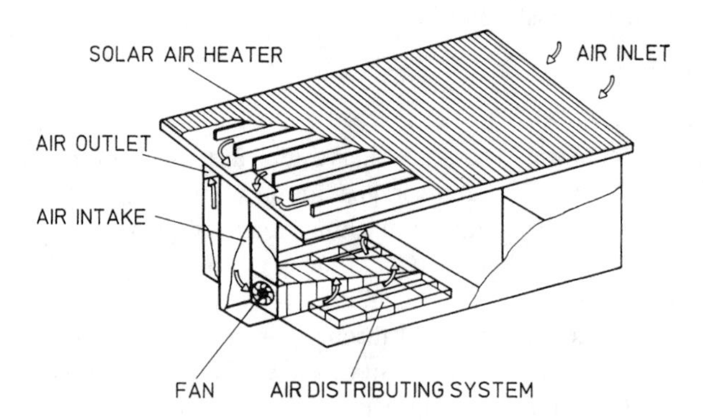


**Figure 2.11. Solar hay dryer with collector incorporated into the roof of the barn [16].**

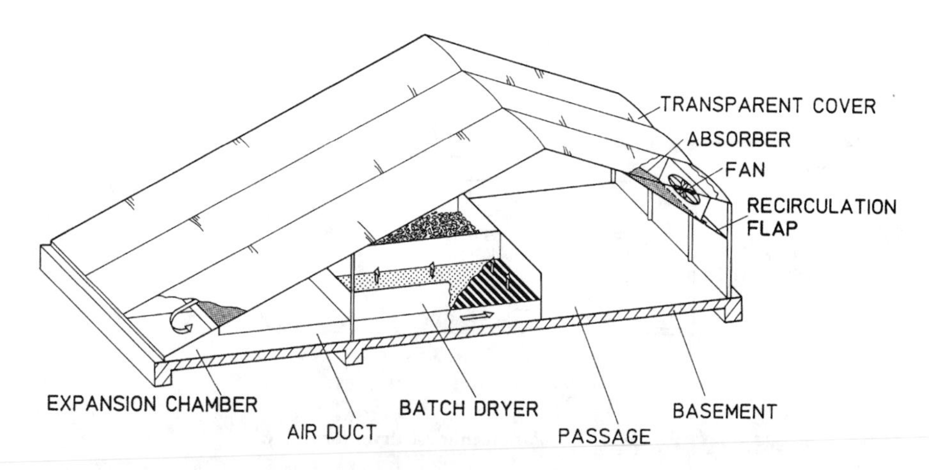


**Figure 2.12. Solar greenhouse dryer for medicinal and aromatic plants [18].**

Another type of greenhouse dryer was developed for drying medicinal and aromatic plants. It is based on a commercially available low-cost plastic-covered greenhouse structure with vertical sidewalls and a span roof [10, 18]. The solar air heater is incorporated into the roof, which extends to the ground on the southern side (Fig. 2.12).

To obtain the desired modular design, the frame of the greenhouse is structured along the ridge into segments. Each segment is equipped with a highly efficient axial flow fan. A batch dryer is incorporated inside. To maximize the use of the solar energy, a flap is installed at the northern sidewall, which allows recirculation of the drying air. This increases the capacity by 30%.

The collector is composed of a highly UV-resistant air-bubble foil as the transparent cover, a black woven fabric as the absorber, and a second air-bubble foil to reduce heat losses at the backside. The foils are fixed with a special fastening system that facilitates the mounting and the replacements of the foils. Because of the strength and durability of the foil, its life span is between 6 and 10 years.

Compared with sophisticated high-temperature belt dryers normally used for drying large quantities of medicinal and aromatic plants, the solar greenhouse dryer saves ~1 kg fuel oil per 1 kg dried drug, aside from reducing the investment significantly. In addition, people can use the structure for storage purposes or as a greenhouse by simply replacing the absorber and the batch dryer. Modification of the construction offers a wide range of additional applications, such as drying fruit, slurry, tobacco, or wood (Figs. 2.13 and 2.14).

### *Developing Countries*

Exposing agricultural products to wind and Sun is the preservation method that has been practiced over the centuries throughout the world. Cereals, legumes, and green forages are dried in the field immediately after harvesting. Fruits, vegetables, spices, and marine products, as well as threshed grains, are spread out in thin layers on the ground or trays, respectively. Other alternatives include hanging the crop underneath a shelter, on trees, or on racks in the field.

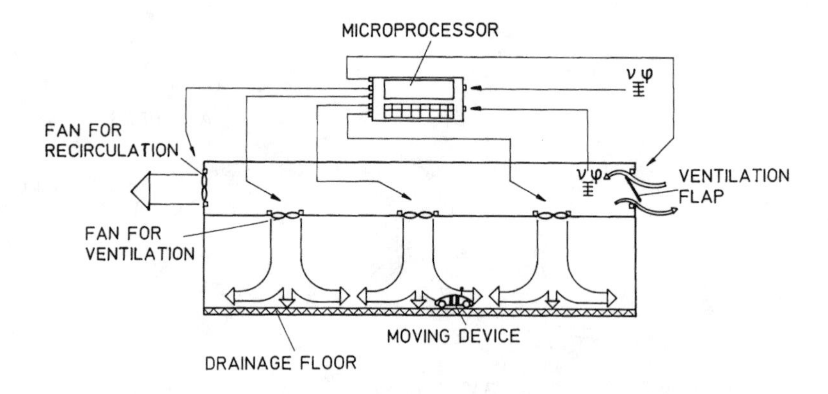


**Figure 2.13. Solar greenhouse dryer for slurry.**

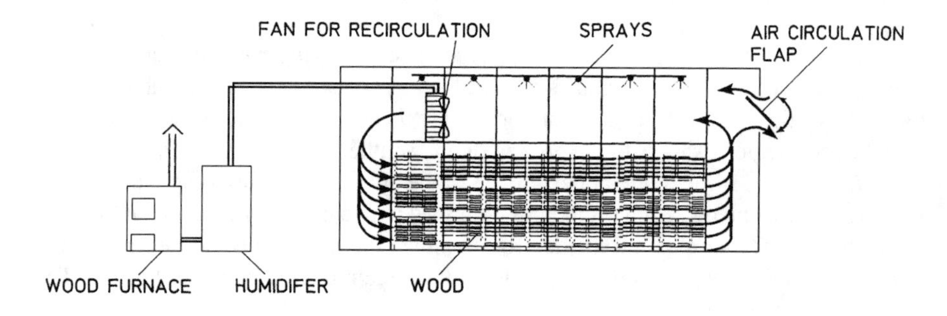


**Figure 2.14. Solar greenhouse dryer for wood.**

However, in tropical and subtropical countries almost all crops are dried by Sun and wind, the only energy sources; only on large private or government estates are mechanical dryers in use [19–23].

*Sun Drying*

During sun drying, heat is transferred by convection from the surrounding air and by absorption of direct and diffuse radiation on the surface of the crop. The converted heat is partly conducted to the interior, increasing the temperature of the crop, and partly used for effecting migration of water and vapor from the interior to the surface. The remaining energy is used for evaporation of the water at the surface or lost to ambient by means of convection and radiation. The evaporated water has to be removed from the surroundings of the crop by natural convection supported by wind forces.

Because of the hygroscopic properties of all agricultural products, during sun drying the crop can either be dried or rewetted. Especially during the night, when the ambient temperature in general is decreasing, causing a simultaneous increase of the humidity,

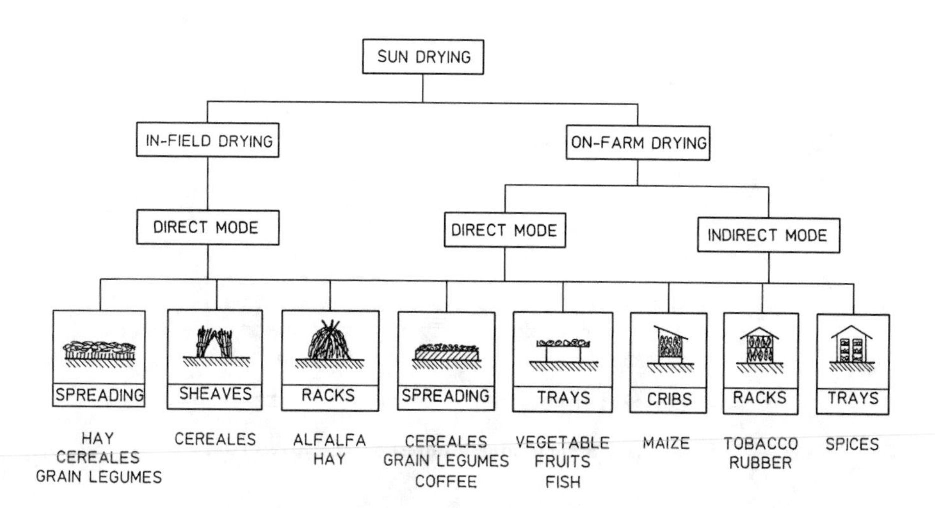


**Figure 2.15. Classification of sun drying methods [10].**

remoistening effects can occur either by condensation of dew or by vapor diffusion caused by osmotic or capillary forces.

A classification of sun drying procedures is shown in Fig. 2.15. In general, the classification can be based on the stage of processing, the location of drying, or exposure to solar radiation.

In tropical and subtropical countries, almost all crops have to be dried, except fruits and vegetables, which are consumed mostly while fresh. Even fish and meat are dried in significant quantities. Spreading the crop in thin layers on mats, trays, or paved grounds and exposing the product to Sun and wind is still the most common drying method used in these countries. Turning the crop in regular, short intervals promotes uniform drying. Storing the crop during the night and rain under a shelter prevents remoistening. Since the drying process is relatively slow, considerable losses occur. Furthermore, insect infestation, enzymatic reactions, micro-organism growth, and mycotoxin development cause a significant reduction in the product quality. Nonuniform or insufficient drying also leads to deterioration of the crop during storage. Serious drying problems occur, especially in humid tropical regions where crops have to be dried during the rainy season.

*Solar Dryers*

In order to ensure a continuous food supply to the growing population and to enable the farmers to produce high-quality marketable products, the development of efficient drying methods is of urgent necessity. Studies have shown that even small and most simple oil-fired batch dryers are not applicable because of the lack of capital and an insufficient supply of energy for operating the dryers. The high-temperature dryers used in industrialized countries are economically feasible in developing countries only if they are used in large plantations and big commercial establishments. To overcome the

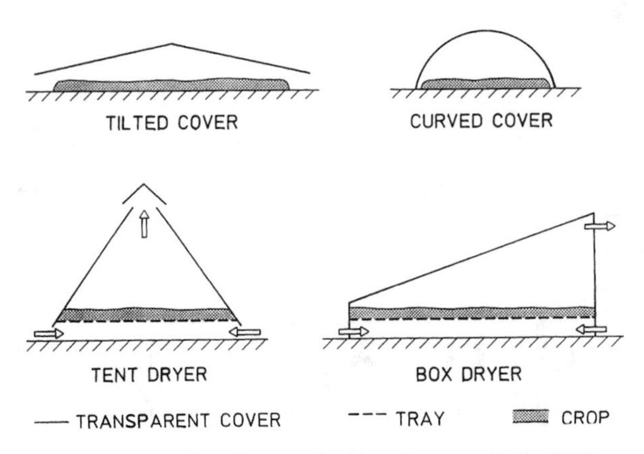


**Figure 2.16. Simple natural convection-type dryers [10].**

existing preservation problems, the introduction of solar dryers seems to be a promising alternative.

Various types of small-scale solar dryers were developed and evaluated for application in tropical and subtropical regions (Fig. 2.16). Considering that a high percentage of farms in these regions are not connected to the public grid, the dryers are designed to use only wind and the Sun as energy sources.

Covering the crop spread out on the ground in a thin layer with a transparent foil is an example of the most simple solar dryer; it is used mainly for drying grapes [13]. For on-farm drying of small quantities of fruits and vegetables, box and tent dryers were developed that can be constructed with locally available materials by the farmers themselves. The transparent cover reduces heat losses and simultaneously gives the product a certain protection from dust and rain. Aeration required for removing the evaporated water is provided by ascending air forces.

Investigations have shown that insect infestations cannot be totally avoided. During long extended periods of adverse weather, the crop deteriorates. Because of the low capacity of the box and the tent dryers, their use is limited to subsistence farmers. Since the application of such dryers makes no considerable contribution to the income of the farm households, the smallholders are hesitant to invest in them. The drying capacity of natural convection-type dryers can be increased by connecting a solar air heater to a drying chamber (Fig. 2.17).

Instead of only the crop's being used as the absorber, solar radiation is further converted into thermal energy in the solar heater. In this dryer, the sloping collector has to be mounted facing south and tilted at an optimum angle, depending on the region and the particular season. The drying air is heated up in the solar air heater and enters at the base of the drying chamber. It then moves upward and passes across the crop, which is spread in thin layers on vertically stacked trays. Air circulation is effected by the ascending air forces. The air flow rate can be increased either by wind coming from the south or by a chimney.

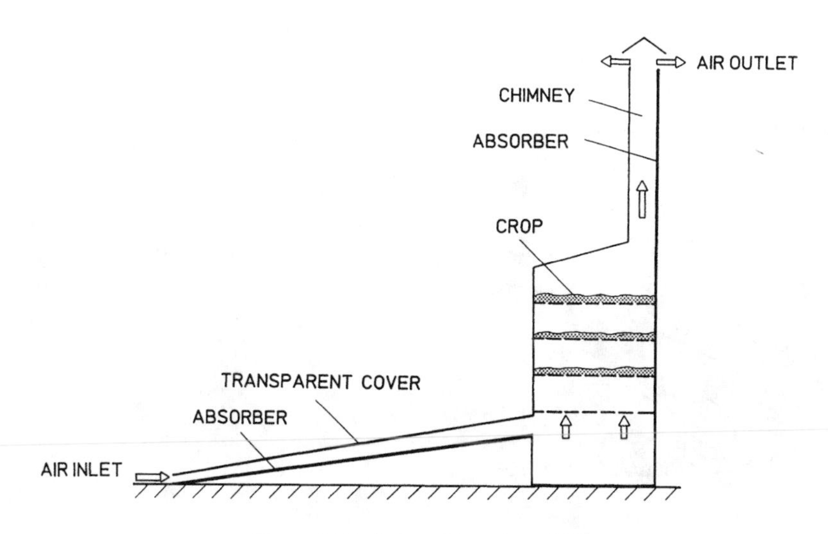


**Figure 2.17. Solar cabinet dryer [23].**

Because of the high air resistance encountered when the air is forced through the crop, only a few trays can be stacked without significantly affecting the air movement. Furthermore, investigations have shown that during the night and cloudy weather, the air circulation breaks down completely. This causes spoilage of the crop due to enzymatic reactions and the growth of micro-organisms. The comparatively high investment, limited capacity, and the risk of crop spoilage during adverse weather conditions have, up to now, prevented the wide acceptance of these dryers.

The highly weather-dependent risk of natural convection-type solar dryers stimulated the development of a solar tunnel in which a fan provides the air flow required for removing the evaporated moisture (Fig. 2.18).

To optimize the use of fluctuating and low-grade solar energy, an extended solar air heater is connected to a tunnel dryer. The solar air heater and the tunnel dryer consist of an insulated floor and a frame along the sides. They are covered with a transparent UV-stabilized plastic foil. The floor of the solar air heater is either painted with black paint or covered with a temperature-resistant black woven fabric. For backside insulation, any kind of locally available material can be used. During drying, the crop is spread out in a thin layer. For easy loading, mixing and unloading of the crop, the solar tunnel dryer is mounted on a substructure, and the cover foil of the tunnel dryer can be rolled up with a water tube. Ambient air is sucked from underneath the dryer by two highly efficient axial fans and is forced between the cover foil and the absorber or crop. The solar air heaters provide the drying air temperatures of 40–80°C required for drying agricultural products [11, 25–27].

Numerous tests in regions with different climatic conditions have shown that fruits, vegetables, cereals, grain legumes, oil seeds, spices, and even fish and meat can be

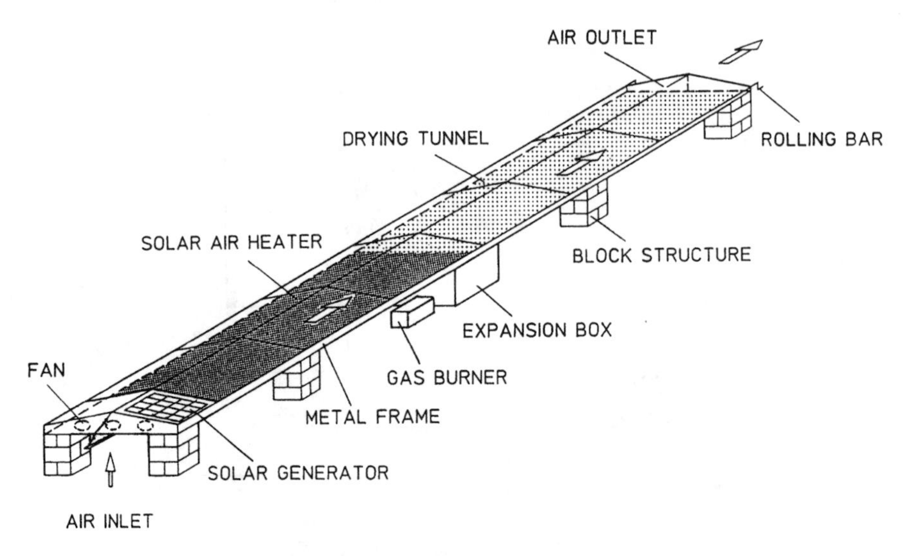


**Figure 2.18. Solar tunnel dryer [10].**

properly dried in the tunnel dryer. In regions with rain during the drying season, the cover has to be tilted to allow draining of the water. Because of the extremely low air resistance when the air is forced over the crop, only 20–30 W are required. This allows a grid-independent operation with only one solar module. The complete protection of the crop during drying, the temperature control by means of the radiation-dependent air flow rate, and mixing the crop in regular intervals result in uniform drying and premium quality. In addition, because of higher drying air temperatures compared with natural sun drying, the drying time can be reduced to almost half. The multipurpose uses, the modular design of the dryer, and local production lead to a payback period of 2–3 years, even when imported solar modules are used. Operating the dryer with electricity from the grid could improve the economic feasibility of its application.

Despite the many advantages of the solar tunnel dryer compared with those of natural sun drying, its adoption can be economically justified only if the farmer gets a higher price for better crop quality.

### 2.3.4 Utilization as Power

*A. Esper, G. Schumm, and W. Mühlbauer*

***Photovoltaic Systems and Components***

The special attraction of photovoltaics compared with that of other power generation technologies lies in the fact that solar radiation is converted directly into electric power by an electronic solid-state process. In general, no moving parts and no specific thermal stresses are involved. Therefore, photovoltaic (PV) systems operate quietly and they can offer extremely high reliability, low maintenance requirements, and a long lifetime. These characteristics are of particular benefit for applications in agriculture and rural

**Table 2.26. PV market shares for main application segments (1990–1994) [28]**

| Application | Megawatt | Percentage |
|---|---|---|
| Grid connected | 50 | 11 |
| Telecommunication | 95 | 21 |
| Other professional | 31 | 7 |
| Water pumping | 54 | 12 |
| Rural domestic electrification | 121 | 27 |
| Consumer applications | 31 | 7 |
| Camping, boating, leisure | 67 | 15 |
| Total | 450 | |

electrification, for which features such as robustness, reliability, simple installation, and simple operation are important.

Because of the nature of the conversion process, one can utilize direct as well as diffuse radiation, which also allows applications in moderate climates with higher fractions of diffuse radiation. Another important advantage of a PV system is its modularity, permitting flexible system sizing for decentral applications down to small load demands.

From small-scale space applications in the 1960's, the commercial PV market has increased to an annual turnover of 90 MW (1996) with a typical annual growth rate of 15% over the past few years. The cost of PV electricity today is still too high for general grid-connected applications, and the application areas at present are dominated by stand-alone systems. These application areas are primarily rural electrification, e.g., water pumping or household electrification, professional applications such as telecommunications or cathodic corrosion protection, and device-integrated consumer applications (Table 2.26).

*Solar Cell*

The smallest independent operational unit of PV systems is the solar cell (Fig. 2.19). The solar cell consists of a specific semiconductor diode, in most cases silicon, with a large aperture area for light absorption. In the conversion process, light is absorbed by the semiconductor, and the absorbed photons generate free charge carriers (electron–hole pairs) that are then separated by the built-in electric field between the $n$- and the $p$-type region. The accumulating charge produces a voltage between the two regions and external contacts, and an electric current can be drawn through an external load.

Depending on the cell efficiency and cell area, the maximum output power for single solar cells is of the order of 1.5 W with typical cell areas of 10 cm $\times$ 10 cm. Output voltages are determined by the electronic properties of the materials involved and are in the range of 0.5–0.7 V. Commercial silicon cells and modules have efficiencies of 12%–16% for the conversion of light into electric power, with polycrystalline cells being typically $\sim$1%–2% lower in efficiency than single-crystalline cells. High-efficiency silicon cells, e.g., for concentrator applications, have been produced with up to 24% efficiency, but they are still at the laboratory stage and restricted to small areas of a few square centimeters. For concentrator applications, gallium arsenide (GaAs) compound semiconductors and related materials are under research, and laboratory efficiencies

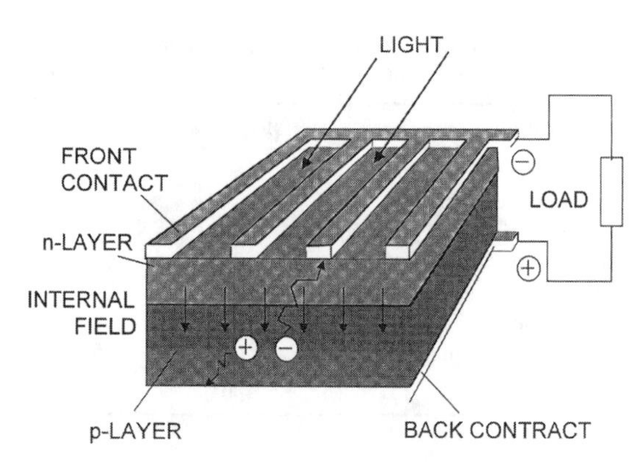


**Figure 2.19. Solar cell.**

above 32% have been achieved. Because of the high cost of GaAs technology, commercial terrestrial applications have not yet emerged.

The production technology for silicon solar cells is based on classical wafer processing in which large monocrystalline or multicrystalline silicon ingots or blocks are formed out of a silicon melt at 1400–1500°C, diced into wafers (0.3-mm thickness and 10 cm × 10 cm area are typical), and then further processed into single solar cells.

To protect the solar cells against the environment, they are usually encapsulated between sheets of glass, metal, or polymer foils. To obtain a desired output power and voltage, a number of cells are typically interconnected into a PV module. Standard flat plate modules with cells in silicon technology dominate the market today. They are designed for voltages in the range of 12–35 V and output powers of 50–200 W, with corresponding module areas of 0.5–2 $m^2$. PV modules may also include rigid frames for better stability and for attaching them to support structures. The vast majority of modules for remote power applications are framed modules with an area of approximately 0.5 m × 0.5 m, an output of ~50 W at 12–15 V, and end-user prices of the order of 300–400 US $.

Although this technology is well developed and has proved reliable, manufacturing costs are still too high for a widespread application in the energy sector. As the cost reduction potential with the classical wafer technology is rather limited, intense research and development has been carried out with alternative materials and production technologies. One strategy for producing flat plate modules that could lead to module costs much lower than with crystalline cells uses thin films (1 $\mu$m,) of materials, such as amorphous silicon (a-Si), copper indium diselenide (CIS), or cadmium telluride (CdTe). The main advantages of thin-film solar modules are (1) lower material consumption and, instead of costly batch processes, (2) utilization of continuous on-line deposition techniques in which the active materials are evaporated or sputtered directly onto large-area glass, metal, or polymer substrates. While present specific manufacturing costs for silicon solar modules range from approximately 3–4 US $/W, it is expected that, with mass production of thin-film modules, costs below 1 US $/W can be achieved as a long-term prospective.

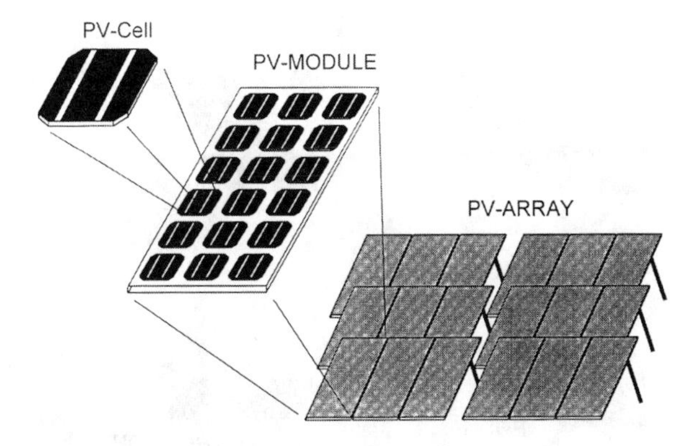


**Figure 2.20. PV-generator.**

*Photovoltaic Generator*

To satisfy a specific power and energy demand by a PV system, a number of solar modules may be electrically interconnected in series and in parallel. The output voltage of the total PV generator is then determined by the number of modules connected in series and the output current by the number of module strings connected in parallel. The size of PV generators may range from single cells with submilliwatt levels (e.g., in consumer products such as calculators) to modules and up to module arrays with many megawatts. The principal structure of a PV generator is illustrated in Fig. 2.20.

The majority of all PV systems installed to date are flat-plate systems with fixed orientation, in which modules are rigidly mounted onto supporting structures. The main advantage of flat-plate systems is that there are no moving parts; therefore the installation can be comparatively simple, and the systems have good reliability and low maintenance requirements. In particular for small-scale systems at remote sites with high-reliability requirements and for which professional maintenance is costly or not available, flat-plate systems are the first choice. But even for large power plants in the past, flat-plate module arrays have been used in most cases.

The cost reduction potential in flat-plate PV systems depends to a large extent on the manufacturing cost of the PV cells and modules, which are the most expensive parts of a PV system. A possible approach to overcome this limitation is the use of tracking and concentrator systems. If PV cells or modules are tracked according to the Sun's movement, an average annual gain in energy harvest, typically between 20% and 35%, can be achieved; the exact value depends on latitude, climatic conditions, and tracking modes [29]. The largest cost reduction potential, however, is associated with additional concentration of the sunlight, because a comparatively expensive solar cell area can be replaced by low-cost standard optical concentrator components such as lenses or mirrors.

Figure 2.21 shows the typical tracking modes applied in field installations. Although the costs associated with the installation and support structures are lower for single-axis tracking compared with the more complicated two-axis tracking, single-axis tracking is an option only for flat-plate modules or for low concentration such as V-trough systems

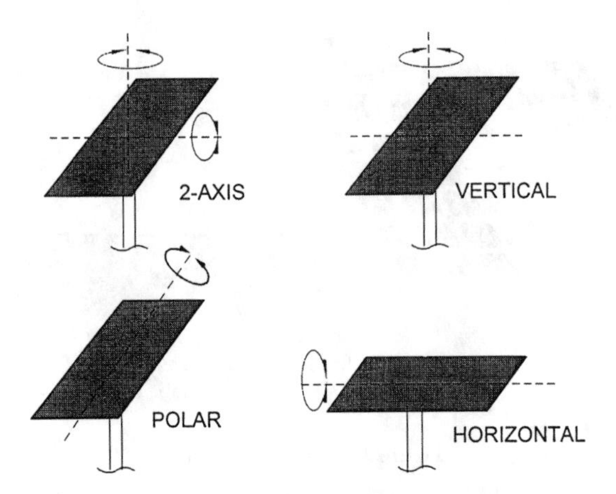


**Figure 2.21. Tracking modes.**

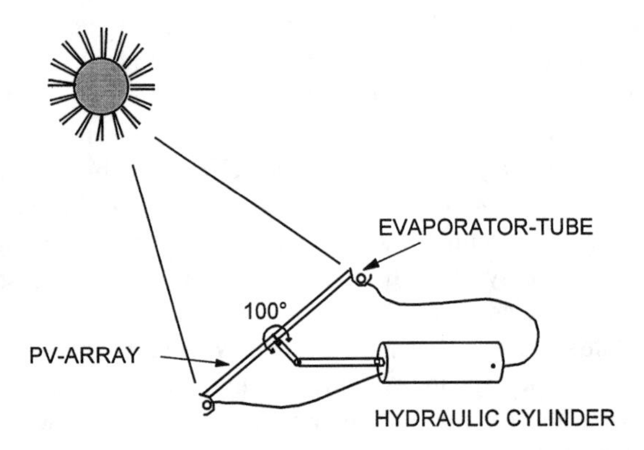


**Figure 2.22. Thermohydraulic tracking principle: Driven by the pressure difference between the two evaporator tubes, the hydraulic cylinder adjusts the PV array until the temperature and the pressure in the two evaporator tubes are equilibrated. This is the case when the Sun is perpendicular to the array.**

in which the losses due to nonideal tracking remain small. Simple passive trackers based on the differential heating of fluids by the Sun and associated piston movements or gravitational forces are popular for these systems (Fig. 2.22).

An example of a simple low-concentration system is the so-called V-trough system with plane-booster side mirrors attached to the modules. These systems have been used already in the early 1980's in large installations (e.g., 6.45-MW grid-connected PV power plant in Carissa Plains, CA) in combination with standard flat-plate modules. However, the PV modules available at that time did not withstand the elevated temperatures and radiation levels they were exposed to in the troughs, and module performance degraded drastically within a few years. In recent years, with better and degradation-resistant module encapsulation materials and techniques, the V-trough systems have regained interest as a low-cost and reliable concentrator option also suitable for decentral applications [30].

With higher concentrations, the requirements for tracking accuracy increase, placing more emphasis on balance-of-system costs for the tracking mechanism and accuracy of support structures and installation procedures. Point- and line-focus systems may be realized with both lens and mirror optics. In contrast to solar thermal systems, in PV systems Fresnel lens optics have been used. In most cases they are made of UV-hardened acrylic, and they have shown good performance so far. A disadvantage of Fresnel lenses, however, is their susceptibility to dust accumulation in the grooves, so they must be protected in a housing, which is a significant cost factor for concentrator modules. At present, parabolic mirror troughs are regaining interest for application in large professional PV concentrator systems with horizontal axis tracking; the installation is similar to that of parabolic trough systems used in solar thermal power plants.

In the past, a number of installations with line- and point-focus systems have been made with system sizes up to a few 100 kW, and today several small firms, mainly in the US, offer concentrators commercially at prices competitive with those of flat-plate installations, but the market for concentrators has remained insignificant up to now. The main disadvantages of concentrator systems are the presence of moving parts and the requirement for direct radiation, which restricts concentrator applications to large and professionally maintained systems in areas with favorable climatic conditions. Nevertheless, as most parts are conventional electromechanical and optical components, concentrators offer a substantial and short-term cost reduction potential already at moderate production capacities.

An important advantage of tracking systems is the constant power output profile produced over the course of the day as the system is always oriented perpendicular to the Sun (Fig. 2.23). In particular for loads connected without a storage battery, such as water pumping or desalination equipment, constant power is desired for running the equipment always at its optimum operating point. For pumping applications, tracking is an option applied widely today, in particular with low-cost passive tracking mechanisms. By use of tracking and V-trough concentration, the savings in required installed PV power can be as much as 50%–60%, leading to an economic advantage of up to 30%–40%, depending on climatic conditions and costs of the tracking structures [27].

*System Configurations*

Depending on the application, a general photovoltaic system may include some of or all the components shown in Fig. 2.24.

***a. PV Generator.*** This consists of a number of modules suitably interconnected, as described above.

***b. Direct Current to Direct Current Converter.*** In PV system design, in most cases the output voltage of the PV generator will be matched with good accuracy to the battery or load voltage. Typical system voltages for small-scale systems are, for example, 12- or 24-V systems. However, as the voltage level of generator and battery is subject to small

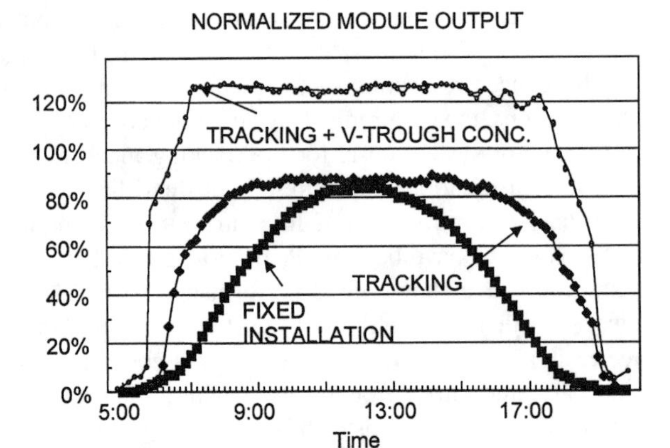


**Figure 2.23. Power output profiles of a PV module or array at a clear day for fixed installation, tracking, and tracking with trough concentration.**

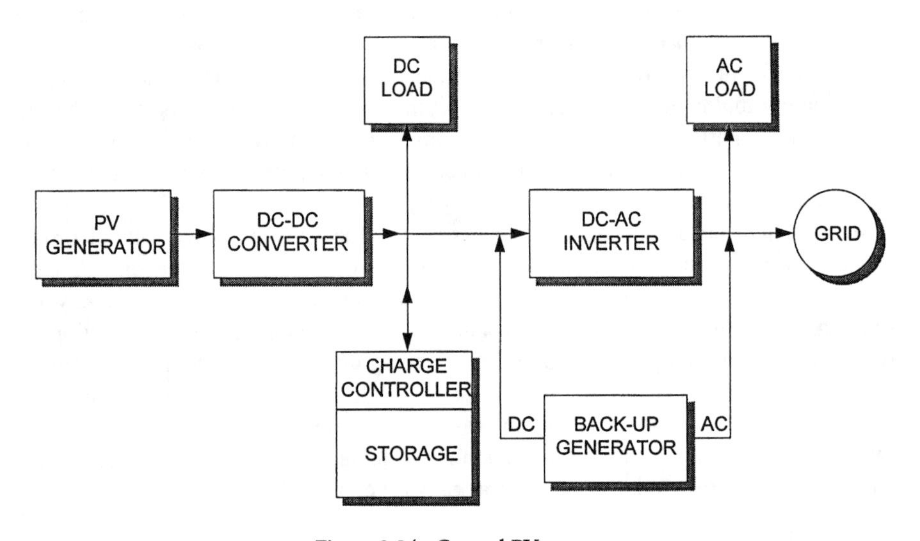


**Figure 2.24. General PV system.**

variations, depending on temperature, charge state, irradiance, etc., in certain cases it may be of advantage to transform the voltage by a dc–dc converter, often in combination with a maximum power tracker, to ensure continuous operation with the highest power output from the generator.

***c. Inverter.*** For PV systems connected to the public electricity grid, an inverter is always required that converts the dc and the voltage produced by the PV generator into an ac with appropriate voltage and frequency levels (e.g., 230 V at 50 Hz). For stand-alone systems only an inverter is required if ac loads are to be operated. This is often the case for larger domestic systems in which a variety of loads are connected. Certain applications such as PV pumping systems with ac motors also require special inverters that are specifically designed for the pumping system and that produce an output of varying frequencies, depending on illumination and corresponding motor speed.

***d. Storage.*** For stand-alone systems, in most cases a storage battery or a backup generator (e.g., diesel) is required for providing power during cloudy and dark periods. There are, however, specific applications for which storage batteries can be omitted. An example is the PV pumping system. Here, the pump is operating whenever there is adequate illumination, and storage is achieved when the pumped water is collected in a tank.

***e. Charge Controller.*** A simple PV system with battery storage may be operated just with direct connection of modules to the battery. This is often the case for small systems or for systems with battery types not sensitive to overcharging. However, with the frequently used lead–acid batteries in which overcharging may result in accelerated deterioration, in most cases it is of advantage to include a charge controller. For small systems up to a few 100 W, charge controllers are usually used in the shunt mode, in which the PV generator is short circuited when the battery is full. For larger systems, charge controllers are usually connected in series with the battery. Charge controllers often include additional functions for operational safety and for system survey and maintenance, such as deep discharge protection of batteries and system status indicators (battery full, empty, charging, etc.).

***Operational Behavior***

*I–V Characteristics*

The operational behavior of solar cells and modules is determined by the nature of the electronic solid-state processes involved. To a good approximation, the solar cell may be described as a semiconductor diode with its corresponding diode $I-V$ (current–voltage) characteristics connected in parallel to a current source, with its current $I_{ph}$ proportional to the irradiance. The resulting $I–V$ curve of such a device is shown in Fig. 2.25. The general shape of the solar cell $I–V$ characteristics is maintained for PV modules and generators as well; the only difference is the scale on the current and voltage axis. The voltage is determined by the number of cells connected in series and the current by the number of modules in parallel.

In practical applications, solar cells and modules in general do not operate under fixed standard conditions. The two most important effects that must be taken into account for PV systems are variations in irradiance and in temperature (Figs. 2.26 and 2.27). Although the main effect of irradiance is a change in electric current proportional to the irradiance, temperature variations produce primarily a shift in voltage. Typical values for the voltage shift are −0.5%/K.

***Maximum Power Point.*** The electric power $P$ that can be extracted from the solar cell or module at an external load (see Fig. 2.19) is given by $P = IV$. It is evident that under short-circuit ($I = I_{ph}$, $U = 0$) or open-circuit conditions ($I = 0$, $U = U_{\infty}$) the extracted power is zero. To draw the maximum available output, the operating point on the $I–V$ curve must be adjusted to the so-called maximum power point (mpp) in which the power

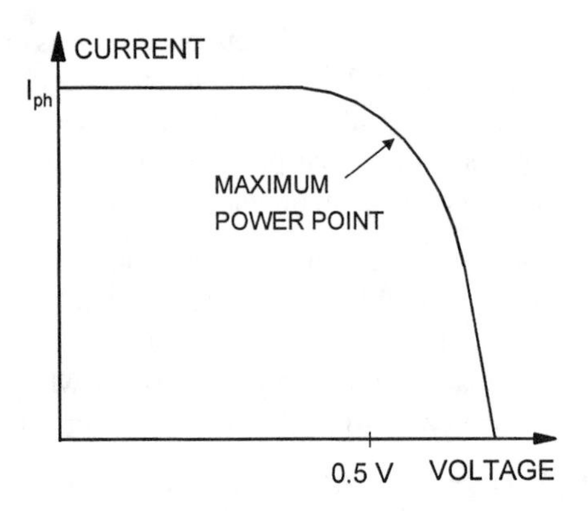


**Figure 2.25. I-V characteristics of solar cells. Also shown is the operating point, at which the cell deliveres the maximum power.**

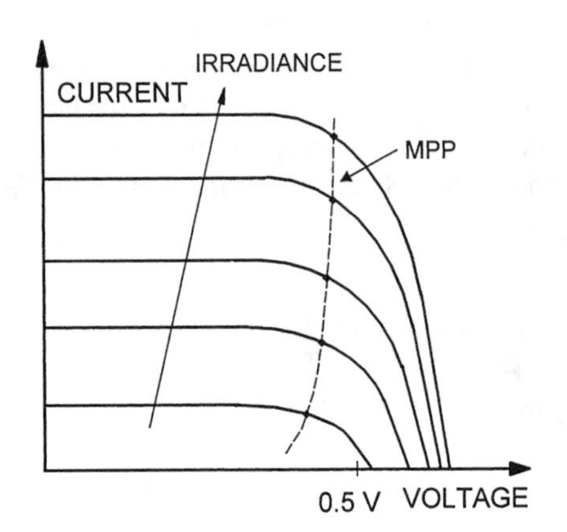


**Figure 2.26. Irradiance dependence of I-V characteristics. Also shown is the change of the mpp with irradiance.**

$P = IV$ is maximized (Fig. 2.25). Depending on the quality of the cells or modules, the mpp current is of the order of 10% less than the short-circuit current, and the mpp voltage is ~20% less than the open-circuit voltage. To operate the module continuously at the mpp for various operating conditions, an electronic load may be used that is continuously adjusted by an internal tracking algorithm. For larger PV systems, mpp trackers are often implemented into the dc–dc converters or dc–ac inverters connected to the output of the PV generator. Figures 2.28 and 2.29 show the range of operating points for a PV generator connected directly to a battery or a load under various illumination conditions. In particular in the case of battery connection, the average loss due to deviations from the mpp is rather small. Taking into account that the electronic mpp trackers have additional

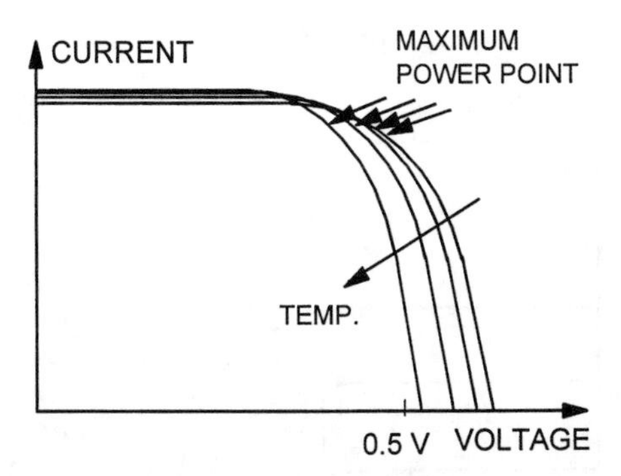


**Figure 2.27. Temperature dependence of IV characteristics. Also shown is the change of the mpp with temperature.**

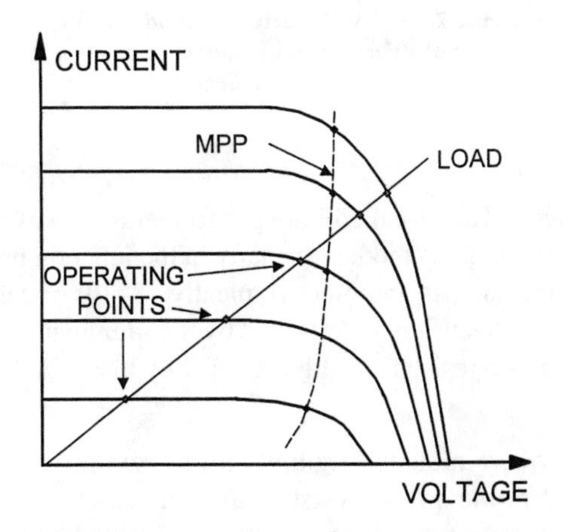


**Figure 2.28. I-V characteristics and working points of PV system with ohmic load for varying irradiance.**

power losses and add cost and complexity to the system, it is often more cost effective, in particular for small systems (up to a few 100 W), to install the systems without mpp tracking devices.

*Partial Shading*

For PV modules and generators, in which many cells are connected in series, an additional problem may occur when modules are inhomogeneously illuminated, for example, because of dirt accumulation or partial shading. An analysis of the electric circuit shows that the shaded cells can act as loads that consume the electric power that is generated by the other illuminated cells. Under unfavorable conditions, the power dissipation at the shaded cells can lead to excessive overheating of the cells. If the cell temperature rises above 85–100°C, the surrounding encapsulation material and eventually the whole

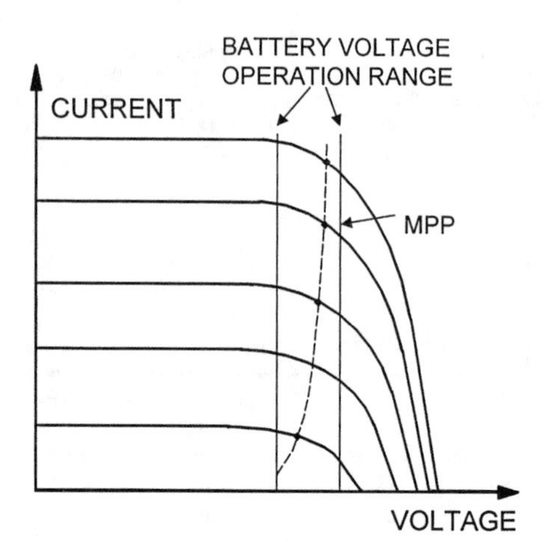


**Figure 2.29. I-V characteristics and working points of PV system with battery storage for varying irradiance.**

module will be damaged. This is called hot-spot formation. To avoid or reduce this effect, it is customary to use, in the module (usually in the junction box), so-called bypass diodes that are connected across a number (typically 18–20) of cells in a string. In the case of shading, the module current determined by the illuminated cells may bypass the shaded cells across the low-resistance path provided by these diodes.

*Orientation*

If modules or arrays are mounted rigidly with a fixed position, the annual energy harvest and generated power profile versus day and season are determined by their orientation. The maximum annual energy harvest is obtained with an inclination of the modules facing the Equator and with a tilt angle corresponding roughly to the geographic latitude. Deviations by 10°–15° in each direction may be tolerated without significant losses being suffered. For specific desired load profiles, orientations may be modified. For example, if the system is to be optimized for highest availability over the year, often a steeper tilt angle is selected. Such an orientation boosts energy harvest during the winter when the Sun's position is low at the expense of summertime. In general, a minimum tilt angle of 15°–20° is maintained to utilize self-cleaning effects by rain and so to avoid excessive dust and dirt accumulation on the module surface.

***Application in Agriculture***

The application of photovoltaics in agriculture depends mainly on the prevailing frame conditons. In this context, significant differences exist between industrialized and developing countries. In industrialized countries, the electrification of farms is widely advanced. In the current organization, farms cannot be operated without access to electricity [31].

Therefore photovoltaics are applied only in cases in which energy must be supplied far away from either the farm or the mains, e.g., for drip irrigation in horticulture, electrified livestock fencing, or a drinking trough on meadows. Prerequisites for the application of photovoltaics in industrialized countries are the investment costs, which should be lower than the cost of a grid connection, an agricultural revenue that will exceed the running costs of the PV system, or laws that will prohibit the use of diesel-driven generators in protected areas [32]. Applications for which energy demand and energy supply are congruent are preferable.

In the developing countries themselves, there is another disparity regarding the electrical energy supply between urban and rural areas that contrasts sharply with the economical importance of most rural areas. While the urban areas are mainly fully electrified, only one third of the more than 2.5 billion people living in the rural areas have grid-based electricity [33, 34].

The current rural energy supply is characterized by supply bottlenecks, blackouts, high costs and allocation disparities, low density of energy demand, extremely low energy consumption, and high technical and nontechnical losses. The revenues from the sale of electricity do not, as a rule, cover more than ~15%–30% of the costs of generation and distribution [35].

The amount of electrical energy required for improving the basic living environment in both the private and the public sectors is only ~300 Wh per person per day. Electricity is needed mainly for house and street lighting, telephone and radio transmitters, health stations, pumping and purification of drinking water, or for powering educational, emergency, and communication equipment such as radio or television [36].

Establishing electrical distribution systems in rural areas of developing countries has reached its technological and economic limitations. As a feasible alternative, more or less decentralized installed, insularly operated diesel power plants have spread rapidly. Fluctuating maintenance costs, particularly of small aggregates, high costs for the diesel oil and spare parts, environmental stress, and risks through transportation and operation have been and continue to be the major problems of these installation.

A World Bank study shows that the hopes of universal electrification through conventional grid services are dim, given the constraints of high capital and operational and maintenance costs. According to the rural energy demand analyses, a combination of several energy technologies in areas with typically low energy demand densities are in many cases the most economical solutions. Therefore the use of photovoltaics is becoming more and more attractive to supply small loads located in remote areas when the conventional approach of connection to the main grid or the use of diesel generators is more expensive. The range under 1 kW of power, which has by far the greatest market potential in sparsely populated rural areas, is the ideal application field for the use of photovoltaics [37].

*Electrical Energy Supply for Rural Households*

Photovoltaics as an energy source for rural areas must be seen in the context that over 90% of the energy consumed is used for cooking [38]. PV systems are not suitable for meeting these needs. 90% of the rural families have a daily electrical energy demand of less than 200 Wh [38]. To meet the electrical energy needs of private households in rural areas, three different technologies can be applied: central station power plants (isolated local grids), battery-charging stations, and so-called solar home systems (SHS's).

***a. Central Power Station.*** Although technically feasible, central power stations with PV generators have up to now not proved to be an economic alternative to a diesel-driven central power station or to a decentralized energy supply. If energy is needed primarily at night, when the output of the solar generator is zero, expensive storage facilities must be included in the system. The conventional 230-V network, which is favorable over a 12- or 24-V network because of the significant lower losses, must be equipped with an inverter to provide the desired 230-V ac. In this case it is usually cheaper to obtain electricity from diesel generators [35]. According to feasibility studies, in the future it will not be economically feasible to utilize PV technology and storage batteries to provide an uninterrupted supply of electric power over a 230-V network unless there is a relatively large, continuous demand for that power.

***b. Battery-Charging Station.*** A battery-charging station is a more decentralized energy supply option than the isolated village grid. Battery-charging stations have typically a generator capacity between 300 and 1000 W. Since they have no power distribution network, individual users have to take their own batteries to the station; the load management is also their responsibility. With just a small number of panels, a technically optimal supply of electric power is ensured, even if demand levels are low. If demand increases, the capacity of the the station can be expanded easily simply by the addition of more panels. The system can also be installed in the immediate vicinity of the users, thus reducing transport costs and time. The closer distance to the consumer is the major advantage of the PV-powered over grid-fed or diesel-powered stations.

Battery-charging stations are an interesting option for areas in which family income is generally low and the major share of the target group cannot expect to be able to afford a SHS, the more expensive alternative in the near future [35].

***c. Solar Home System.*** The SHS represent the most decentralized mode of energy supply. If produces just enough energy to meet the basic energy needs of a typical rural household for the operation of lamps and home entertainment equipment. SHS's normally do not compete against other electrical installations; they replace kerosene, candles, dry-cell batteries, and other charging options for automotive batteries.

SHS's usally consist of one or two solar modules with a power output of ~50 W, mounting structure, lead–acid batteries with ~50–100-A h capacity, a charge/discharge controller, and one or two fluorescent lamps of 10 to 15 W each (Fig. 2.30). The module converts solar radiation directly into electricity, which is then stored in a battery to make it available during off-sunshine periods. To enanble the installation of a reliable SHS, the charge regulator must be equipped with high- and low-voltage disconnects, a LED indicator of the battery charge, protection from moisture and corrosion, protection from short circuiting and reverse polarity. Furthermore, fluorescent lamps with a well-designed ballast have to be used to prevent excessive power consumption and short lifetime [39].

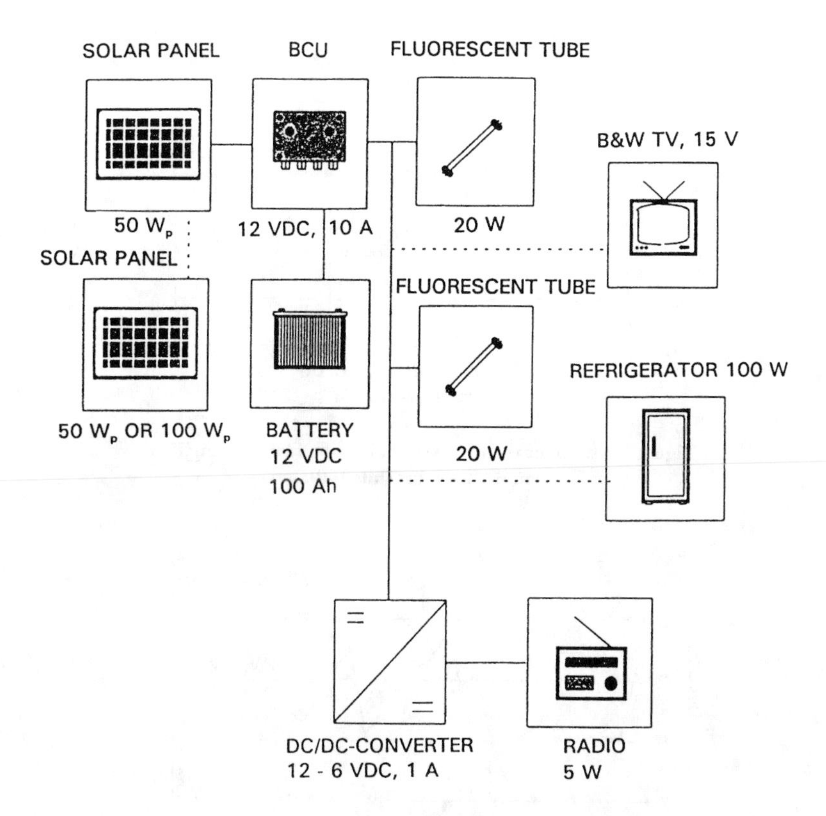


**Figure 2.30. System diagram for a Solar Home System [35].**

A typical daily load profile of a family living in a rural village in Chile that uses three 11-W lamps and a radio or a TV is shown in Fig. 2.31.

Up to now the load demand of most families in the rural areas is considerably lower than the power supply of a SHS (Fig. 2.32) [40]. This reveals the potential use of a SHS for labor-intensive operations such as milling or pumping.

In the past few years, considerable attempts were made to introduce SHS's. Up to now more than 300,000 SHS's that were imported, as well as locally produced ones, were installed to meet the basic energy needs of the rural population. In many developing countries, the local production of components or the whole system was initiated.

Experiences showed that the introduction of these systems had a significant impact on the social life of the users. Living conditions in the evening are made easier, working hours are extended, and access to information and entertainment by radio or TV is made possible. In this respect, the use of SHS's plays a big role in improving the living standard of rural people and somehow prevents the migration problem from the rural to the urban areas.

*Photovoltaic Pump System*

For the rural population it is of vital importance to have pure drinking water as well as water for animals and for irrigation of crops. But approximately two thirds of the

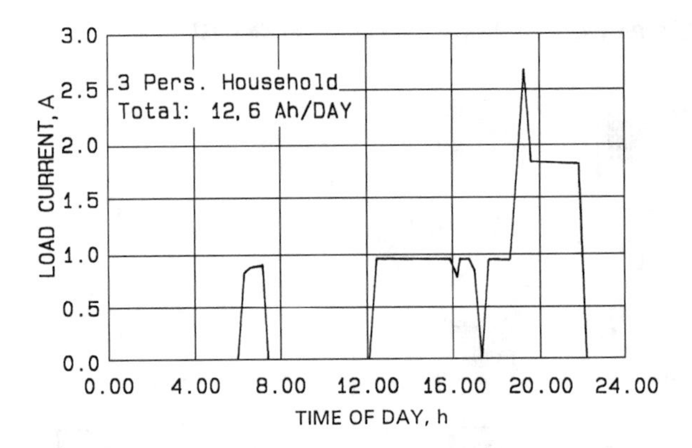


**Figure 2.31. Typical daily load profile of a family living in a rural village in Chile [40].**

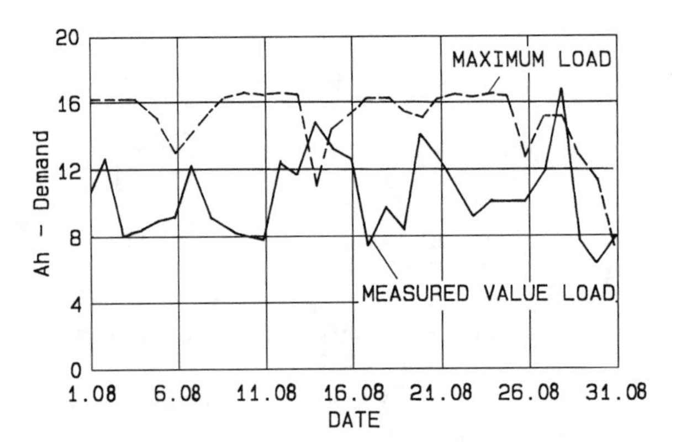


**Figure 2.32. Daily course of load demand and power supply [40].**

world's population have to survive without adequate water supplies—in most cases in rural areas without a power grid. Diesel water pumps are expensive to run and require fuel and regular maintenance [35].

Pumping of water in remote and isolated areas is today one of the most promising fields of PV application. More than 10,000 PV pump (PVP) systems have been installed worldwide. A wide range of application for autonomous PVP systems exists: village drinking-water supply, drinking-water supply for livestock, or small-scale irrigation [41].

In arid and humid regions, contaminated surface or shallow water is often used for domestic purposes. The use of inferior-quality water in these areas is one of the major causes of illness. At a daily domestic water requirement per capita of 10–30 l, a village of 1000 inhabitants needs a minimum amount of 10 to 30 m$^3$ of water every day [42].

This daily water demand remains almost constant. To satisfy these requirements, water has to be pumped from deep wells with pumping heads of 20 to 50 m; the daily energy requirement ranges from 0.5 to 4 kWh.

The daily drinking-water demand of animals depends mainly on the species. Poultry has a daily requirement of 0.2 l while dairy requires 70–100 l per day. In intensive animal production systems, the daily demand is almost constant while in extensive production systems the drinking-water demand occurs only during dry periods. For example, a herd of cattle with 100 animals requires 2–4 $m^3$ per day. To pump water from a depth of 20–40 m requires energy from 0.1 to 0.4 kWh.

Irrigation is an important input to secure production of different agricultural products. It is a prerequisite to increase crop yields and the number of harvests per year. Furthermore, it allows continuous crop production in areas with insufficient precipitation. The water requirement of 100 $m^3$/ha is comparatively high and depends mainly on the type of crop, the stage of vegetation, precipitation, and the losses of water along the distribution systems. Distribution losses range from 15% for drip irrigation to 50% for furrow irrigation. Normally surface water is used for irrigation purposes. The pumping heads usually range from 0.5 to 5 m. The daily energy requirement for irrigating 1 ha ranges from 0.14 to 1.4 kWh. In contrast to drinking water, irrigation-water demand varies directly with irradiation.

Solar pumping systems compete with both hand pumps or animal-driven pumps and with motor-driven pumps. Hand pumps, because of their low power output of 20 W for 6 continuous working hours, can provide 0.12 kWh. On the other hand, animal-driven pumps require power gears and their use is limited to certain regions. Considering a net available power of 150 W when donkeys are used or up to 600 W when a pair of oxen is used, at 4–6 working hours, at least 0.6 to 2.4 kWh are provided per day (Table 2.27).

Deep wells that are used to satisfy larger water daily demands require motor-driven pumps. However, aside from the considerably high investment cost, they require maintenance and repair as well as diesel fuel that, if ever available, is expensive in rural areas.

To close the gap between small manual or animal-driven pumps and motor pumps, PVP's were developed. PVP's consist of a solar generator, an electric motor, a water pump, and a water storage tank (Fig. 2.33).

**Table 2.27. Average power output of manual operation, draft animals, and one-wheel tractor**

| Energy Source | Average Power Output (kW) |
|---|---|
| Human (manual operation) | 0.01–0.08 |
| Donkey | 0.1–0.2 |
| Cattle | 0.2–0.4 |
| Ox | 0.3–0.5 |
| Mule | 0.3–0.6 |
| Camel | 0.4–0.7 |
| Horse | 0.4–1.0 |
| Buffalo | 0.4–1.0 |
| One-wheel tractor | 10–15 |

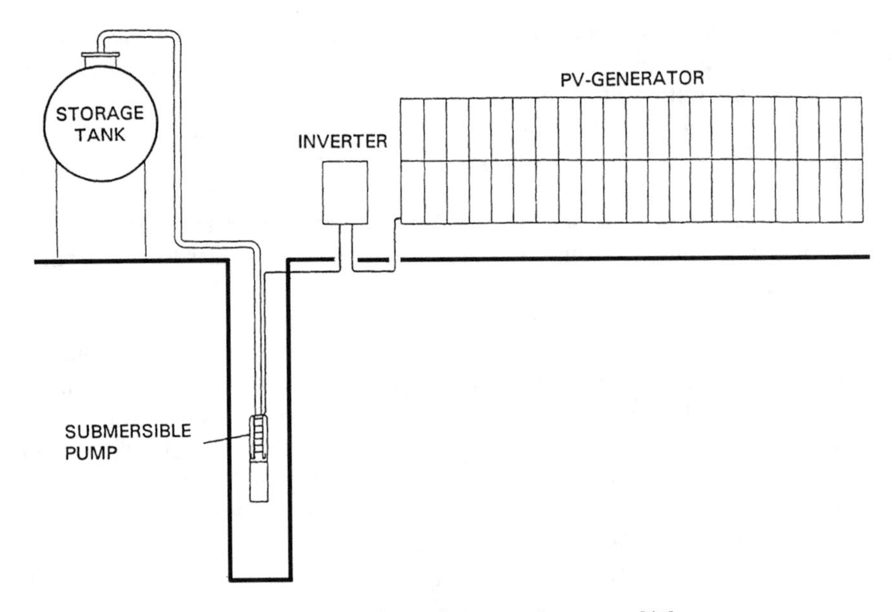


**Figure 2.33. Photovoltaic pumping system [43].**

The interaction between insolation, module performance, delivery head, water level dynamics, and water volume delivered is shown in Fig. 2.34.

Since PV modules produce dc electricity, the use of dc motors seems logical. However, even if dc motors have higher potential efficiency, they contain brushes that need regular replacement and therefore are not recommended for submersion. The dc motors that are used in only the lower power range (<1 kW) have to be installed on top of the well and connected to the pump by a shaft. On the other hand, brushless, maintenance-free, and electronically commutated motors are offered in only the low power range and have the disadvantage of having significantly lower efficiency because of the losses of the electronic control system. As an alternative, ac motors can be used for swim pumps in combination with dug wells or open water. In larger installations (1–5 kW), three-phase ac motors are recommended in combination with submersible multistage centrifugal pumps. The required ac power is delivered by an inverter. An inverter with mpp tracking has an efficiency range of 90%–95%. In general, batteries are not used in PV pumping systems because water stored in tanks is generally more economical.

It must be emphasized that the designs of electric motors and pumps are intended for grid operation under constant power supply. The major optimization criteria for these units are high reliability, high capacity in relation to scale, minimum costs, and high efficiency under normal operating conditions. The utilization of solar energy for pumping water requires a special adaptation of the pumps and motors based on the specific properties of the solar radiation. As a result of the fluctuation of solar radiation, part load is the most probable operating condition for solar pumping system. Figure 2.35 shows the influence of the load ratio on the efficiency of electric motors and pumps. As

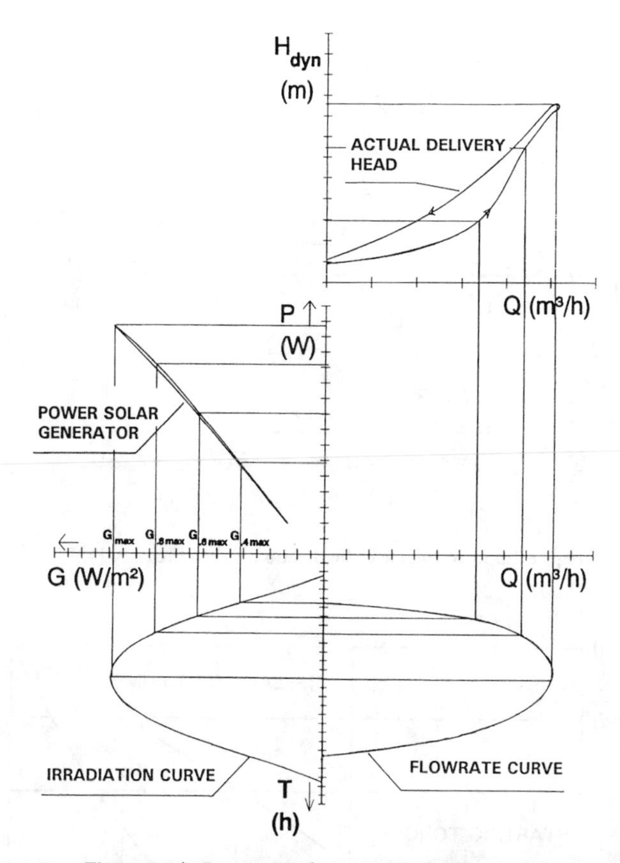


**Figure 2.34. Interaction between insolation, module performance delivery head, water level dynamics and water volume delivered [44].**

indicated, the start-up torque of the pumps influences the system efficiency significantly (Fig. 2.36).

Piston pumps, for example, require a relatively high starting torque. On the other hand, hydrodynamic pumps could start at even low insolation. However, it requires a much higher minimum rotating speed to pump the water at a certain pumping head. This means that it requires much higher insolation when the pump is in operation.

Since solar panels are expensive, system optimization is urgently needed to reduce the panel area. Because of the low efficiency of the solar modules (7%–10%), low threshold, and low part-load efficiency of motor and pump, the daily system efficiency reaches 2% to 3%. Even under steady-state conditions and at full power operation, the system efficiency reaches only 5%.

PV pumping systems have been successfully tested in many developing countries and are commercially produced. Because of the high investment and the low system

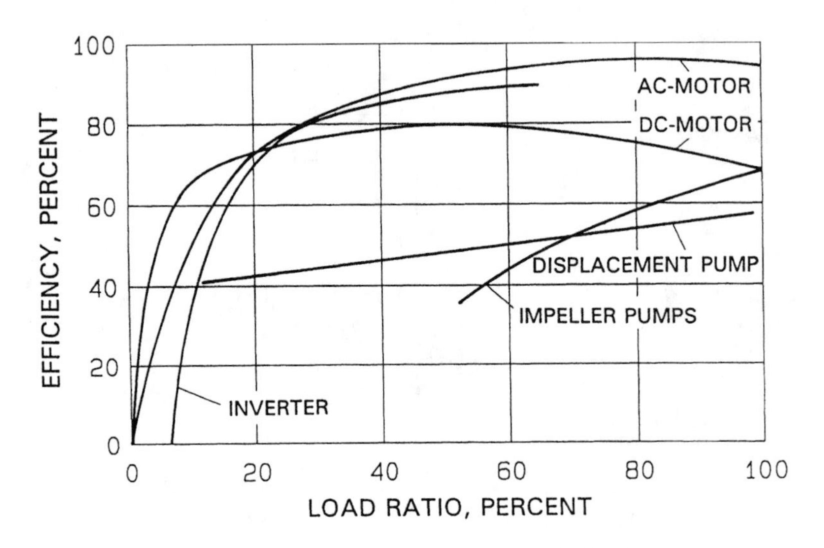


**Figure 2.35. Influence of load ratio on the efficiency [42].**

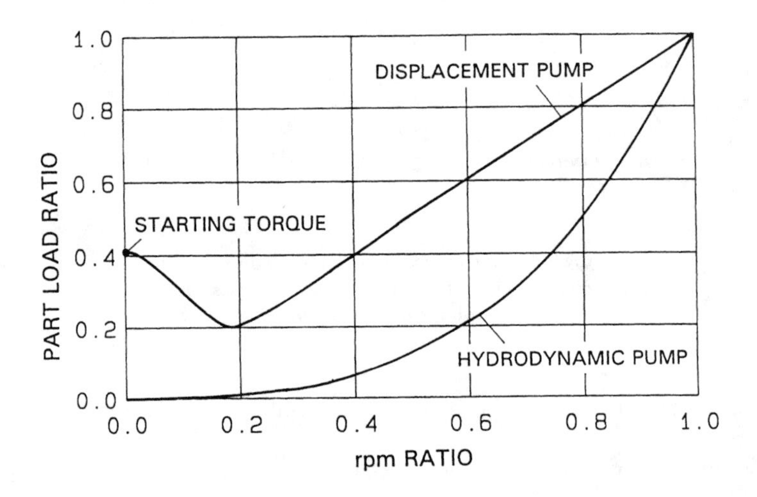


**Figure 2.36. Start up torque of pumps [42].**

efficiency, PV pumping systems are feasible only in remote areas with sufficient solar radiation (>4.5 kWh/$m^2$). Furthermore, proper matching between energy supply and water demand is required.

*Photovoltaic Drives*

In tropical and subtropical countries, the use of machinery in agriculture and commerce that leads to a significant improvement in the labor conditions depends on the availibility of electical energy. Up to now in rural areas the power required for driving small-scale agricultural machinery is borne either by manpower, draft animals, or fossil fuel for power generators. The load range lower than 500 W, which on the one hand can hardly be provided by manpower and on the other hand is rather insufficiently served by internal-combustion engines, seems to be particularly apt for the application of PV energy sources [45]. Its specific benifit compared with that of conventional engines is due to the high costs for the engines in the low load range. The use of diesel generators is beset with many problems such as meeting the maintenance requirements of the generators, a lack of a properly coordinated fuel supply, and a lack of adequately trained technologists to mantain the generator. Besides the effect of releasing women from hard physical labor, the use of appropriate machinery allows stable product quality in the case of mills or even inferior products in the case of dryers and will shorten labor time significantly.

The mechanization of farm labor, which is already in progress, could be guided toward more efficiency on both microeconomic and macroeconomic grounds if PV drives were supplied to substitute for combustion engines.

PV drives will be used only if they can compete with conventional systems with respect to investment cost, repayment, and capital appropriations. The main cost factor of a PV drive is still the PV system, particularly the solar modules. Therefore, before a PV system is developed, the power requirement of the equipment itself has to be minimized.

PV-driven small-scale agricultural machinery, such as rice hullers, oil mills, and graters are currently not available on the market, and even in experimental stages they have been only rarely investigated. A prototype of a grain mill had been optimized in terms of the geometry of the grinder and the material composition of the mill stone in order to enable the use of photovoltaics for driving the motor and to ensure a high product quality [43]. The favorable geometry of the grinder not only reduces the energy requirement of the motor but also enables a gentle milling. The PV-driven grain mill has already proven its technical performance and reliability in operation within practical tests [46]. A PV module with a power of 50 W was directly coupled to the motor. The daily amount of grain ground to a finely crushed grain results in 15 kg of wheat, 12 kg of millet or 10 kg of maize. An economical feasibility study showed that the initial costs of the newly developed grain mill are significantly lower than the installation costs of a small combustion engine.

Because of a comprehensive optimization concept, a PV-driven solar tunnel dryer was developed and introduced in more than 35 countries worldwide [26, 27, 47].

In a solar generator, different modes for driving the drying process can be distinguished. On the one hand, the continuous operation of the fan during day and night is analogous to conventional high-temperature drying; on the other hand, the direct drive of the fan depends on the solar radiation (Fig. 2.37).

The solar tunnel dryer can be operated with one PV module only in the case of direct coupling of the solar generator and the dc motor, when continuous operation of the fan is not required. The solar generator is installed at the inlet of the solar air heater. This enables cooling of the solar generator by ambient air that is forced underneath the back of the module (Fig. 2.38). Despite comparatively low efficency, car cooling fans with dc motors are the cheapest and most adequate solution since they are available in developing countries at low prices.

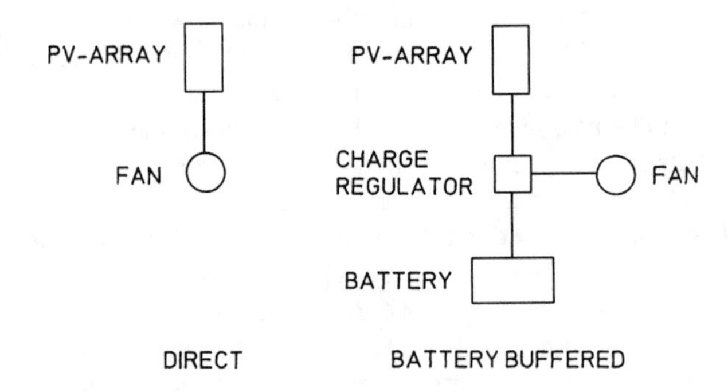


**Figure 2.37. Photovoltaic drives [47].**

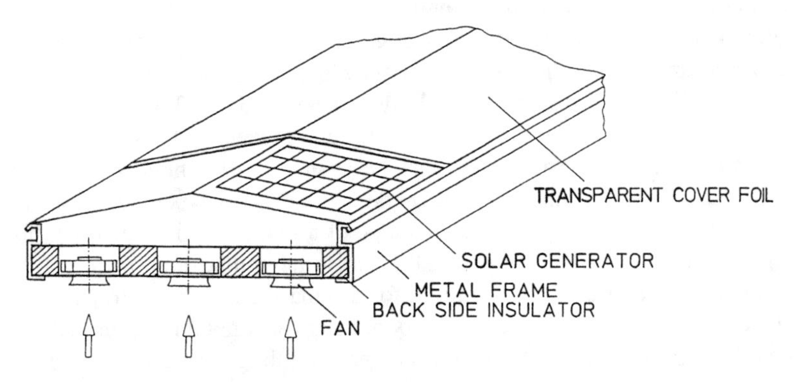


**Figure 2.38. Solar air heater with axial flow fans and integrated solar generator [26].**

Continuous operation of the fan is required only in the case of perishables such as grapes or cocoa in the first phase of the drying process. In this case the PV system must be equipped with battery storage for providing the energy required for ventilation during the night or adverse weather. The main advantage of the battery-buffered system is that the operating point of the fan will vary within only a small range according to the fluctuations of the voltage and the current, which depend on the stage of charge of the battery system. The dimensions of the fan used in this type of system can be measured analogous to those of the grid-connected one. The disadvantages of the battery-buffered systems are that the a larger PV array and a comparatively expensive battery system are needed. As experience showed, in tropical and subtropical countries, because of the poor manteinance and the prevailing weather conditions, a short life span of the battery system can be expected.

Using a battery-buffered system that provides a nearly constant air flow, which results in a temperature profile at the outlet of the solar air heater, is similar to the course of the global solar radiation. This means that in the morning and the late afternoon the drying air temperature is near the ambient temperature, which is not sufficient to accelerate the

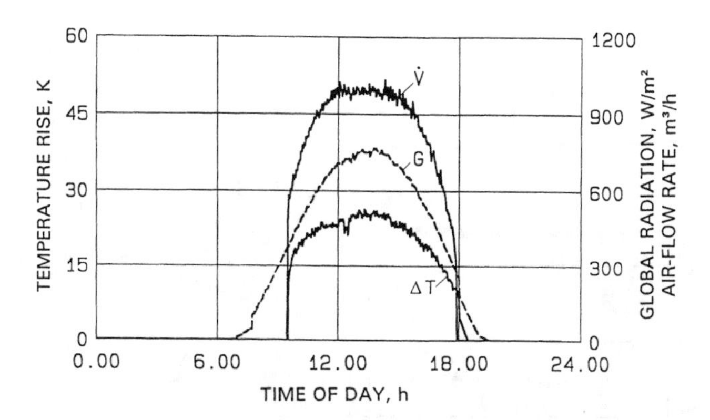


**Figure 2.39. Course of temperature rise and air flow rate in a solar tunnel dryer equipped with a direct coupled PV-driven fan [26].**

drying process. At noon at high solar radiation the drying air temperature exceeds the permissible temperature. The requirement of an almost constant drying air temperature can be considerably better matched by a PV-operated fan (Fig. 2.39).

The temperature of the drying air is automatically controlled by the irradiation when a PV drive is used. During periods of low radiation, the air flow rate results in a comparatively high temperature rise. High radiation causes a high air flow rate, which results in a relatively low temperature increase [47]. The fans started and stopped at the global solar radiation threshold value of 106 and 102 W/m$^2$, respectively (Fig. 2.40). The higher threshold value for starting the fan is due to the higher torque required for starting the fans [27].

*Aeration/Ventilation*

The use of photovoltaics for the ventilation of greenhouses or the aeration of fish ponds will gain more importance, since, in these applications, maximum energy demand and energy supply occur during the summer.

To enable or even to increase the production of animals or plants, greenhouses or stables must be kept at a moderate temperature level.

Intensive aquaculture is being applied more because of the continuously increasing demand of protein supply of the worldwide growing population. Aquaculture requires artifical aeration for keeping the oxygen content of the water between 2 and 5 mg/l. Surface aeration with systems like paddle wheels or gyroscopic aerators is a favorable technique as it gives a better performance in shallow ponds common in aquaculture than does blowing air through the water. For this application the use of photovoltaics seems to be an interesting option, since aeration is mainly necessary during the summer, although aeration has to be done during the night. During the day, the water plants themselves can produce enough oxygen. Therefore all systems have to be equipped with a battery buffer. The size of the system is mainly influenced by the water temperature and number of fish in the pond.

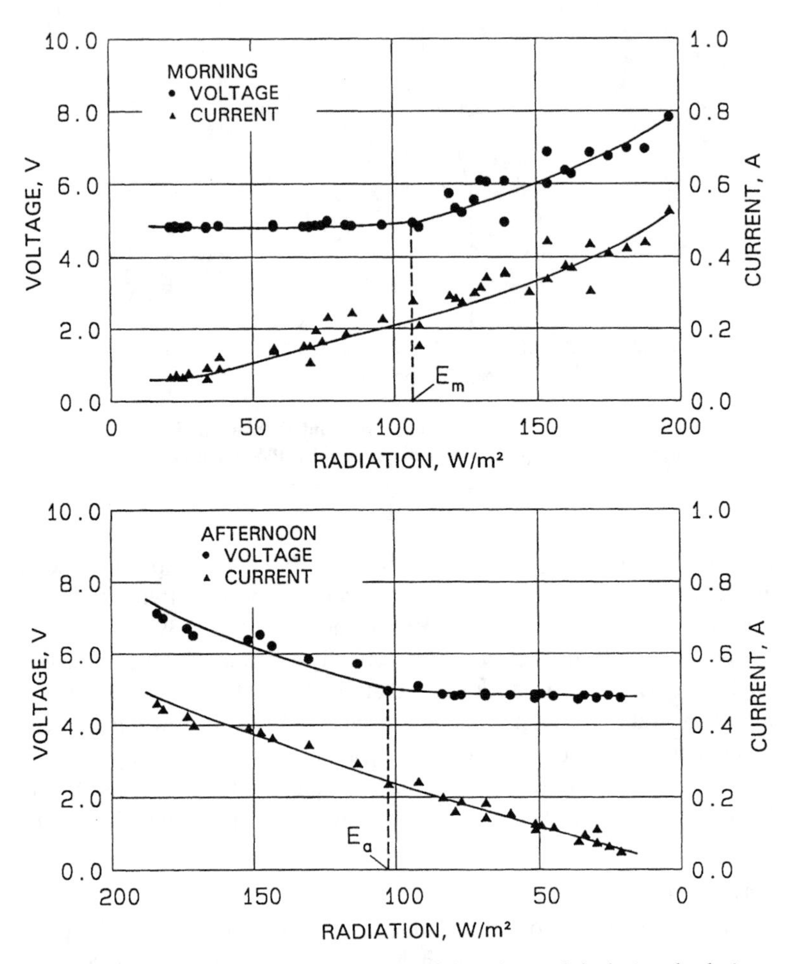


**Figure 2.40. Current and voltage output of the solar module during the drying process [27].**

# References


1. Thekaekara, M. P. and A. J. Drummond. 1971. Standard values for the solar constant and its spectral components. *Nat. Phys. Sci.* 229(6):6–12.
2. Duffy, J. A. and W. A. Beckmann. 1980. *Solar Energy of Thermal Processes*. New York: Wiley.
3. Halacy Jr., D. S. 1980. Solar energy and biosphere. *Solar Energy Technology Handbook Part A*. Engineering Fundamentals. Eds. Dickinson W. C. and P. N. Cheremisinoff. New York: Marcel Dekker.
4. Garg, H. P. 1982. Treatise on solar energy. *Fundamentals of Solar Energy*, Vol. 1. New York: Wiley.

5. Kleemann, M. and M. Meliss. 1988. *Regenerative Energiequellen*. Berlin: Springer-Verlag.
6. Liu, B. Y. H. and R. C. Jordan. 1960. The interrelationship and characteristic distribution of direct, diffuse and total solar radiation. Solar Energy 4(3): pp. 3.
7. Stoy, B. 1980. *Wunschenergie Sonne*. Heidelberg: Energie-Verlag.
8. Mühlbauer, W. 1979. Drying of agricultural products with solar energy in the Federal Republic of Germany. *German-Greek Solar Conference*: 1–18.
9. Linckh, G. 1993. Thermodynamische Optimierung von Luftkollektoren für solare Trocknungsanlagen. *Forschungsbericht Agrartechnik des Arbeitskreises Forschung und Lehre der Max-Eyth-Gesellschaft*, No. 207. Frankfurt: Max-Eyth-Gesellschaft.
10. Mühlbauer, W. and A. Esper. 1995. Solar air heaters-state of the art and future perspective. *Proceedings of the ISES Solar World Congress*: pp. 310.
11. Mühlbauer, W., J. Müller, and A. Esper. 1997. *Sun and Solar Crop Drying*. New York: Marcel Dekker.
12. Mühlbauer, W., A. Esper, and J. Müller. 1993. Solar energy in agriculture. *Proceedings of ISES Solar World Congress* 8:13–27.
13. Mühlbauer, W., W. Hofacker, H. M. Müller, and M. Thaler. 1981. Die Kaltlufttrocknung von Weizen unter energetischem und mikrobiologischem Aspekt. Grundlagen Landtechnik 31(5):145–154.
14. Mühlbauer, W., G. Maier, T. Berggötz, A. Esper, G. R. Quick, and F. Mazaredo. 1992. Low-temperature drying of paddy under humid tropical conditions. *Agricultural Mechanization in Asia, Africa and Latin America* 23(4):33–41.
15. Eckström, N., R. Henriksson, and G. Gustavsson. 1984. The application of solar collectors for drying of grain and hay. *Proceedings of UNESCO/FAO-Working Group Meeting Solar Drying*: 31–36.
16. Pfister, T. 1986. *Heubelüftung mit Sonnenkollektoren*, p. 33. Location: Flawil.
17. Chang, H. S. 1978. Solar energy utilization in a greenhouse solar drying system. *Agricultural Mechanization in Asia*: 11–16.
18. Müller, J., G. Reisinger, J. Kisgeci, E. Kotta, M. Tesic, and W. Mühlbauer 1989. Development of a greenhouse-type solar dryer for medicinal plants and herbs. *Solar Wind Technol.* 6:523–530.
19. Malik, M. A. S. 1980. *Solar Crop Drying*. vol. 1–5. Washington DC: World Bank.
20. Szulmayer, W. 1973. Thermodynamics of sundrying. In *Sun in the Service of Mankind*, Paper V24. UNESCO Conference.
21. Mühlbauer, W. 1983. Drying of agricultural products with solar energy, Bulletin 3. Rome: Food Agriculture Organization.
22. Kilkis, B. 1981. Solar energy assisted crop and fruit drying systems: theory and applications. *Energy Conservation and Use of Renewable Energy in the Bioindustries*: p. 307. Oxford: Pergamon.
23. Brace Research Institute. *A Survey of Agricultural dryers*. C 19, p. 151. Quebec: McGill University. (Canada).
24. Eissen, W. 1983. Trocknung von Trauben mit Solarenergie. *Forschungsbericht Agrartechnik des Arbeitskreises Forschung und Lehre der Max-Eyth-Gesellschaft*, No. 85. Frankfurt: Max-Eyth-Gesellschaft.

25. Esper, A. 1995. Solarer Tunneltrockner mit photovoltaischem Antriebssystem. *Forschungsbericht Agrartechnik des Arbeitskreises Forschung und Lehre der Max-Eyth-Gesellschaft*, No. 264. Frankfurt: Max-Eyth-Gesellschaft.
26. Esper, A. and W. Mühlbauer. 1996. Solar tunnel dryer for fruits. *Plant Res. Dev.* 44:61–80.
27. Schirmer, P., S. Janjai, A. Esper, R. Smitabhindu, and W. Mühlbauer. 1996. Experimental investigation of the performance of the solar tunnel dryer for drying bananas. *Renewable Energy* 7(2):119–126.
28. Tiberi, U. and J. Bonda. 1995. Photovoltaics in 2010. Summary Report compiled by the European PV Ind. Assoc., for the Directorate General for Energy DG XVII, European Commission.
29. Nann, S. 1990. Potentials for tracking pv systems and v-troughs in moderate climates. Solar Energy 45:385–393.
30. Klotz, F. H. 1995. PV systems with v-trough concentration and passive tracking—concept and economic potential in europe. *13th European PV Solar Energy Conf. Proc.*: 1060–1064.
31. Ratschow, J.-P. 1996. Strom aus erneuerbaren Energien aus der Sicht der Landwirtschaft. KTBL-Arbeitspapier 235: *Energieversorgung und Landwirtschaft*. KTBL-Schriften, Vertrieb im Landwirtschaftsverlag GmbH Münster-Hiltrup.
32. Von Oheimb, R. a. M. Strippel. 1994. Photovoltaik-Versorgungen hofferner Ställe und Einrichtungen. KTBL-Arbeitspapier 208:*Photovoltaik-Anwendungen im Agrarbereich*. KTBL-Schriften, Vertrieb im Landwirtschaftsverlag GmbH Münster-Hiltrup.
33. De Groot, P. 1996. A photovoltaic project in rural africa: a case study. *Proc. of the World Renewable Energy Congress*: 163.
34. Williams, N. 1994. Financing small photovoltaic applications. *Proc. of the International Conference on Solar Electricity: Photovoltaics and Wind*: 477.
35. Biermann, E., F. Corvinus, T. C. Herberg, and H. Höfling. 1992. *Basic Electrification for Rural Households*. Eschborn: German Agency for Technical Cooperation (GTZ) GmbH.
36. Ramakumar, R. 1983. Renewable energy sources and developing countries. IEEE *Trans. Power Appar. Syst.*, PAS-102:502–510.
37. Biermann, E. 1992. Sample fact sheet for solar electricity in developing countries. *Plant Res. Dev.* 36:85–89.
38. Haars, K. 1991. Vom Kraftwerk zur individuellen Stromversorgung. *Energiewirtschaftliche Tagesfragen* 41:53–57.
39. Aulich, H. A. 1996. Small economical PV power generation systems to provide lighting, communication and water supply to rural areas. *Proc. of the World Renewable Energy Congress*: 44.
40. Sapiain, R., R. Schmidt, R. Ovalle, and A. Torres. 1996. Ländliche Basiselektrifizierung im Norden Chiles. *Proceedings of 11th Symposium Photovoltaische Solarenergie*: 261.
41. Muhaidat, A. 1993. Decentralized photovoltaic water pumping systems. Amman Jordan: Royal Scientific Society.

42. Mühlbauer, W., A. Esper, and J. Müller. 1993. Solar energy in agriculture. *Proceedings of ISES Solar World Congress: Biomass, Agriculture, Wind* 8:13–27.
43. *PV for Pumping Systems*. 1993. Eschborn: GTZ, Energy Division. 4.
44. Cunow, E., S. Makukatin, M. Theissen and H. Aulich. 1994. The CLISS project: a large scale application of photovoltaics in West-Africa. *Proceedings of the World Renewable Energy Congress*.
45. Esper, A., O. Hensel, R. A. Mayer, and W. Mühlbauer. 1992. Photovoltaic driven systems for agricultural machinery. *Proceedings of the International Seminar on the Commercialization of Solar and Wind Energy Technologies*: 377–392.
46. Hensel, O. and A. Esper. 1995. Photovoltaisch betriebene Getreidemühle für die ländliche Bevölkerung in Entwicklungsländern. *Proc. 19th Conference of CIGR*, Section IV.
47. Esper, A., O. Hensel, and W. Mühlbauer. 1996. PV-driven solar tunnel dryer. *Proceedings of the Agricultural Engineering Conference*.

### 2.3.5 Greenhouse

*Y. Hashimoto and H. Nishina*

***Greenhouse System***

The basic components of greenhouses are the covering material and the structural material. The most important component is the covering material, which envelops air and creates quite a different environment from that of the outside [1, 2].

It was considered that the reason why the temperature inside the greenhouses increased was due to the mechanism called the mousetrap theory. As for greenhouses with glass as the covering material, it was considered that solar radiation (short-wave radiation with the wave region of 300–3000 nm) came into the greenhouse and that the radiation emitted from the interiors such as soil, plants, and heating system (long-wave radiation with the wave region of 3000–80,000 nm), was trapped by the glass. Glass is almost transparent for short-wave radiation and opaque for long-wave radiation. It was considered that the energy accumulated inside the greenhouse caused the temperature increase. This phenomena is called the greenhouse effect. Nevertheless, it must be noted that the mechanism mentioned above is misleading for the temperature increase inside the greenhouse. The true mechanism is due to the enveloping of air and the decrease in air exchange between the inside and the outside of the greenhouse.

***Energy and Mass Transfer in Greenhouses***

Energy transfers inside and outside the greenhouse occur in the form of heat conduction, sensible heat transfer, latent heat transfer, absorption of solar radiation, and long-wave radiative exchange. Figure 2.41 shows heat conduction, sensible heat transfer, and latent heat transfer (water vapor transfer) among the various components and to the outside air. Figure 2.42 shows the absorption of solar radiation and the long-wave radiative exchange.

As for heat conduction, it is enough to consider the heat conduction in the soil. A part of solar radiation is absorbed by the soil surface, moves into the soil by heat conduction,

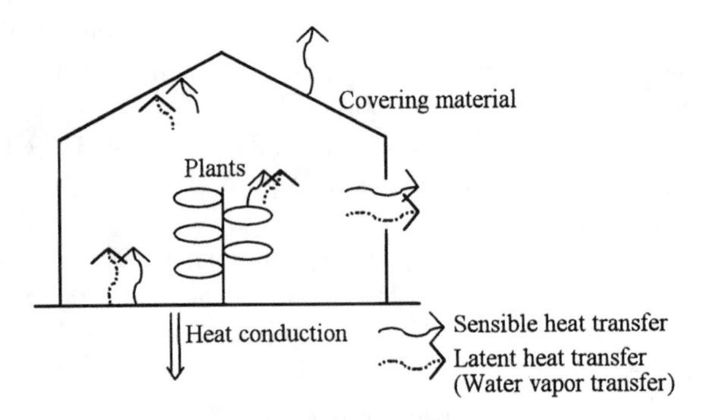


**Figure 2.41. Heat conduction, sensible heat transfer, and latent heat transfer inside and outside the greenhouse.**

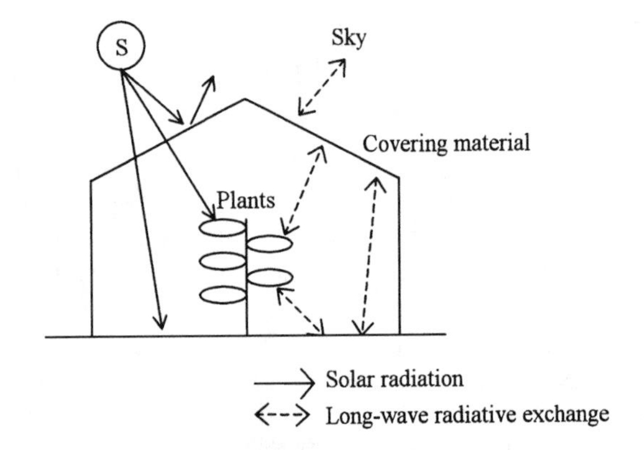


**Figure 2.42. Solar radiation and long-wave radiative exchange inside and outside the greenhouse.**

and is stored in the soil during the day. The stored heat returns into the greenhouse by heat conduction and contributes to keeping the inside temperature at night. Heat conductivity and specific heat of the soil affect the inside temperature and the heating load at night.

Sensible heat transfer must be considered among the various surfaces and the air: the soil surface and the inside air, the plant surface and the inside air, the inner surface of the covering material and the inside air, and the outer surface of the covering material and the outside air. In the case with a thermal screen, sensible heat transfer between the inner surface of the covering material and the inside air is divided into three parts: the inner surface of the thermal screen and the inside air, the outer surface of the thermal screen and the air between the thermal screen and the covering material, and the inner surface of the covering material and the air between the thermal screen and the covering material. Furthermore, air exchange by ventilation or air leakage causes sensible heat transfer.

Latent heat transfer, which follows evaporation or condensation of water, occurs between the soil surface and the inside air (evaporation from the soil surface), between the plant surface and the inside air (transpiration from the plants), and between the inner surface of the covering material and the inside air (condensation at night). Water vapor transfer by ventilation or air leakage can be considered as latent heat transfer.

A part of solar radiation is reflected to the outside by the covering material. Another part is absorbed by the covering material and the structural material. The remaining part comes into the greenhouse and is absorbed by the plants and the soil surface. The ratio of the solar radiation absorbed by the plants or by the soil surface depends on the size of the plants.

Long-wave radiative exchange occurs between the soil surface and the plant surface, between the soil surface and the inner surface of the covering material, between the plant surface and the inner surface of the covering material, and between the outer surface of the covering material and the sky.

As for mass transfer, it is enough to consider the transfer of water vapor and carbon dioxide. Water vapor transfer occurs from the soil surface to the inside air (evaporation from the soil surface), from the plant surface to the inside air (transpiration from the plants), from the inside air to the inner surface of the covering material (condensation at night), and from the inside air to the outside air (ventilation or air leakage). Carbon dioxide transfer occurs in relation to photosynthesis and respiration of the plants and ventilation or air leakage.

***Solar Energy Collection and Storage System***

There are basically two types of collection of solar energy: an active type and a passive type [3]. In the active type, solar collectors that are independent of the greenhouse are used.

Nevertheless, the greenhouse itself can also be utilized as a solar collector because the greenhouse is designed for maximizing solar energy gain. In the passive type, excess heat inside the greenhouse is transferred with the inside air to the heat storage system and is stored in the heat storage medium during the day. The stored heat is used to satisfy the heating load of the greenhouse at night. In order to get more excess heat, it is desirable to heighten the set-point temperature of ventilation during the day, but this may cause problems, such as plant disease.

In addition, excess heat during the day includes latent heat in the form of water vapor, which makes heat transfer into the heat storage medium more effective through the process of water condensation on the heat exchange surface. Nevertheless, the condensed water may evaporate and make the inside humidity higher at night, which does not have a good effect on plant growth and disease. Therefore it is desirable to remove the condensed water before the beginning of heat release.

Passive solar greenhouses can be grouped according to the characteristics of the heat storage system [4]. Water, latent heat storage material, soil, and rock have been utilized as the heat storage medium.

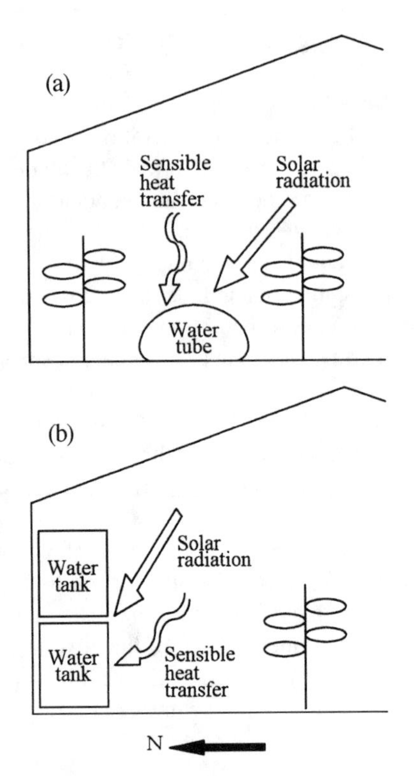


**Figure 2.43. Schematic diagram of (a) water tube system, (b) water tank system.**

*Water*

The heat storage systems that utilize water as the heat storage medium can be grouped into the three types: a water tube, a water tank, and a water storage unit with a fan for forced convection.

Figure 2.43(a) shows the schematic diagram of a water tube system, in which plastic bags filled with water are laid on the soil surface along the pathway between the rows of the plants. Figure 2.43(b) shows the water tank system, in which water tanks are placed in front of the north side of the greenhouse. In the water tube system and the water tank system, the water stores solar energy by direct absorption of the incident solar radiation and natural convection (sensible heat transfer from the inside air) during the day. At night, the stored heat is returned to inside the greenhouse by natural convection and long-wave radiation.

Figure 2.44 shows an example of a water storage unit with a fan [5]. The water storage unit is composed of a fan, air ducts, insulating material, and so on. In this system, the air is circulated between inside the greenhouse and in the water storage unit by the fan. The water stores solar energy from the air by forced convection during the day and releases the heat to the air by forced convection at night.

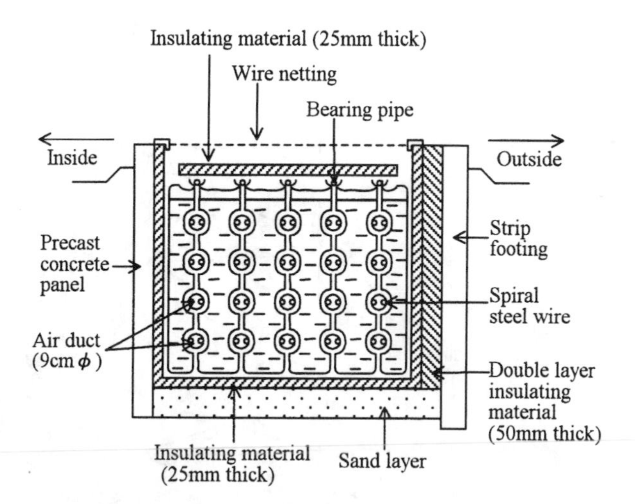


**Figure 2.44. Example of water storage unit (cross section).**

*Latent Heat Storage Material*

Latent heat storage material (phase-change material) is an alternative heat storage medium. Materials such as $CaCl_2\ 6H_2O$, with a melting temperature of 29°C, and $Na_2SO_4\ 10H_2O$, with a melting point of 32°C, have been examined and improved for the use of greenhouse solar heating.

Hot and humid air inside the greenhouse is circulated through the latent heat storage unit during the day. Heat is absorbed by the latent heat storage material and stored for later use. The material changes phase during this process. At night, cold air inside the greenhouse is circulated through the latent heat storage unit and is heated before returning into the greenhouse. The latent heat storage material then returns to its solid phase.

Figure 2.45 shows an example of a latent heat storage unit [6]. The latent heat storage material is composed of $0.4\ Na_2SO_4\ 10H_2O/0.6\ CO(NH_2)_2$ and has a melting point of 15–19°C, a freezing point of 13–17°C, and a heat of fusion of 167 kJ/kg (234 MJ/m$^3$). Batches of the latent heat storage material weighing 7.5 kg each are encapsulated in bags made of aluminum-laminated polyethylene film. A heat storage box consists of 20 bags that are hung at intervals of 4 cm. Three boxes are lined up in the direction of the air flow and a ventilating fan is attached, which makes the total heat storage unit. The total weight of the latent heat storage material is 450 kg and the total heat storage capacity is 84.78 MJ.

During the monitoring of the system [7], the inside temperature was maintained 8°C higher than the minimum outside air temperature. The mean air temperature at night was 9.5°C and the saving ratio of oil consumption was 66%.

*Soil*

When plastic or aluminum pipes that are buried in the soil are used, the excess heat in the air is transferred from inside the greenhouse, through the buried pipes, to the soil.

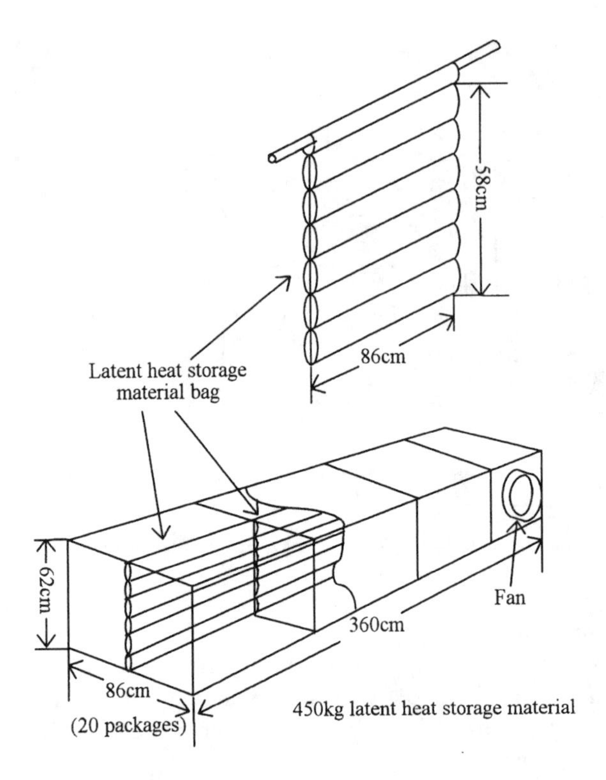


**Figure 2.45. Example of latent heat storage unit.**

The pipes usually run along the length of the greenhouse with air inlet and outlet on opposing sides. Figure 2.46 shows the schematic diagram of this system.

During the day, hot and humid air is drawn from the upper part inside the greenhouse and is driven through the buried pipes. The temperature of the soil in the immediate vicinity of the pipes increases. At night, cold air from inside the greenhouse is circulated again through the buried pipes. Heat is transferred from the soil to the air and then back to inside the greenhouse. Heat recovery is also achieved by passive heat conduction through the soil and then radiation and convection from the soil surface to the air. The water condensed within the buried pipes is either drained away or evaporated by the forced-air flow. In summer, the same system can be used for cooling the air inside the greenhouse.

***Related Equipment***

*Introduction to the Control for Greenhouses*

It may be evident that the greenhouse, including the solar plant factory, is expected as the most advanced system in plant production in the 21st century. In the aspect of agricultural engineering, the control systems are important, and there have been two approaches to the environmental control of the greenhouse: one is analytical method [8–10], and

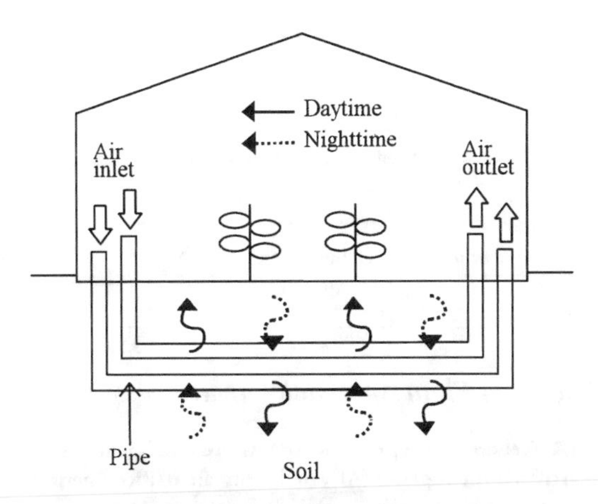


**Figure 2.46. Schematic diagram of heat storage system with buried pipes.**

the other is a more flexible method that includes optimal strategy [11, 12]. As the latter also aims at obtaining both the strategy based on the plant responses, in other words, the speaking plant approach (SPA) [13] and the strategy of the horticultural operation based on the knowledge of horticultural experts [14], it needs several computers with each computer used for data acquisition, decision of optimal strategy, inference of the expert's answer, and actual real-time control of environmental factors [15]. Further, it includes the robot as well within the aspect of intelligent environmental control [16]. To realize intelligent control, including both environmental control and control of the automated machinery or robot (as in seeding, transplanting, ridge-widening, and harvesting stages), remarkable progress has been made on the computer-integrated system from the point of view of both hardware in the computer network and software in the algorithm [17].

*Speaking Plant Approach as Intelligent Control*

***What is the Speaking Plant Approach?*** Plants, including both crops and fruits after harvesting, exhibit physiological responses as a reaction to changes in the environment; sometimes these are called environmental stresses. These responses are so complicated that a mathematical model of them is not effective, even if the modeling may be possible. Mostly we can observe only some tendencies. Sometimes an expert farmer has an optimum strategy based on these observations. As a more scientific methodology, measurement techniques for these physiological responses instead of mere observations should be used for better environmental control. That is the so-called SPA to environmental control. It depends not on a mathematical optimization but on a decision based on inference [13].

At present well-advanced artificial intelligence technology. Furthermore, the measurement of some aspects of plant response to environmental stresses appears feasible. Therefore it can be noted that intelligent control of a greenhouse should be examined

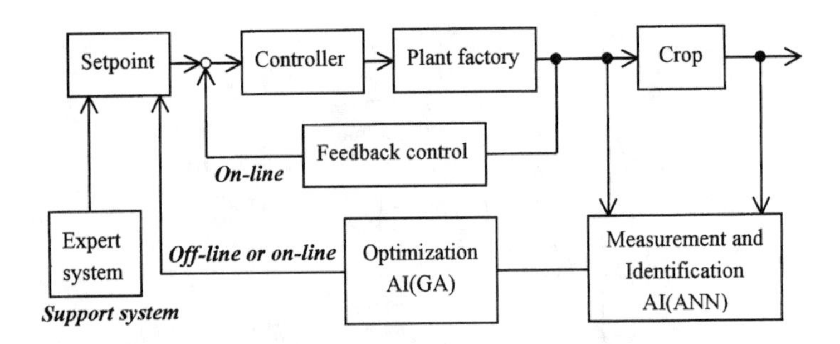


**Figure 2.47. General concept of the SPA as the intelligent control system based on artificial intelligence (AI), including an artificial neural network (ANN) and a genetic algorithm (GA).**

based on this SPA [18]. Furthermore, it should be applied to postharvest processes [19] as well as cultivating processes.

Figure 2.47 shows the block diagram of the SPA based on intelligent control. The control system is composed of two subsystems: one is the expert system that is the decision support system for an optimal set point in the longer term of the cultivating process; the other is the intelligent control system that is also composed of two subsystems, both of which are described as on-line and off-line or on-line. The on-line subsystem is the feedback loop to keep temperature, humidity, and $CO_2$ at the set points. The set point is generally adjusted by operators who usually decide the optimum based on their experience, although they sometimes take account of their observation of the crops. The proposed system, on the contrary, includes off-line or on-line, in which the physiological processes of the crops are identified by an artificial neural network, and then the optimal strategy is obtained through the model simulation with a genetic algorithm. The responses to environmental stresses are identified from the relation between the actual environmental condition and the measured data under this condition. In this case, artificial neural networks and genetic algorithms are more effective than an operator.

*Informatics as Diagnosis of the Growing Crops*

It may be reasonable that, when the computer-integrated system is used, three-dimensional shape recognition is available for the diagnosis of the growing crops in the greenhouse. Also, for greenhouse automation, a computer network based on an integrated services digital network is effective. It should be noted that a virtual factory would be introduced to the greenhouse by remote diagnosis [20].

## References

1. Takakura, T. 1993. *Climate under Cover*, p. 155. Dordrecht, The Netherlands: Kluwer Academic.
2. Bot, G. P. A. 1993. Physical modelling of greenhouse climate. *The Computerized*

*Greenhouse*, eds. Hashimoto, Y., G. P. A. Bot, W. Day, H.-J. Tantau, and H. Nonami, pp. 51–73. New York: Academic.
3. Albright, L. D. 1991. Production solar greenhouses. *Solar Energy in Agriculture*, ed. Parker, B. F., pp. 213–231. Amsterdam: Elsevier.
4. Santamouris, M., C. A. Balaras, E. Dascalaki, and M. Vallindras. 1994. Passive solar agricultural greenhouse: a worldwide classification and evaluation of technologies and systems used for heating purposes. *Solar Energy* 53:411–426.
5. Kozai, T., K. Shida, and I. Watanabe. 1986. Thermal performance of a solar greenhouse with water tanks for heat storage and heat exchange. *J. Agric. Eng. Res.* 33:141–153.
6. Machida, Y., Y. Kudoh, and T. Takeda. 1985. PCMs of Glauber's salt-based eutectic mixtures for greenhouse heating: *Proc. International Symposium on Thermal Application of Solar Energy* (Japan Solar Energy Society): 537–541.
7. Nishina, H. and T. Takakura. 1985. Solar-heating greenhouses by means of latent heat storage units: *Proc. International Symposium on Thermal Application of Solar Energy* (Japan Solar Energy Society): 543–545.
8. Udink ten Cate, A. J. 1983. Modelling and control of greenhouse climates, p. 159, Thesis, Wageningen Agricultural University.
9. van Henten, E. J. 1994. Greenhouse climate management—An optimal control approach, p. 329, Thesis, Wageningen Agricultural University.
10. Chalabi, Z. and W. Zhou. 1996. Optimal control of greenhouse heating: *Proc. 13th IFAC World Congress, B*: 393–398.
11. Tantau, H.-J. 1993. Optimal control for plant production in greenhouses. *The Computerized Greenhouse*, eds. Hashimoto, Y., G. P. A. Bot, W. Day, H.-J. Tantau, and H. Nonami, pp. 139–152. New York: Academic.
12. Challa, H. and G. van Straten. 1993. Optimal diurnal climate control in greenhouse as related to greenhouse management and crop requirements. *The Computerized Greenhouse*, eds. Hashimoto, Y., G. P. A. Bot, W. Day, H.-J. Tantau, and H. Nonami, pp. 119–137. New York: Academic.
13. Hashimoto, Y. 1989. Recent strategies of optimal growth regulation by the speaking plant concept. *Acta Hortic.* 260:115–121.
14. Hashimoto, Y. and K. Hatou. 1992. Knowledge based computer integrated plant factory: *Proc. IFAC Workshop on Expert Systems in Agriculture*: 9–12.
15. Hatou, K., H. Nishina, and Y. Hashimoto. 1990. Computer integrated agricultural production: *Proc. 11th IFAC World Congress* 11:306–310.
16. Hashimoto, Y. 1993. Computer integrated system for the cultivating process in agriculture and horticulture-approach to "Intelligent Plant Factory". *The Computerized Greenhouse*, eds. Hashimoto, Y., G. P. A. Bot, W. Day, H.-J. Tantau, and H. Nonami, pp. 175–196. New York: Academic.
17. Morimoto, T., T. Torii, and Y. Hashimoto. 1995. Optimal control of physiological processes of plants in a green plant factory. *Control Eng. Prac.* 3:505–511.
18. Morimoto, T., K. Hatou, and Y. Hashimoto. 1994. Speaking plant approach for environment control based on artificial intelligence: *Proc. CIGR 12th World Congress* 1:190–197.

19. De Baerdemaeker, J. and Y. Hashimoto. 1994. Speaking fruit approach to the intelligent control of the storage system: *Proc. CIGR 12th World Congress* 2:1493–1500.
20. Hatou, K., H. Matsuura, T. Sugiyama, and Y. Hashimoto. 1996. Range image analysis for the greenhouse automation in intelligent plant factory: *Proc. 13th IFAC World Congress, B*: 459–464.

## 2.4 Wind Energy

*R. N. Clark*

### 2.4.1 Overview

The use of windmills as an energy source has its roots in antiquity. Archeological and written records have shown that early forms of windmills were used by the Chinese, Egyptians, Persians, and Babylonians. Many European documents after the 13th century refer to various windmill designs; however, by the 1700's, two basic windmill designs had developed, the post mill and the tower mill [1]. The post mill consisted of a large house and a tail pole that was used to turn the structure and bring the rotor into the wind. The tower mill was more like the modern windmill in that it consisted of a rotor and tail mounted on a fixed tower. Because of the ease of orientation, the tower mills were used in locations that did not have a significant prevailing wind direction.

The rotors or sails of early windmills were constructed primarily of wood, reeds, and canvas. Many different concepts and designs were used, including springs and shutters to increase or decrease the sail area. The sails on larger mills were ~12 m long and 3 m wide, providing a peak power of 30 kW [1]. These European designs were duplicated in other parts of the world as colonists moved to other continents, but these machines did not provide the flexibility needed to withstand the fickle weather in many arid and semiarid regions. In the United States, Daniel Holladay began making wind machines in 1857 that were self-regulating by using paddle-shaped blades that would pivot, or feather, as the wind-speed increased. Other designs used a solid wheel assembly and a side vane to turn the rotor out of the wind as the velocity increased. These designs evolved into the multibladed farm windmill that used a reciprocating-type pump. These systems worked well to lift water from deep wells, and windmills as large as 6 m in diameter were commonplace along the railroads and provided water for steam engines. Enclosed gears, metal wheels, and towers improved the systems, allowing them to run smoothly in light winds. T. Lindsey Baker's *A Field Guide to American Windmills* [2] is an excellent resource on the design, construction, and operation of farm windmills made between 1875 and 1975.

By 1930, wind power was also used to generate electricity. These electrical generating systems were quite different from the multiple-bladed water pumpers in that they usually had only two or three blades and rotated at a much higher speed. The rotor was connected without gears to a direct current generator with an output of 6, 12, 24, or 32 V. The electricity was used to charge batteries for later use. The electricity from the batteries was used to power a radio and two or three light bulbs. Between 1931 and 1957, tens of thousands of these electrical units were sold until they became uneconomical and inconvenient when inexpensive electricity became available from rural electric cooperatives. Most wind-electric generators were removed after the installation of electric power lines by the Rural Electrification Administration.

However, windmills never entirely disappeared; water pumping units remained in the western United States to provide livestock with water in remote areas. Several designs from the United States were adapted to locations in other parts of the world, and companies in Argentina, Australia, and South Africa became the leading manufacturers of water-pumping windmills. Small electrical generating wind machines were never produced worldwide like the mechanical water-pumping units; however, today China is the largest producer of small electric generators.

### 2.4.2 Wind Characteristics

***Global Circulation***

Global wind circulation is driven by solar radiation and the rotation of the Earth and the atmosphere. Seasonal variations are due to the tilt of the Earth's axis relative to the plane of the Earth's movement around the Sun. Since solar radiation is greater per unit area near the equator, heat flows from the regions near the equator toward the poles. Because of the rotation of the Earth and the conservation of momentum, a circulation pattern is established that produces the predominate surface winds know as easterlies or westerlies and are the northeast trade winds in the northern hemisphere and the southeast trade winds in the southern hemisphere.

Superimposed on this circulation is the migration of cyclones and anticyclones across the midlatitudes, which disrupt the general flow. The jet streams also influence the surface winds. Local winds are due to local pressure differences and are influenced by the topography of mountains, valleys, etc. The diurnal (24-h) variation is due to temperature differences caused by daytime heating and nighttime cooling. Coastal breezes are caused by temperature differences between land and sea.

To select a suitable site for harnessing the wind, you must become familiar with its patterns. Wind is best described by Park [3] as "a fickle servant," and "It may not be available when you need it, and you can be overwhelmed by its abundance when you don't." The nature and the magnitude of wind characteristics are important in determining the economic feasibility of a particular wind turbine at a given site. Reliable procedures for comparing the wind characteristics at potential wind turbine sites are essential.

***Energy***

The wind contains kinetic energy that is harvested by a rotor or wheel and transferred to a rotating shaft. Energy in the shaft is then used to pump water directly, drive an electric generator, or produce heat. As a result, the energy in the wind is normally converted to either mechanical, electrical, or heat energy. Power is frequently used in describing the performance of wind machines and is a measure of the energy extracted during a specific period of time. The theoretical power in a wind stream is determined by

$$P = 1/2\rho A V^3, \tag{2.5}$$

where $P$ is the power in watts, $\rho$ is the air density in kilograms per cubic meter, $A$ is the cross-sectional area of the intercepted wind stream (rotor area) in square meters, and $V$ is the wind-speed in meters per second. Actual power harvested by a wind turbine is less than the theoretical because of power losses in the system. The efficiency or power coefficient ($C_p$) is determined by

$$C_p = \frac{\text{power delivered}}{\text{theoretical power}}. \tag{2.6}$$

The maximum power coefficient is 0.59, as determined by the Betz limit, and a good efficient rotor would average 0.30.

***Wind Speed***

Historically wind-speeds have been measured near airports to assist aircraft in landing and take-offs and near evaporation pans to relate varying evaporation rates. Agriculturalists have known the importance of wind-speed measurements in predicting evapotranspiration rates, field drying rates, and loading on farm structures for many years. For all these applications, a simple daily wind run total or an instantaneous peak value was usually adequate to predict the expected outcome. However, because wind power is dependent on the cube of the wind-speed, a wind-speed distribution is needed. Figure 2.48 illustrates an actual wind-speed distribution for Bushland, TX, showing the number of hours or percentage of time that a given wind-speed could be expected. Researchers have examined several mathematical distributions to determine which best represents actual wind-speed distribution. The Rayleigh distribution was selected as providing the simplest, best approximation of wind-speed characteristics and is compared in Fig. 2.48 with an actual distribution.

The Rayleigh wind-speed distribution gives a ratio of time the wind blows within a given wind-speed band, a probability interval of $V$ to $V + dv$, and the total time under

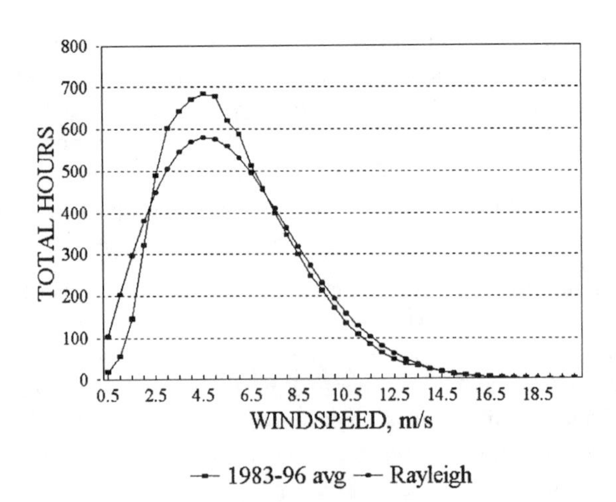


**Figure 2.48. Average wind-speed distribution and Rayleigh predicted wind-speed distribution for Bushland, TX.**

consideration. The Rayleigh is a Weibull distribution with a constant of 2 and depends only on mean wind-speed. It is defined by

$$F(V)\,dv = (\pi/2)(V/\nabla^2)\exp[-(\pi/4)(V/\nabla)^2]\,dv, \tag{2.7}$$

where $F(V)$ is the Rayleigh frequency distribution as a function of $V$, $\nabla$ is the mean wind-speed, $V$ is the instantaneous wind-speed, and $dv$ is the wind-speed probability interval. In most cases, the Rayleigh distribution predicts the wind characteristic with less than a 10% error [4].

Because local winds are influenced by temperature and pressure differences, it is important to determine the prevailing wind direction. Often each season will have its own prevailing wind direction, and in mountain valleys there is usually a diurnal change in wind direction. Wind direction is normally up the mountain from noon to sunset because of the daytime heating, and then reverses direction and flows downward from evening to sunrise [5]. Wind-direction determinations are much easier in noncomplex terrain, but are important when you are selecting a site for your wind turbine.

***Wind-Speed Measurements***

Accurate wind-speed measurements are needed for wind-turbine siting and predicting energy capture. The most common way of determining the wind-speed is to use an instrument called an anemometer. The most popular anemometer is the cup type, consisting of small cups mounted on a rotating shaft. Many have a small DC generator to produce a direct, readable voltage that is proportional to wind-speed. The physical size and the configuration of anemometers have been as varied as the manufacturers, with little standardization.

Anemometers used to measure wind-speeds near evaporation pans have usually been at a height of 0.5 m and have been of the wind totalizing type, indicating the number of kilometers of wind passing the anemometer per day. Other wind measurements for predicting evapotranspiration have usually been taken at 2 m, with few measurements made above 3 m. In the past, anemometers at airports have been placed in the most convenient places, such as on top of buildings or along the runway. The measurement heights have not been standardized, and historical records include many different heights, even at the same airport. Standard anemometer heights have been used more in the areas of environmental quality measurements than in any other area. Standard anemometer heights for nuclear and coal-fired electrical generating plants are 10 and 40 m, respectively. Before reported wind-speed data are used for predicting energy capture, a determination of how the data were collected (recording, frequency, averaging time, etc.) is needed along with the height of the anemometer since they all affect the outcome.

Many studies have been made with recorded wind data from government weather services, government agricultural and energy departments, and other government and private agencies. Figure 2.49 is the result of a study by the U.S. Department of Energy [6]. This map presents wind data as wind-power classes rather than actual wind-speeds.

Because of the complex nature of wind flow patterns at any site, it is recommended that actual wind-speed data be collected for at least 1 year for each site. The anemometer should be placed at the hub height of the planned machine to eliminate inherent errors in predicting wind-speed at different heights. The anemometer should also be capable of

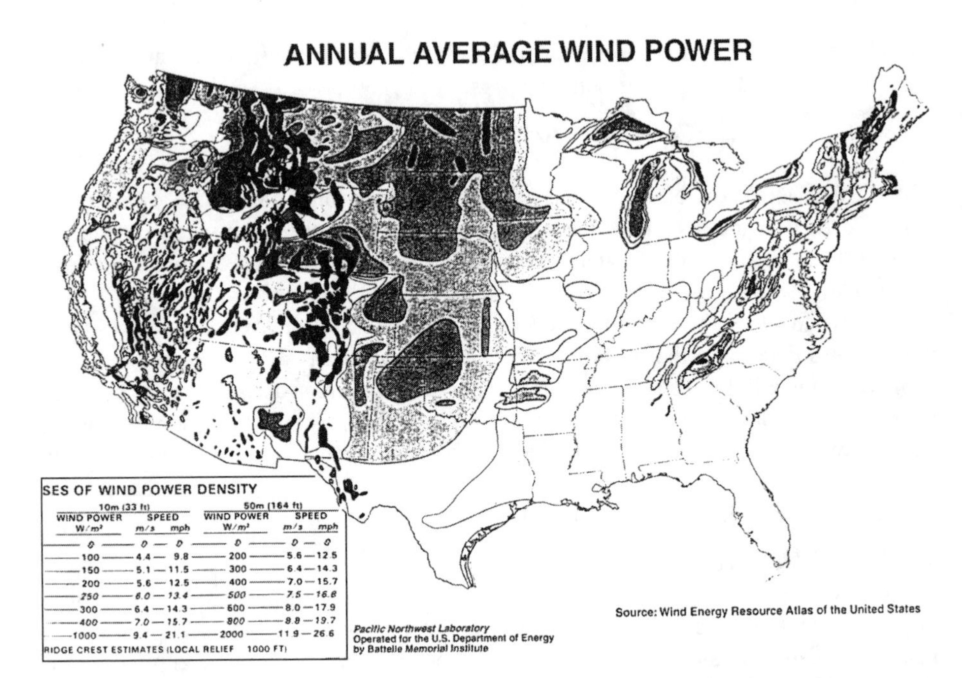


**Figure 2.49. Estimates of average annual wind power for the continental United States.**

supplying data to a recorder at least every 5 min. It is important to determine the variation in the wind-speed since the energy produced is a function of the wind-speed cubed.

***Wind Shear***

As the wind passes over the Earth's surface, wind close to the surface is retarded because of the roughness caused by trees, buildings, and crops. The extent to which wind speed is decreased near the Earth's surface and the resulting variation with height is called wind shear. The height at which free wind (100% of potential) is available depends on the surface terrain (Fig. 2.50). The free wind height over smooth terrain (water) could be as low as 40 m, but when determined over complex terrain (large cities), the height could be as high as 500 m. The extent of wind shear is also influenced by temperature and humidity. Cold, damp air has a larger shear than warm, dry air.

Meteorologists have suggested several equations for adjusting the wind speed measured at one height to represent the wind speed at another height. The most common equation is

$$V_2 = V_1(h_2/h_1)^N, \quad (2.8)$$

where $V_2$ is the wind speed at the new height ($h_2$) and $V_1$ is the known wind speed at height ($h_1$). Detailed measurements have shown that this equation does not always predict the correct wind speed at the new height because of variations in moisture, pressure, and temperature profiles as well as differing wind profiles. It is intended to represent average or long-term conditions.

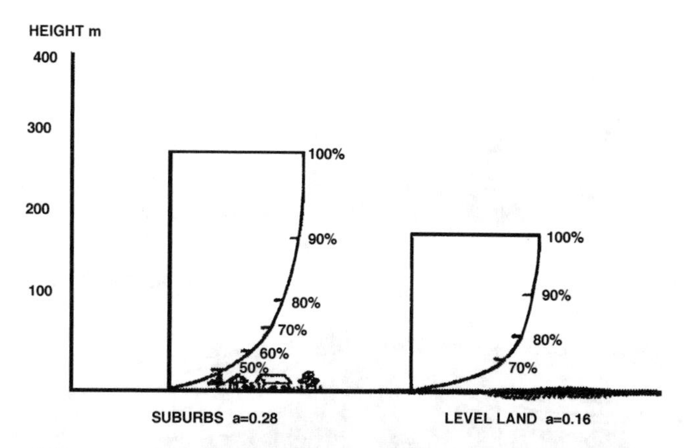


**Figure 2.50. Wind-speed profiles over flat, open country and urban areas or large farmsteads.**

### 2.4.3 Site Analysis and Selection

***Wind Characteristics***

The most important decision concerning the use of a wind–energy conversion system is the selection of the site. Because of the high initial cost of wind systems, it is imperative that the best site be used rather than just a good one. Three main questions should be answered when a site analysis is made:

(1) Is there sufficient wind for the machine to produce usable power at least 50% of the time?

(2) What effects will surface terrain have on the wind profile?

(3) What barriers might affect the free flow of the wind?

The speed and the duration of the wind are determined from the wind-speed distribution that can be calculated by a Rayleigh distribution or by actual measurements. It is important that an analytical determination be made because most people tend to overestimate their wind resource. Only sites that have an annual average wind-speed in excess of 5 m/s at a 10-m height should be considered for wind power.

***Surface Roughness***

The roughness of the terrain determines the amount of wind-speed reduction that occurs near the surface. Selecting a good site in flat terrain is much easier than in hilly or mountainous terrain. Figure 2.51 shows the effect of hills on the transition height and its downwind result. Other large natural changes in terrain have similar effects and should be considered in the site selection process. Often an increase in tower height will provide the needed change to place the rotor in the free wind, rather than moving the entire unit to another location. As a general rule, wind turbines should be 20 to 30 m above ground level and at least 10 m above surrounding obstacles.

Determining the effect of surface terrain in hilly or complex terrain requires much work and analysis. Wind profiles over hills provide areas of acceleration on the windward face and areas of turbulence and decreased flow on the leeward side. The shape of the hill determines the magnitude of each of the phenomena. Normally the top and the upper

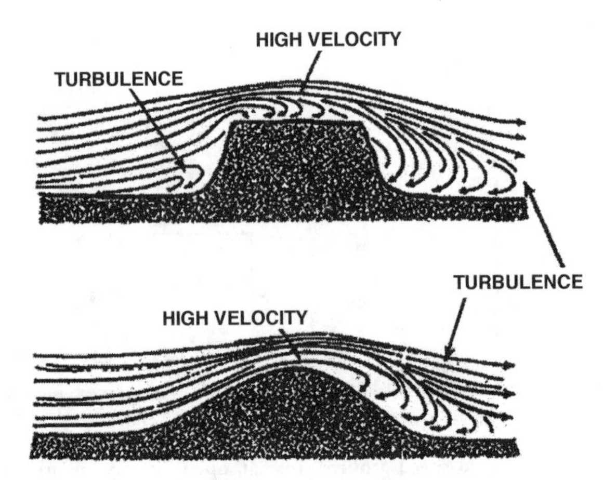


**Figure 2.51. Schematic wind-flow patterns over hills showing areas of turbulence.**

half of hills on the windward face or where the prevailing wind is tangent to the site are considered good sites because these are points of wind acceleration. The leeward sides and lower portions of the hills should be avoided because of reduced speeds and turbulence. Also, larger wind turbines should be placed where the entire rotor sees wind of a similar characteristic to avoid uneven loading on the rotor. Flows through valleys, canyons, basins, and gaps each create a special situation that requires extra evaluations when one is selecting a site [5].

When a potential site is first evaluated, it is useful to observe the vegetation. Vegetation is often deformed by high average winds and can be helpful in remote areas to evaluate the site before anemometry is installed. Hewson *et al.* [5] developed four easily observed deformities of trees for estimating wind-speed. They are:

- Brushing (3–5 m/s): Branches and twigs bend downwind like the hair of a pelt that has been brushed in one direction only. This deformity can be observed in deciduous trees after their leaves have fallen. It is the most sensitive indicator of light winds.
- Flagging (4–6 m/s): Branches stream downwind, and the upwind branches are short or have been stripped away.
- Throwing (6–8 m/s): A tree is wind thrown when the main trunk and the branches bend away from the prevailing wind.
- Carpeting (8–9 m/s): This deformity occurs because the winds are so strong that every twig reaching more than several inches above the ground is killed, allowing the carpet to extend far downwind.

The availability of Graphical Information Systems for personal computers can prove invaluable in a first analysis or screening of suitable sites for wind turbines. The sites with the highest potential could be classified by closeness to utility lines, access to existing transportation roads, elevations that are conducive to wind potential, and positioning of potential wind plants on terrain to take advantage of the prevailing wind patterns [7].

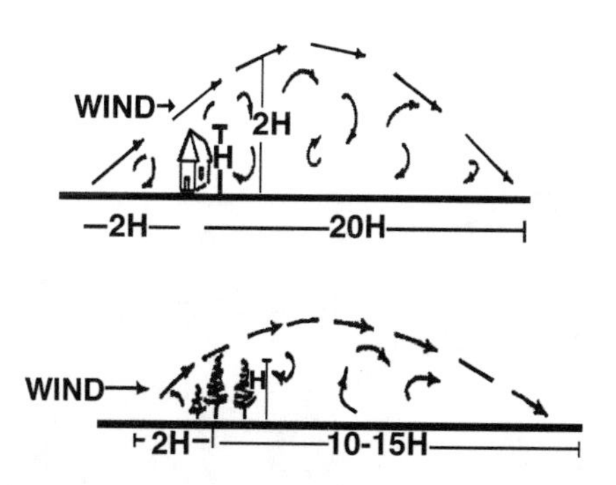


**Figure 2.52. Zones of disturbance and turbulence around buildings and windbreaks.**

***Manmade Barriers***

Buildings and trees hinder wind flow and not only cause lower wind-speeds in their wake, but also create turbulence. The reduced wind-speed in the vicinity of such an obstruction will affect the power that can be produced by a wind generator. Also, the turbulence caused by the wind swirling around obstructions will shorten the life span of the unit. Small wind turbines; however, are best located near the point where the energy is to be used since the cost of transmission lines from a good site 1 km away may exceed the cumulative value of the conventional energy displaced.

A site must be far enough upwind or downwind to avoid the disturbed wind flow around an obstruction. Buildings produce effects on wind flow that are different from those trees and bushes. The turbulent zone over buildings and trees is shown in Fig. 2.52. Turbulence is the result of many factors, but is primarily due to the height of the building or tree ($H$). Note that even the flow of air downwind from the building is affected. A machine should be located away from the building by a distance of $2H$ upwind and $10H$ downwind.

Shelterbelts or windbreaks have an even greater effect on wind flow than do buildings. The kinds of trees in the windbreak, spacing, and height are the controlling factors. For a tower equal in height to that of the trees, power in the wind is reduced 80% at a distance of $10H$ downwind from the windbreak. A general rule of thumb is to erect a tower tall enough to clear the highest obstruction by 10 m and to site it at least 90 m from the nearest trees or buildings [5].

### 2.4.4 Types of Wind Machines

Any device that converts the kinetic energy in the wind to a usable form of energy is called a wind machine. Many types of wind machines have been devised, and there are as many patents on wind machines as on almost any other type of device. Basically, all wind machines remove kinetic energy from the wind by slowing it down and converting this energy to mechanical energy transmitted by a rotating shaft. Two basic types of machines have evolved and are classified as drag and lift types.

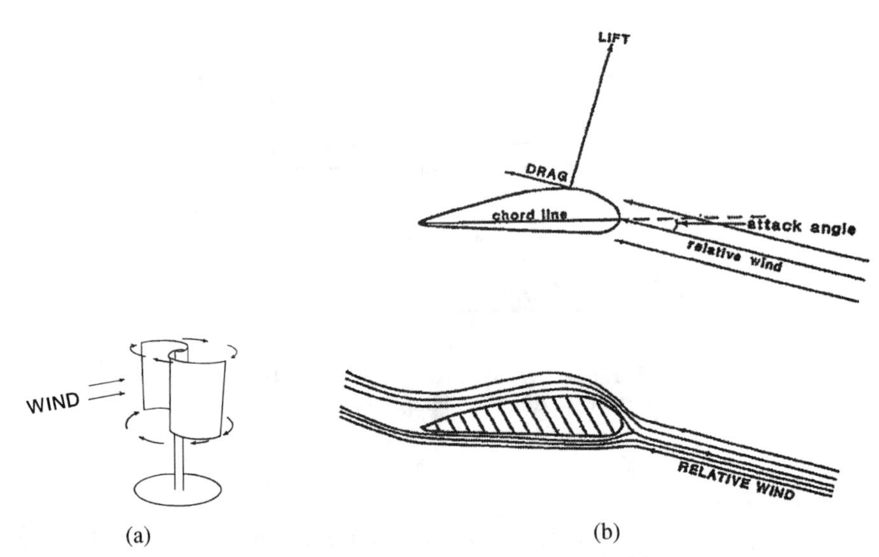


(a) (b)

**Figure 2.53. Wind machines: (a) drag type, (b) lift-type showing lift angle and drag force.**

***Drag Devices***

These simple machines have flat, curved, or cup-shaped blades made of wood, steel, or other materials, the most common example being the American farm windmill. Another drag device is the cup anemometer used to measure wind speed. In both examples, the wind pushes on the blade or cup, forcing the rotor to turn about its axis [Fig. 2.53(a)]. This axis can be horizontal (parallel to the ground) as on the farm windmill or it can be vertical (perpendicular to the ground) as on the cup anemometer.

Drag devices characteristically produce a high starting torque and are well suited to pumping water in low volumes. Experimenters have tried several approaches to improving their performance. One popular example is the S or Savonius rotor, often made from split oil drums. The farm windmill and similar drag devices are inherently limited in the amount of energy they can extract from the wind. At best, only a third of the energy available can be captured by such wind machines [3].

***Lift Devices***

Lift devices use slender airfoils for blades rather than plates or cups. When the wind strikes one of these blades, it flows over and around the blade, creating lift. This aerodynamic lift pulls the blade about the axis of rotation, causing the rotor to spin by using the same lift principle that supports an airplane wing in flight [Fig. 2.53(b)].

In general, wind turbines that use lift have only a few (two, three, or four) blades, in contrast to the multiple blades of drag devices. At first it seems mysterious that a windmill with only two blades can operate more efficiently than one with a large number of blades. A modern wind turbine that uses airfoils can convert twice as much wind energy to useful work as a drag device of the same rotor area. This is one of the reasons why the multibladed windmill has not been adapted to generate electricity.

**Figure 2.54. Water-pumping windmills are used to provide water for livestock across much of the Southwest.**

### *Generic Types*

Four generic types of wind machines are discussed to provide a better understanding of components and operating theory.

#### *Multibladed Farm Windmill*

The multibladed farm windmill was developed in the mid-1800's and was designed primarily for pumping water from wells (Fig. 2.54). The units manufactured today have changed little from the ones made in 1930 when an enclosed, oil-lubricated gearbox was introduced. Rotor diameters range from 2 to 6 m, and units contain 16 to 18 curve-shaped metal blades. Air flow through the rotor causes a cascading effect, thus enhancing the drag and increasing rotor efficiency to 30%. The desirable features of this multibladed rotor are:

- high starting torque,
- simple design and construction,
- simple control requirements,
- durability,

and the undesirable features are that this rotor:

- exerts a high-rotor-drag loads on a tower,
- is not readily adaptable to loads other than water pumping [3].

There are many units operating throughout the world, and several new manufacturers have started production since 1980. This is a good example of a complete system manufactured to perform a particular task.

*Savonius Rotor*

The Savonius rotor, or S rotor, was invented in the early 1920's and has received considerable attention because of its simple construction. Figure 2.55 shows the arrangement of the curved surfaces. The desirable features of the high-drag S rotor are:

- easy manufacture,
- production of a high starting torque,

and the undesirable features are:

- difficulty of control because controls to limit rotational speed in high winds have not been devised,
- poor use of materials because it presents a small frontal area for a fixed amount of construction materials [3].

This unit is best suited for pumping water, driving compressors or pond agitators, and other direct loads. Because of its slow rotational speed and control problems, it is not well suited to electrical generation.

*Horizontal Propeller Types*

These lift-type machines are the most common wind machines used for electrical generation and normally have two or three blades (Fig. 2.56). Most machines are designed to produce a maximum efficiency of 40% to 45% at a tip–speed ratio of 5 to 6. The tip–speed ratio is the ratio of the linear speed of the blade tip to the corresponding wind speed. Because of this higher operating speed, these units are almost always used for generation of electricity. The desirable features of these high-speed, propeller-type rotors are:

- slender blades that use less material for the same power output,
- higher rotational speeds that reduce gearbox requirements,
- lower tower loads,
- larger diameter and high power levels that are more easily obtained.

The undesirable features are:

- production of a low starting torque,

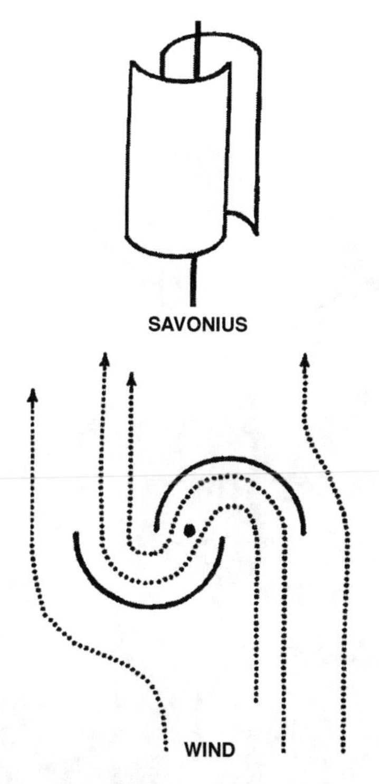


**Figure 2.55. Savonius or S rotor has several curved vanes to catch the wind.**

- blades that require good aerodynamic design,
- possible flutter and vibration problems.

An important feature of horizontal-axis machines is the manner in which they respond to changes in wind direction. Many small wind generators use a tail vane to keep the rotor perpendicular to the wind. Medium-sized wind generators, and even some of the smaller ones, operate by placing their rotors downwind of the tower and omitting the tail vane. The blades are angled slightly downwind (coned), making the turbine respond to changes in wind direction much like those with tail vanes. Larger units use a motor-driven gear to hold the rotor into the wind. This is why horizontal-axis wind turbines are also classified as upwind or downwind.

The major components of a horizontal-axis downwind machine are shown in Fig. 2.57. Early manufacturers were expending a large percentage of their design efforts into perfecting the blades, but soon learned that bearings, gearboxes, and brakes require special attention too. The large number of operating hours per year and vibratory fatigue loads require specially designed components.

Electricity is transferred from the generator through slip rings or twist cables to the ground. For the larger machines, twist cables have proven to be more reliable and cost

**Figure 2.56. 15-m-diameter wind turbine with a 50-kW induction generator.**

effective. Because propeller machines have a low starting torque, they are usually not started under load, thus requiring an electronic control system for activating the load when the correct rotational speed is reached. In a similar manner, the control system must maintain a reasonable rotational speed in high winds. Therefore, the control system becomes an important part of the wind machine and often determines the success of a particular design.

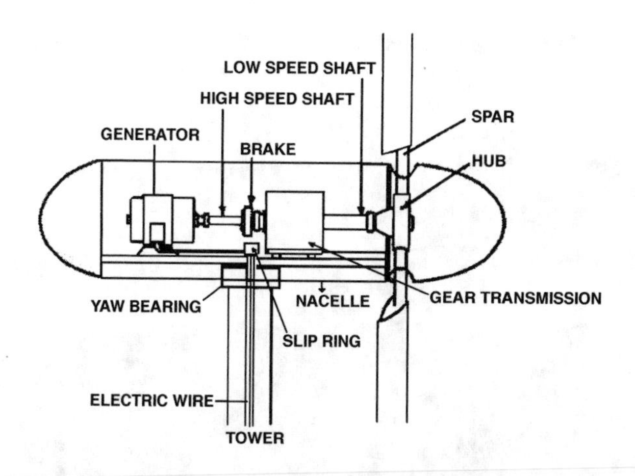


**Figure 2.57. Components of a typical horizontal-axis wind turbine with an electric generator.**

*Darrieus*

The Darrieus rotor was developed by the Frenchman G. J. M. Darrieus in the early 1920's. The Darrieus is classed as a vertical-axis unit, but differs from the Savonius in that it operates on aerodynamic lift principles. It has a tip–speed ratio of 6 to 7 and a rotor efficiency of 40% to 45%. The desirable features of Darrieus or vertical-axis wind turbines are:

- Low material usage for a high power output,
- high-maintenance items at ground level,
- acceptance of wind from any direction
- use of mechanical power at ground level,

and the undesirable features are:

- a rotor that is not normally self-starting,
- the requirement of a larger land area to accommodate guy cables,
- need of a good rotor control system to prevent rotor overspeed.

A Darrieus unit is shown in Fig. 2.58 with its two symmetrical, airfoil-shaped blades. Recent data have indicated that large vertical-axis units may be more efficient and aerodynamically stable than horizontal-axis units because each blade increment on the vertical axis operates in the same wind-shear region, whereas on a horizontal-axis unit, each blade increment rotates through several wind-shear regions, causing extreme changes in blade loading per revolution. Unequal blade loading is a major problem on horizontal-axis machines in excess of a 50-m diameter.

Vertical-axis machines normally have curved blades attached at each end to reduce the stresses in the blades. However, straight-bladed units have been built and tested, some with fixed blades and others with articulating blades that allow the machines to operate with variable pitch. Because of the large number of moving parts, these units have not been reliable. Vertical-axis turbine designs have been largely passed over by commercial manufacturers.

**Figure 2.58. Darrieus or vertical-axis wind turbine used to generate electricity.**

Wind machines come in many shapes and sizes, but all include the basic components of a rotor, power transmission, and control system. Regardless of the shape, power output is determined primarily by the rotor-swept area. A machine should be selected from a proven manufacturer (one that has sold at least 100 units), and should be large enough to provide the power required.

### 2.4.5 Water Pumping with Wind Power

Water pumping was one of the first uses of wind power. Records show that the Persians and the Egyptians used wind power to pump water as early as 1200 BC. The Dutch used windmills to reclaim land from the sea in the 17th and the 18th centuries. Much of the Great Plains region of the United States and the Outback of Australia were not occupied until windmills were used to provide a dependable water supply. In recent times, windmills have been used to replace hand pumps or animal-driven pumps.

Over 1000 known windmill manufacturers have produced water-pumping windmills through the years, with many of them out of the business by 1920 [2]. By the 1970's, only a handful of manufacturers remained, and they produced only a few machines per year.

The desired volume of water to be pumped and the water source have dictated the type of pump used and the pump has dictated the design of the windmill. Bucket and chain pumps have been used with small sail-wing-type windmills because there was a need for a slow-moving rotating shaft. The low-lift, high-flow pumps needed to move large volumes of water to reclaim land resulted in a screw-type pump driven by a large geared rotor. And finally, the deep wells of the Great Plains required a low-volume, high-lift pump and a windmill with a high starting torque and reciprocating motion. All mechanically driven windmills were designed to operate with a particular pump and to provide water from a particular source.

***Mechanical Water Pumping***

Almost all mechanical windmills made today are used for pumping water from wells, and the volume is usually sufficient for livestock water, domestic uses, or small-scale irrigation. Normal flow rates vary from a few liters per minute to 50 to 75 L/min. They start pumping water at a wind speed of approximately 3.5 m/s and reach a peak flow at 9 to 10 m/s, depending on the furling spring tension [8].

These windmills have been improved by use of all-steel rotors and enclosing the gears in an oil bath. Except for these two modifications, little has changed on these windmills since the early 1900's. Similarly, the pumps have not changed significantly. Urethane cups have replaced leather cups in the single-acting piston pump to provide longer life.

Windmills in current production have rotor diameters of 1.83 m (6 ft.), 2.44 m (8 ft.), 3.05 m (10 ft.), 3.66 m (12 ft.), 4.27 m (14 ft.), and 4.88 m (16 ft.), with the 2.44-, 3.05-, and 3.66-m diameters the most popular models. Nearly all units are mounted on 10-m-high steel towers. The horizontal-axis windmills are backgeared and have a gear ratio of 3.3 to 1. All gears operate through an oil reservoir for lubrication. For overspeed control, the rotor turns sideways out of the wind, which is called furling. The wind speed at which furling begins can be adjusted by changing the spring tension between the gearbox and the tail vane. The maximum pump speed is also adjusted by changing the spring tension acting on the tail vane. The installation manual reads "The speed of the windmill is regulated by hooking the end of the regulating spring in the various holes provided for this purpose in the vane stem horizontal flange. There are five holes. Increase tension of the spring to increase speed of the wheel. Decrease tension of spring to decrease speed of the wheel. Approximate maximum speed is 32 strokes of pump rod per minute" [9].

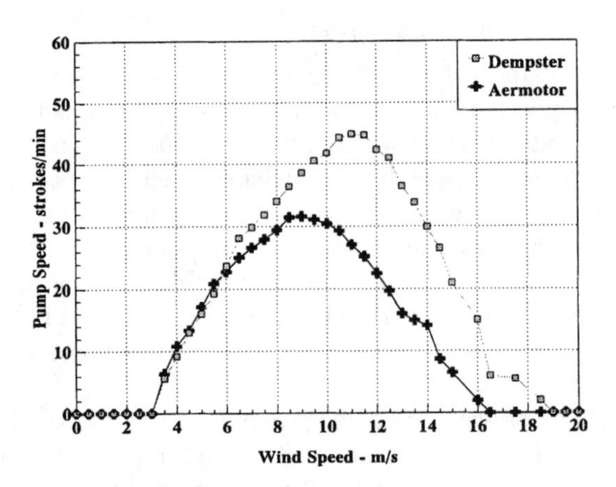


**Figure 2.59. Comparison of pump speed for Aermotor and Dempster windmills with the furling spring in the third adjustment hole and a pumping lift of 20 m.**

With a fixed pump stroke length and pump diameter, stroke speed of the single-acting piston pump is the only variable to change the rate of pumping as wind-speed changes. The flow rate or volume of water pumped is proportional to the pump speed. Because of the influence of pump speed on the overall performance of the windmill, the adjustment of pump speed (rotor wheel speed) becomes the single most important setup and maintenance factor with mechanical windmills. Figure 2.59 shows the influence of spring tension on pump speed for two different manufacturers of windmills. The springs were set at the same hole, but produced different tensions, which produced speeds of 30 strokes/min for an Aermotor and 44 strokes/min for the Dempster. (The mention of trade or manufacturer names is made for information only and does not imply an endorsement, recommendation, or exclusion by the U.S. Department of Agriculture, Agricultural Research Service.) Because of differences in spring tensions, each windmill needs to be checked to determine whether it is operating within the suggested speed range.

The pump used with almost all mechanical windmills is a single-acting piston pump with two valves. One valve functions as a check valve and is positioned at the bottom of the pump cylinder, and the other valve is attached to the pump rod and oscillates up and down with the action of the windmill rotor. Pumps can be purchased in several sizes, which in turn will produce different volumes of water. Table 2.28 contains pumping capacities for various pump diameters and the recommended size of windmill to use for various pumping depths. Table 2.28 was reproduced from product literature distributed by the Aermotor Corporation and has been used for over 75 years to match pumps and rotor sizes. Although many pump sizes are available, most pumps purchased correspond to the available pipe sizes. Normally a pump is selected that is slightly smaller than the standard pipe; for example, a 47-mm (1 7/8-in.) pump is used with a 51-mm (2-in.) drop pipe. This is done so that the pump valves can be pulled out for replacing the lifting cups without removing the drop pipe. This becomes important when wells are over 20 m deep.

Table 2.28. Pumping capacities[a]

| Diameter of Cylinder (mm) | Capacity (L/h) | | Total Pumping Lift (m) | | | | | |
|---|---|---|---|---|---|---|---|---|
| | | | Rotor Diameter (m) | | | | | |
| | 1.83 | 2.4–4.9 | 1.83 | 2.44 | 3.05 | 3.66 | 4.27 | 4.88 |
| 44 | 386 | 568 | 40 | 56 | 85 | 128 | 183 | 305 |
| 48 | 473 | 681 | 36 | 53 | 79 | 119 | 179 | 280 |
| 51 | 492 | 719 | 29 | 43 | 66 | 97 | 140 | 229 |
| 57 | 681 | 984 | 23 | 34 | 52 | 76 | 110 | 180 |
| 63 | 852 | 1230 | 20 | 28 | 43 | 64 | 91 | 149 |
| 70 | 1003 | 1457 | 17 | 24 | 36 | 55 | 79 | 129 |
| 76 | 1211 | 1779 | 14 | 21 | 30 | 47 | 67 | 110 |
| 83 | — | 2082 | — | — | 27 | 40 | 56 | 93 |
| 90 | 1665 | 2422 | 11 | 15 | 23 | 35 | 49 | 81 |
| 95 | — | 2763 | — | — | 20 | 30 | 43 | 70 |
| 102 | 2157 | 3141 | 8 | 12 | 18 | 26 | 38 | 61 |
| 114 | 2744 | 3974 | 6 | 9 | 14 | 21 | 30 | 49 |
| 127 | 3406 | 4920 | 5 | 8 | 11 | 17 | 24 | 40 |
| 152 | — | 7096 | — | 5 | 8 | 12 | 17 | 26 |
| 178 | — | 9651 | — | — | 6 | 8 | 12 | 20 |
| 203 | — | 12,490 | — | — | 4 | 7 | 9 | 15 |

[a] Adapted from a chart provided by the Aermotor Company: "Capacities are approximations based on the mill set at the long stroke and operating in a 7–9 m/s wind."

As a result of the rotor speed control, windmills reach their maximum rotor speed and pumping rate at a wind speed of 8–10 m/s and almost totally stop when the wind speed reaches 18 m/s. A typical flow-rate curve is shown in Fig. 2.60 for a mechanical windmill with a 50-mm-diameter pump and lifting water 21 m. For this case, water began flowing when the wind-speed exceeded 2.5 m/s and reached its peak flow rate of 11 L/min at 9-m/s wind speed. As the wind speed increased above 10 m/s, the flow rate decreased until it totally stopped at 18 m/s. The shape of the curve in Fig. 2.60 changes very little as long as the maximum pump speed of 30 to 35 strokes/min is not exceeded, and it is not significantly influenced by pump diameter or pumping depth. Similar data for a 70-mm pump lifting water 30 m show that the maximum pumping rate increases to 23 L/min at 9-m/s wind speed and the starting wind-speed is increased to 3 m/s [10].

The volume of water pumped can be predicted with a pump curve and a wind-speed histogram. Since most water-use plans are based on daily water usage, I suggest that water volumes be estimated on a daily basis. Figure 2.61 shows the average water pumped per day based on monthly wind-speed histograms. This representation is suggested when the month-to-month variation in wind-speed is significant. The most water pumped per day was in May, with the average being 10,450 L/day, and the least pumped was in August, with the average being 8700 L/day. The yearly average was 9628 L/day. Most mechanical systems using single-action piston pumps do not have a large difference between high wind-speed months and low wind-speed months because the mechanical units reach their

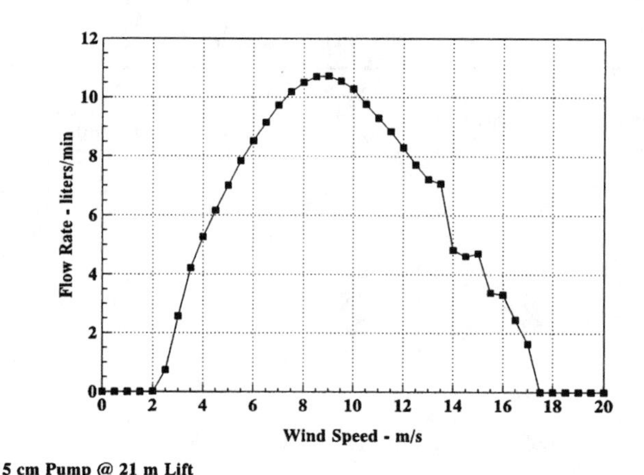


**Figure 2.60. Rate of water pumped by a mechanical, multibladed windmill with a piston pump.**

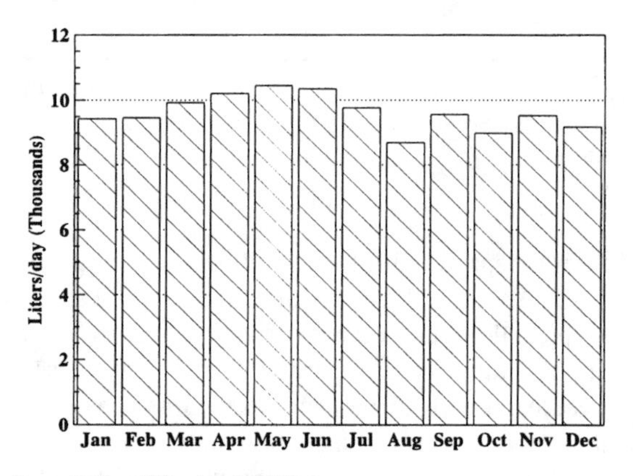


**Figure 2.61. Average daily water volume pumped by a mechanical, multibladed windmill.**

maximum pumping rate at 8–10-m/s wind-speed. Therefore higher winds speeds do not significantly contribute to more water pumped [11].

Any windmill that will provide an average of 8000 L/day or more will provide sufficient water for a herd of 100 head of cattle or 700 sheep or goats. It is recommended that each head of cattle have available 50 to 60 L/day of water and 10 to 12 L/day for each sheep or goat. Additional water is needed for evaporation and other water losses from the storage tank. I suggest that a storage tank with a 4–5-day water supply should be provided. This volume of water provides for days when insufficient wind is available and those times needed for maintenance and repairs.

### *Electrical Water Pumping*

A wind-electric water-pumping system consists of a wind-turbine generator connected directly to a standard three-phase induction motor driving either a centrifugal or submersible pump. This system was developed by the United States Department of Agriculture, Agricultural Research Service to operate independently of any electric utility as a stand-alone unit. It was designed to provide water at any remote location where water was needed for domestic uses, livestock, or irrigation. Also, it was designed to meet the demands for a low-maintenance, high-reliability, renewable-energy-powered water-pumping system that could be easily transported anywhere in the world [12].

The wind turbine used in a wind-electric water-pumping system uses electric generators that are direct-driven, permanent-magnet generators with three-phase, 240-V, AC nominal output. The generators produce a frequency and a voltage that are proportional to the rotational speed of the rotor. The key to operating an electric motor at variable frequency and variable voltage is to let both the frequency and the voltage vary in the same ratio. Electric motors do not consume additional current as long as the voltage/frequency ratio does not vary from the design ratio. A typical 220-V, three-phase electric motor will have a voltage/frequency of 3.6 and it will operate between 3.0 and 4.0 without overheating.

The permanent-magnet generators are wound to produce a frequency of 30 Hz and a voltage of 100 V at a wind speed of 4.5 m/s, which becomes the starting point for using wind-electric water pumpers. As wind speed increases, voltage and frequency ramp up proportionally until the mechanical overspeed control takes over at a wind speed of approximately 14 m/s (Fig. 2.62). Throughout the range of wind speeds between 4.5 and 14 m/s, the voltage/frequency ratio ranges between 3.2 and 4.0. This relationship can be maintained with a simple controller that senses the frequency or the voltage to determine the initial point of connection between the generator and the electric motor. Normally a small amount of capacitance is added to help boost the voltage at the higher wind speeds

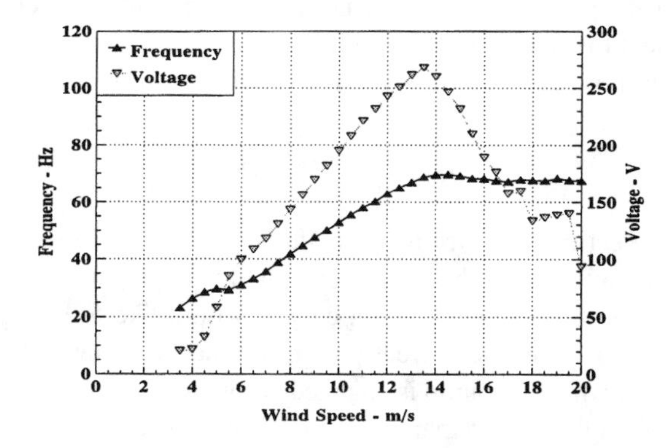


**Figure 2.62. Measured frequency and voltage from a 1500-W wind turbine for each wind-speed bin for a pumping depth of 45 m.**

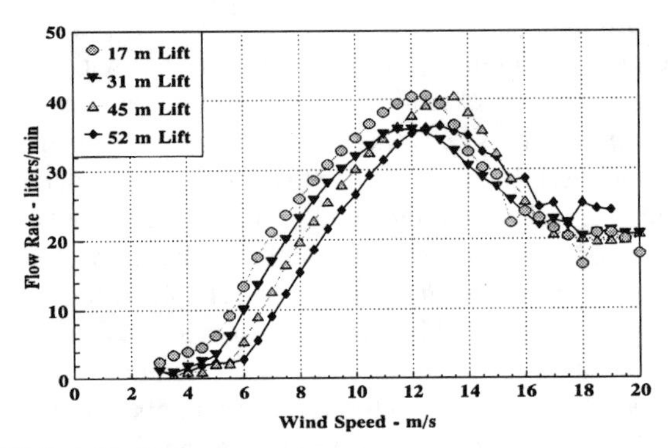


**Figure 2.63. Water flow rates for four pumping depths with a submersible pump and a 1500-W wind turbine.**

[13]. A new controller has been designed by the U.S. Department of Agriculture, Agricultural Research Service that adds the capacity to monitor the frequency, voltage, and voltage/frequency ratio. If any of the three parameters fall outside the preset conditions, the controller reacts by either adding more capacitance or disconnecting the motor load [14].

Typical pump curves for a wind-electric water-pumping system are shown in Fig. 2.63. There are four curves presented that indicate the differences in pumping rates as affected by the pumping head. It is recommended that pumps be selected based on the pumping head and the size of the wind turbine selected. A major advantage of the wind-electric pumping system is that the pump can be selected somewhat independently of the wind turbine used. For each wind turbine, there is a preferred electric motor size, that is, a 1.1-kW motor is used with a 1.5-kW wind turbine or a 0.75-kW motor is used with a 1.0-kW wind turbine [15]. Manufacturers or their representatives are well versed in the match of the electric motor to the wind-turbine generator that is best. However, in most cases, there are several pumps that could be used with a particular size electric motor. For the pump, you select the one that will provide water from the required pumping depth (head) of your well. It is suggested that you use the following equation to determine the selection pumping head:

$$H_s = H_a(f_{60}/f_c)^2, \tag{2.9}$$

where $H_s$ is the selection head from the manufacturer's literature, $H_a$ is the actual pumping head of the well, $f_{60}$ is the frequency at which the manufacturer's curves were determined, and $f_c$ is the frequency at which you would like to start pumping water. Using this equation to select the correct pump provides sufficient lift capacity of the pump to obtain water when the pump is engaged at low wind speeds of 4–5 m/s because the pump is turning slightly more than half of its design speed.

The average daily water volume pumped for a 1.5-kW wind-electric pumping system lifting water 17 m is shown in Fig. 2.64. The highest daily average water pumped was in May at 23,490 L/day, with little difference among the months of March, April, or May,

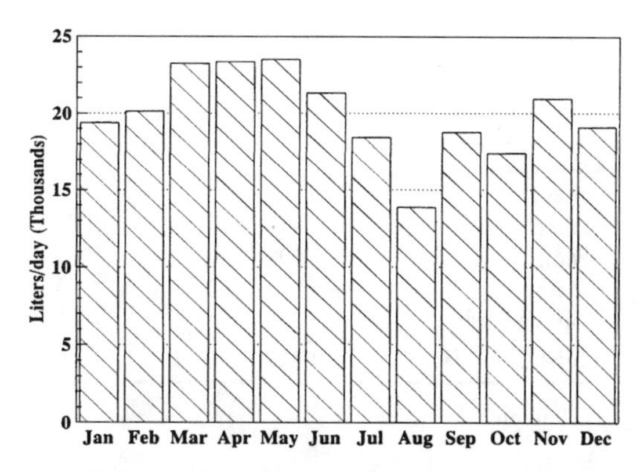


**Figure 2.64. Average daily water volume pumped for each month calculated with 10-year wind-speed histograms from Bushland, TX.**

and it dropped to 13,860 L/day in August. Because the wind-electric pumping systems utilize wind speeds up to 14–16 m/s, the influence of the months with either high or low wind speeds is clearly seen. This system averaged 19,930 L/day for the year or enough for ~200 head of cattle or 1400 sheep.

Wind-electric pumping systems offer several advantages over mechanical systems: (1) flexibility in locating the wind turbine some distance from the pump, (2) the ability to pump larger water volumes than mechanical wind pumps, (3) lower maintenance requirements of advanced small wind turbines compared with mechanical wind pumps or diesels, (4) higher wind-to-water average operating efficiency compared with that of mechanical wind pumps [9%–12% compared with 6%–8%], and (5) ability to provide integrated community water and household electricity. The disadvantages of the wind-electric pumping system are that (1) it is a new technology, requiring a different type of installation and new job skills training, and (2) there are few dealers and manufacturers.

***Economics of Wind-Powered Water Pumping***

Mechanical windmills have been used for years to pump water, the technology is mature, and all costs are well defined. Wind-electric water-pumping systems are new and are still being refined and the costs are not as well known, but the advantages of the wind-electric systems warrant their being considered for new installations and many replacement installations if most of the system is replaced. Figure 2.65 is a comparison of the daily water volumes between mechanical and electric-water-pumping systems lifting water 45 m. The electrical system averages 12,534 L/day, and the mechanical system averages 8600 L/day with very little differences in volume pumped in August. For the year, the electric system exceeds the mechanical system by almost 4000 L/day or ~45% more water. These data clearly show that electrical wind pumps operate better than mechanical systems when the average wind speed is above 5 m/s and operate about the same as mechanical systems when the wind speed is between 4 and 5 m/s [10].

**Table 2.29. Catalog list prices of various components used in a wind-powered water-pumping system[a]**

| Component | Mechanical (US $) | Electrical (US $) |
|---|---|---|
| Wind turbine | 3275 | 3195 |
| Tower | 1800 | 1576 |
| Pump control | — | 475 |
| Pump | 194 | 381 |
| Motor | — | 283 |
| Pipe and rods | 1050 | 300 |
| Total Cost (US$) | 6319 | 6210 |

[a] Pumping depth is 45 m.

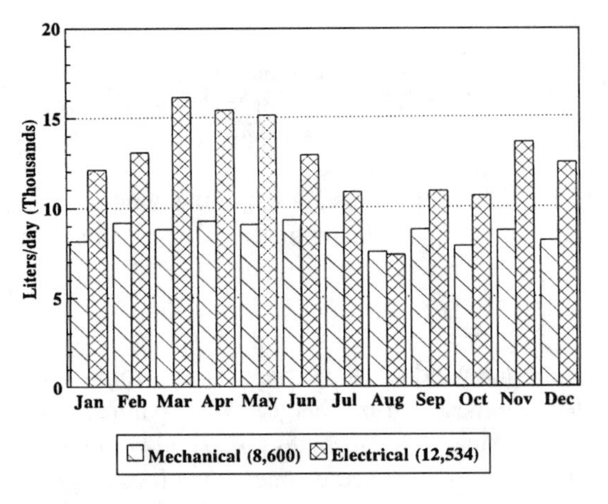


**Figure 2.65. Comparison of the average daily water pumped between a mechanical windmill and a wind-electric water-pumping system. The pumping lift was 45 m.**

The cost of the wind turbines and towers of both the mechanical and the electrical systems are almost the same (Table 2.29). The significant differences in cost are in the pumps and piping. For the comparison in Table 2.29, steel pipe and rods are used for the mechanical system, and polyethylene pipe is used for the electrical system. Although the electrical system requires a controller and a more expensive pump and electric motor, these costs are more than offset by the higher cost of the steel pipe. The overall system costs are almost identical for the two systems.

## References

1. Golding, E. W. 1955. *The Generation of Electricity By Wind Power*, p. 332. London: E. and F. N. Spon.
2. Baker, T. L. 1985. *A Field Guide to American Windmills*, p. 516. Norman, OK: University of Oklahoma.

3. Park, J. 1981. *The Wind Power Book*, p. 253. Palo Alto, CA: Cheshire.
4. Clark, R. N. 1980. Data requirements for wind energy. ASAE Paper No. 80-4517. St. Joseph, MI: ASME.
5. Wegley, H. L., J. V. Ramsdell, M. M. Orgill, and R. L. Drake. 1980. *A Siting Handbook for Small Wind Energy Conversion Systems*. Report PNL-2521, Rev. 1. Washington, DC: U.S. Department of Energy.
6. Elliott, D. L., C. G. Holladay, W. R. Barchet, H. P. Foote, and W. F. Sandusky. 1986. *Wind Energy Resource Atlas of the United States*. Report DOE/CH 10093-4. Washington, DC: U.S. Department of Energy, Solar Energy Research Institute.
7. Nelson, V., E. H. Gilmore, and K. Starcher. 1993. *Introduction to Wind Energy*. Report 93-3. Canyon, TX: Alternative Energy Institute, West Texas A&M University.
8. Kamand, F. Z., R. N. Clark, and R. G. Davis. 1988. Performance evaluation of multibladed water pumping windmills: *Proc. of Windpower'88*: 281–289.
9. *Dempster Windmill Assembly Instructions*, Dempster Industries Inc., Beatrice, NE.
10. Vick, B. D. and R. N. Clark. 1997. Performance and economic comparison of a mechanical windmill to a wind-electric water pumping system. ASAE Paper No. 974001. St. Joseph, MI: ASAE.
11. Clark, R. N. and K. E. Mulh. 1992. Water pumping for livestock: *Proc. Windpower'92*: 284–290.
12. Clark, R. N. 1994. Wind-electric water pumping systems for rural domestic and livestock water: *Proc. of European Wind Energy Assoc*: 768–777.
13. Clark, R. N. and S. Ling. 1996. A smart controller for wind-electric water pumping systems. *Wind Energy 1996*, pp. 204–208. New York: ASME.
14. Ling, S. and R. N. Clark. 1997. The role of capacitance in a wind-electric water pumping system: *Proc. of Windpower'97*: 601–609.
15. Clark, R. N. and B. D. Vick, 1995. Determining the proper motor size for two wind turbines used in water pumping. *Wind Energy 1995*, ASME SED, vol. 16, 65–72. New York: ASME.

## 2.5 Hydraulic Energy

*V. Schnitzer*

### 2.5.1 Water-Power Development

Of all power sources, besides human and animal power, water power is the oldest and most important in the development of agricultural processing and industrial development. Deriving from early application in the flour mills, the power source had been utilized for other processes. The names of sawmills, grinding mills, hammer mills, and even steel mills indicate the role of this power source in our civilization. The rapid progress of electrification was then possible with the introduction of engines and turbines driven by fossil fuels; water power, independent in its location, still has a big share- and the biggest in renewable energy in today's energy demand.

In spite of fast expansion of the electric grids, individual power sources in remote and agriculturally dominated areas still play the main role and foster development.

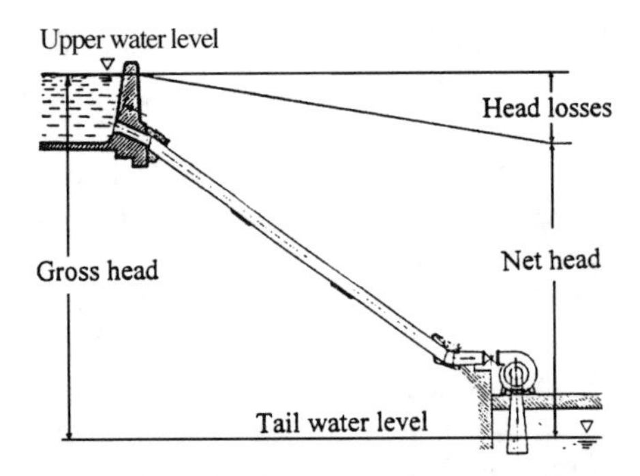


**Figure 2.66. Section of medium head plant.**

The focus of the overview given here is on the microsize to small-size hydropower range; more details are given in special literature [1–3].

***Principles***

Wherever a drop or fall of water can be developed, either by diversion before waterfalls, cascades, or slopes or by dams and barriers, the difference created between the water levels can be utilized to generate power (Fig. 2.66). For the calculation of power $P$ (in kilowatts) to be generated, the following formula, deriving from the hydrodynamic laws, applies:

$$P = QH\gamma\eta.$$

- Flow ($Q$) refers to the discharge, the share of the flow of the river passing through the turbine (in cubic meters per second).
- Head ($H$) refers to the net head between the upper and the lower water levels (in meters).
- $\gamma$ stands for the acceleration due to gravity ($\gamma = 9.81$ kN/m$^3$).
- The overall efficiency ($\eta$) of turbine and electricity generation is assumed to be ~70% for the power estimate.

The estimation formula may further be simplified by making $\gamma = 10$ instead of 9.81, deriving

$$P = QH0.7.$$

For calculation of the annual power production, the total working hours of 7000–8000 h is multiplied by the mean power produced [Fig. 2.67(c)].

***Hydraulic Resources***

Water-power potential depends on the hydrogeographic conditions of the area. Whereas the flow available depends on the rainfall rate and the catchment area, the head depends on the topography with the slopes available. [Fig. 2.67(a)]. Mountainous regions with regular rainfalls are therefore well suited for hydropower development.

The annual rainfall and seasonal behavior [hydrograph, 2.67(b)] is evaluated in the flow-duration curve, which shows the hydropower potential of a river catchment area [Fig. 2.67(c)].

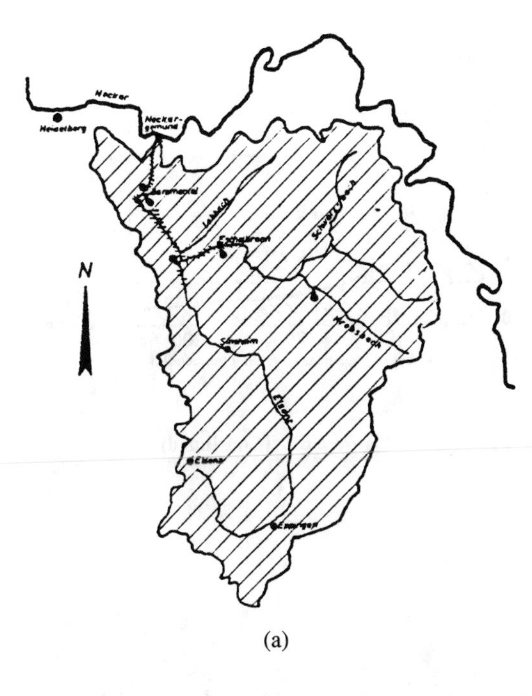


(a)

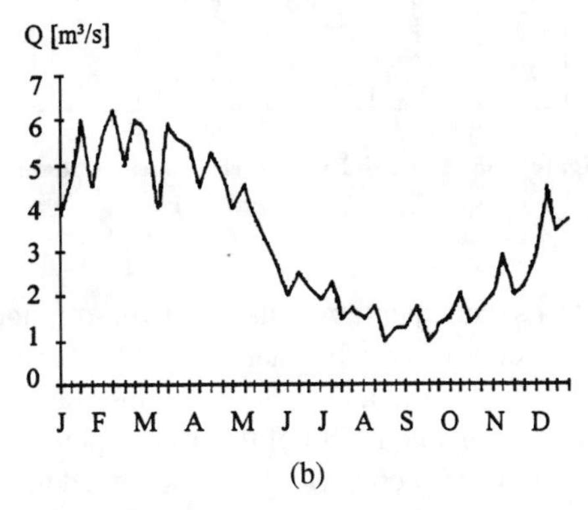


(b)

**Figure 2.67. (a) Example of catchment area of river basin, (b)annual hydrograph, (c) flow-duration curve and annual power production.**

Depending on the characteristics of the area, the flow that can be utilized can be determined. In general, for a firm power required, the water flow should be ensured in the dry season. Storage dams and reservoirs collect surplus flow to distribute it for power generation in a time of less flow or peak power demand.

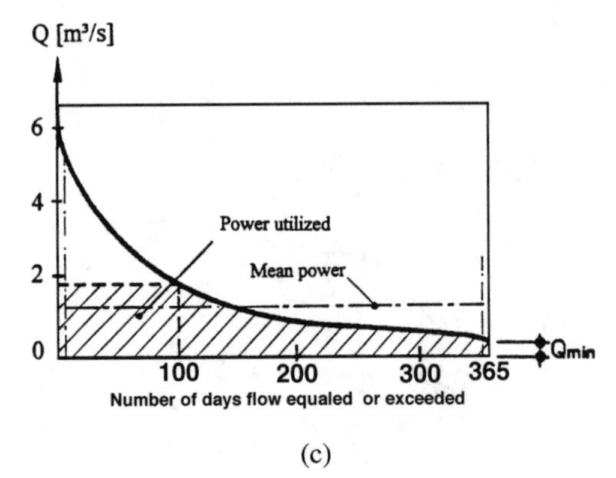


(c)

**Figure 2.67. (Continued)**

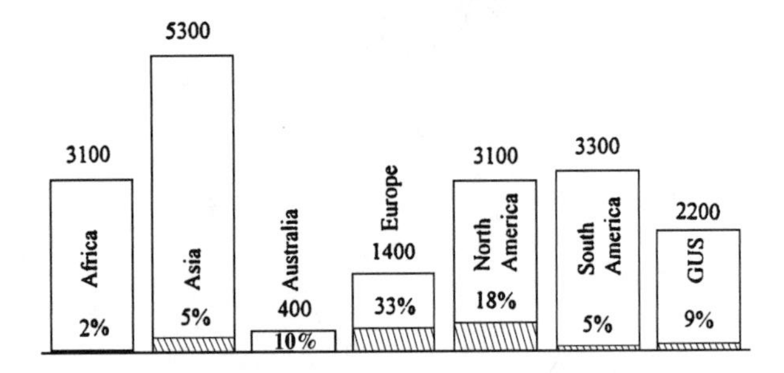


**Figure 2.68. Potential and developed water power.**

***Potential***

In a river system with its catchment area, its water run-off, and a drop in the water level, a theoretical potential of power can be established. On average, less than one fourth of this potential power is usable, and a capacity to be installed in power usually rates below one third or even less than one fourth of this usable potential capacity.

Only highly developed industrial countries have succeeded in developing a high rate of their potential in the past. There is, however, a vast potential in Africa, Asia, and South America still to be developed. (Fig. 2.68).

Most of the potential can be developed in big hydropower stations and will serve the national grids. A relatively small amount can be developed in smaller hydropower stations, which play an important role in decentralized rural development. As for smaller communities that dominate in the rural areas, the smallest potential plays a key role.

A classification based on the installed capacity distinguishes between micro, mini, and small water power according to the following ratings: micro is up to 100 kW, mini is 101–2000 kW, and small is 2001–15,000 kW, with a unit size from 1001 to 5000 kW.

As micro and mini hydropower stations up to 2000 kW play a major role in agricultural development areas, the following information refers mainly to these areas. The micro range, which substitutes diesel engines in remote locations, is essential for economic development as it leads to saving expensive fossil fuels and rendering systems self-sufficient.

### 2.5.2 Plant Layout and Civil Works

***Water-Power-Plant Layout***

Depending on the topographic conditions and the fall available for utilization, the design has many variations for reaching an optimal solution. The basic components, already typical for a traditional mill, are those shown in Fig. 2.69.

*Weir, Diversion, and Intake Structures*

At the beginning of waterfalls, cascades, or steep slopes, a weir is constructed to permit the diversion of water at an intake structure.

*Water Conduit*

Water is passed along contour lines in open canals as far as possible and collected in a forebay, from where water is directed to the power plant.

*Penstock/Pressure Pipeline*

As penstock constructions are expensive, the penstock follows the shortest distance to the power house. In difficult terrains or steep slopes, the penstock is let directly from the intake to the power house.

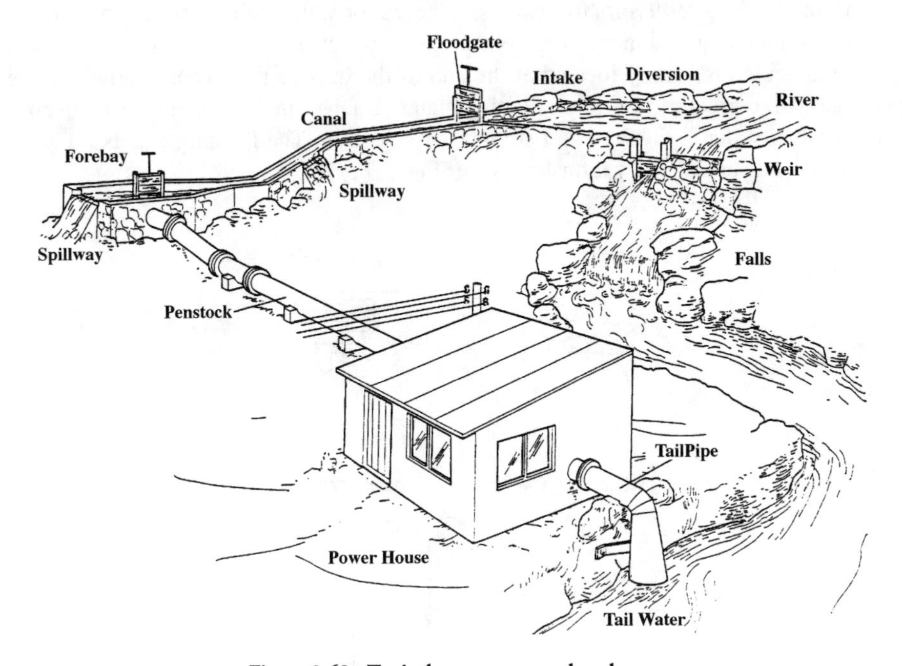


**Figure 2.69. Typical water-power-plant layout.**

*Power House*

Located in a flood-secure place, preferably on a rock with the shortest distance for the tail water back to the river, it has to be easily accessible for construction or for the public (in the case of a mill).

Low headwater power plants may have the turbines installed directly in the weir or barrier or may use a power canal, diverting the water to reach the maximum fall.

***Civil Works Components***

As civil works take up the highest portion of cost of a water-power project, the hydrogeographic evaluation and component design require much attention [3–5].

*Weir, Diversion, and Intake Structure*

Weirs or low dams serve to gain head in the river and enable controlled water diversion to the power canal or to the turbines directly.

Simply structured temporary weirs made locally are often found in remote hilly areas with a tradition of irrigation and water mills. They consist of compiled rubble, sometimes reinforced with wooden structures or improved by gabion support (Fig. 2.70).

- Preference is given to fixed-crest weirs either built in either natural stone work or concrete, possible also in gabion construction. Weirs that involve gates or, barrages or movable weirs are required in bigger plants in combination with flood control.
- Gravity weirs with the height of one third of the base width are automatically stable. Intake structures divert an adequate quantity of water shaped to divert the bed load and solid matter.
- Dams creating reservoirs for water storage, as for water supply, irrigation, or flood control are required in multipurpose projects, which are usually bigger.

The diversion is ideally located at the end of the outside river bend; variations are possible with corrections to divert floating material. Diverting groins improve the conditions when water is diverted at a straight river section. The flushing canals for solid matter may be controlled by a flushing gate (Fig. 2.71) [6].

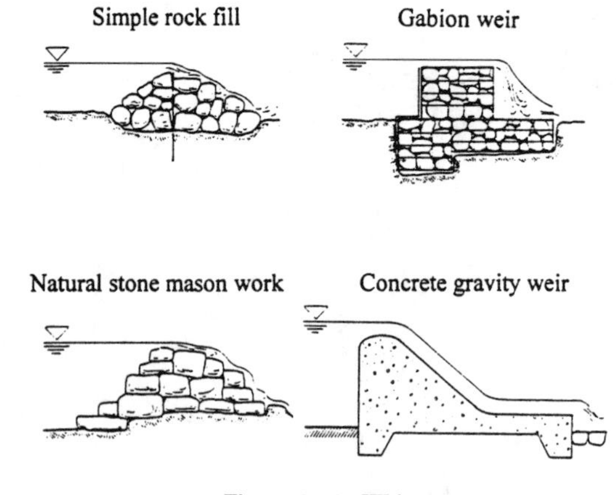


**Figure 2.70. Weirs.**

Optimal abstraction behind bend

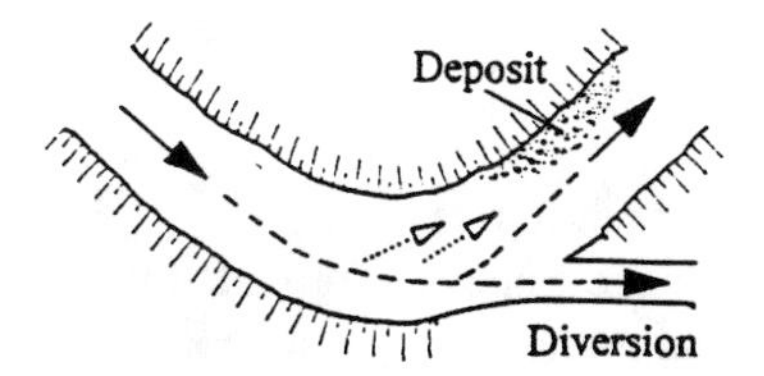


Side intake corrected with groin

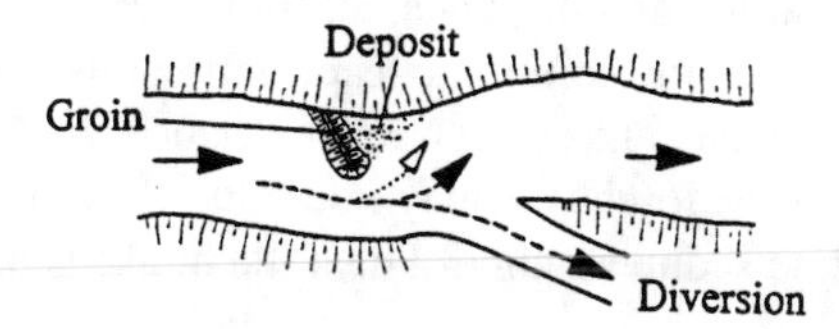


Structure of side intake with weir and flushing canal

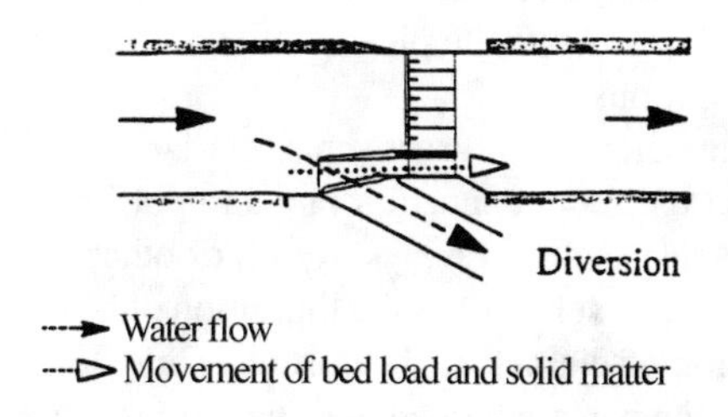


**Figure 2.71. River-diversion options.**

*Canals and Forebays*

Canal designs are determined by the topographical conditions; however, various options in design should be left open.

The flow in the canal is a function of the velocity, the roughness of the surface, and the cross sectional profile. The velocity of water should be kept above a minimum value to prevent sedimentation and aquatic plant growth, but below a maximum value to prevent erosion of unlined canals.

| Material | Sand | Loam | Clay | Masonry | Concrete |
|---|---|---|---|---|---|
| Velocity | 0.3 | 0.5 | 0.8 | 1.5 | 2.0 |
| Maximum | 0.5 | 0.8 | 2.0 | 2.0 | 2.2 |

Natural canals are excavated with the required cross section and slope; for improvement they may be sealed with clay or mortar mixture or lined with brick or stone masonry work.

Concrete lining and concrete constructions are higher in cost; however, they permit higher velocities and are often used to cover critical portions of the canal, especially where porous and unstable soils prevail.

Hydraulic calculations for canals are well documented for fair accuracy, especially in literature for irrigation works design [5].

Crossings, ducts, and drainage canals are provided where obstacles occur along the canal alignment. Spillways along the canal permit controlled overflow whenever excessive flow occurs. They are arranged near the intake, spaced along the canal, and one is required near the end.

The forebay, the basin at the end of the canal, fulfills various functions: It provides water storage for the operation of the turbine, especially at the start-up phase, contains water control for the turbine inlet, and protects against debris. As the water velocity slows down when the water enters the forebay, it may also fulfill the function of a settling basin, especially when no separate sediment control near the intake is provided.

*Penstock*

From the forebay, water is conveyed through a pipe or penstock to the turbine. In high head plants the penstock may be the most important and cost-intensive component whereas in medium or low head plants the pipe solution is less problematic and can be covered by familiar pipe solutions.

Pipe length, material, diameter, and pressure characterize the penstock. Although the penstock alignment aims at a direct connection in a straight line, the nature of terrain, suitability of ground, obstacles such as ravines, rocks, or other structures influence the layout. The criterion for diameter selection is the limitation of head loss, which increases drastically with the reduction of diameter. The cost of energy loss due to the friction is compared with the cost for the penstock in an optimization process (Fig. 2.72).

As the penstock has to withstand the maximum occurring pressure, considering also dynamic effects such as water hammer and surges, the material and the wall thickness of the pipe are selected accordingly.

From the great variety of material available for pipes, such as polyvinyl chloride, high-density polyethylene, prestressed concrete, fiber cement, spun ductile iron, glass-reinforced plastic, and steel, steel is still the most common material used. Its strength

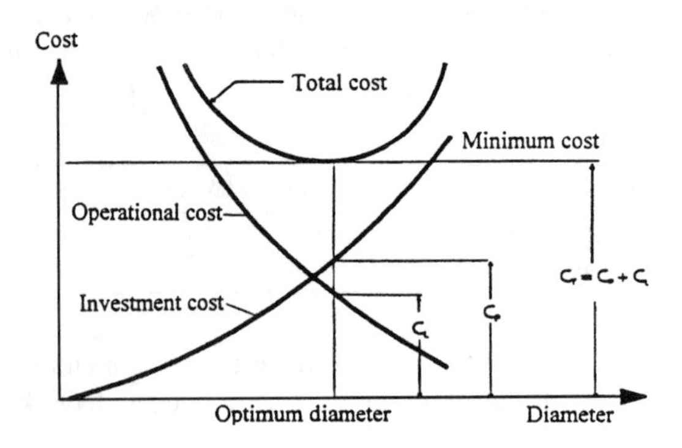


**Figure 2.72. Optimization of penstock diameter.**

properties, availability, and local adaptation still secure its dominating role. It may be supplied in welded pipes or spiral welded pipes, or it may be rolled and welded locally. When the steel is laid above the ground, the jointing, anchoring, and crossing of ducts are less problematic. Plastic pipes are more frequently used in medium and low head schemes, when laid in the ground, because of their advantages such as light weight, easy handling, and corrosion resistance, as well as price—if locally available.

Penstock flow is controlled by valves; air valves are installed at the high points of the pipeline. To avoid excessive pressure caused by water hammer, special calculations may lead to special pressure-relief solutions (i.e., surge tanks, relief valves).

*Power House*

The shape and the size of the power house depend mainly on the type of turbine and the space requirement for auxiliary equipment such as electric gear, hoist, and service facilities.

Water mills for direct power use accommodate the mill or working machines in additional space adequate for their function.

Sufficient elevation is required for flood protection. Solid masonry work or concrete on good foundations are required against the errosive impact of water.

Depending on the discharge arrangement of the turbine, a tail pipe or draft tube may lead directly into the tail water or may discharge water through a diffusor bend of increasing size to recover the velocity energy in the discharged water.

Superstructures of power houses and mills vary according to the size of plant, the local building techniques, and the means of the owner.

### 2.5.3 Plant Equipment

***Water Turbines***

The potential energy of water is converted into mechanical power in a rotating machine, the turbine. The designs of turbines depend on the head and flow and are classified into two main types:

Impulse turbines are driven by the water velocity from a water jet directed on the runner. The runner operates above the tail water level. Impulse turbines, namely Pelton and Turgo, are applied in high and medium head plants, whereas the cross-flow turbine covers the ranges from low to medium heads.

The reaction turbines convert the energy in a runner fully immersed in water within a casing. Guide vanes and runner profiles are shaped for optimal energy conversion in the flow. The runners may be located above the tail water level, converting the remaining energy in a conic tailpipe or diffusor.

From the great variety of turbines available, only an overview is given here. Competent and experienced firms have developed standard programs for a wide range of solutions to suit specific sites and conditions for micro, mini, and small hydropower (Fig. 2.73).

The Francis turbine, with its wide range of applications, was commonly used in the early years of turbine applications; it has now been replaced, to a wide extent, by the Kaplan turbine, with its variable propeller blades. If both the propeller blades and guide vanes are adjustable, the double regulated Kaplan turbine can operate over a wide flow range with good efficiency.

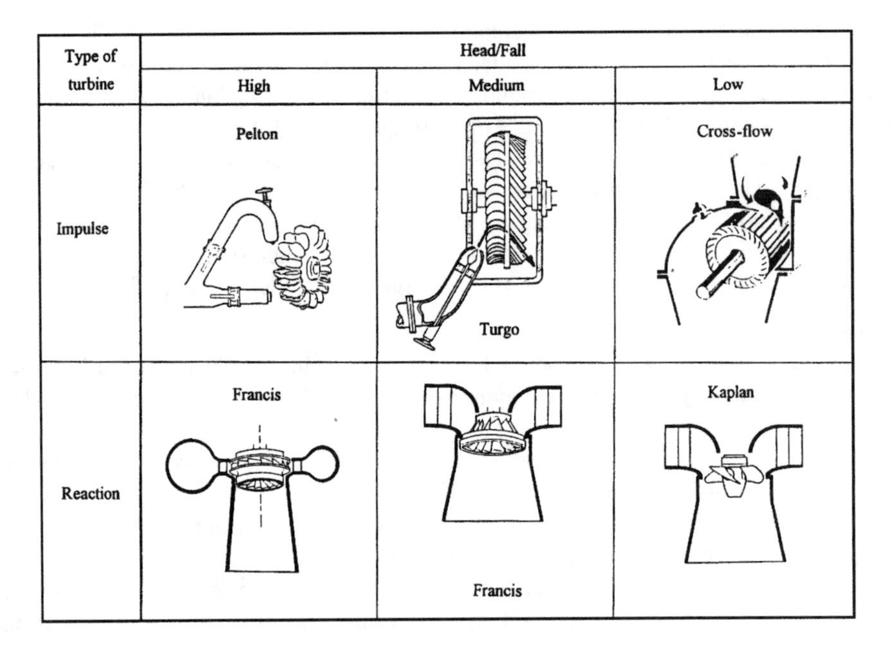


**Figure 2.73. Classification of turbines.**

Installed in vertical, horizontal, and inclined positions, it meets especially the low head range of river run-off and by-pass canal plants.

The cross-flow turbine is also referred to as the Banki, Mitchell, or Ossberger turbine. The water passes through the drumlike radial runner; a regulating guide vane enables the turbine to cope with the wide flow range with fairly good efficiency.

As it has spread worldwide with its good operating range, it has also been the model that was first manufactured in developing countries and has established its position as a good standard in some regions.

Standard centrifugal pumps operate as turbines when the flow is passing in reverse. They are limited to applications with rather constant flow and are therefore ideal for energy recovery from pipelines and pressure systems. Because of their simplicity, rigidity, and low price, they have also been introduced as a power drive for mills with constant power demand.

Typical efficiency curves are summarized in Fig. 2.74.

### *Water Wheels*

Horizontal water wheels have been known since ancient times and are still in use to drive small flour mills in remote high mountains [7]. Improvements were successfully introduced to increase efficiency and extend the service range [8] (Fig. 2.75).

Overshot, breastshot, and undershot water wheels have been developed for powerful drives in the process of industrialization [9] (Fig. 2.76). They reached their optimum design in the first third of this century; their decline began with the introduction of the

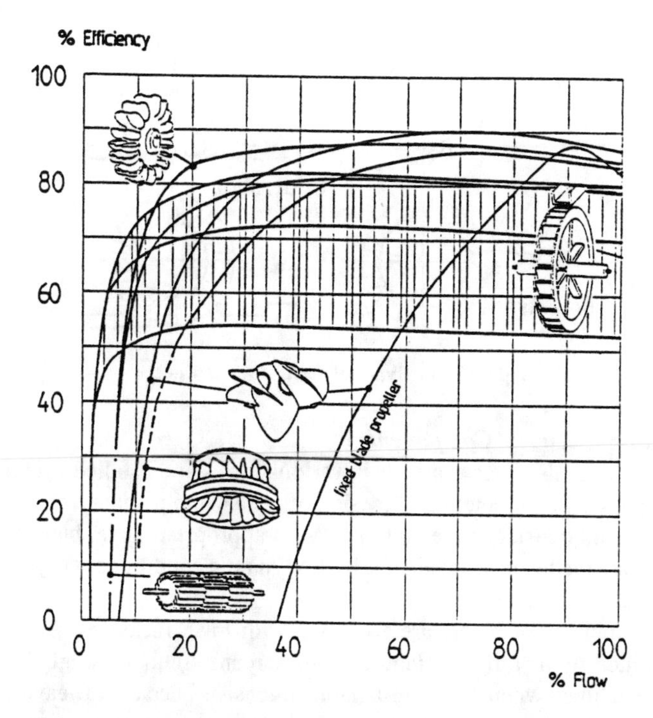


Figure 2.74. Typical efficiency curves of turbines and water wheels.

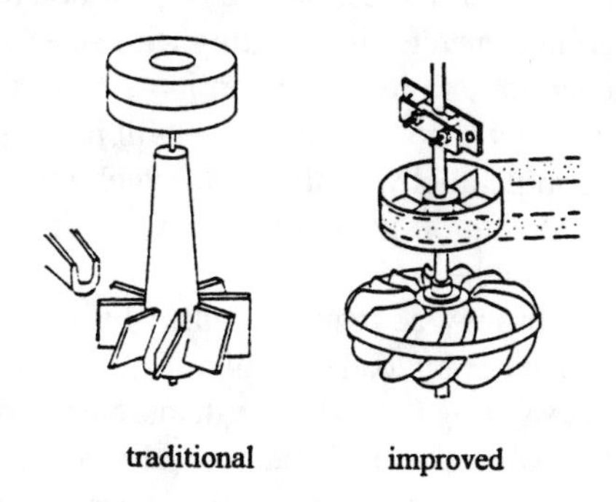


Figure 2.75. Horizontal water wheels.

modern turbines. Most of the existing water wheels were replaced by turbines. Many plants nowadays have been reactivated with modern elements for power transmission and electricity generation.

In spite of a good part-load behavior (Fig. 2.74), the low speed of the wheel remains critical for power transmission [10].

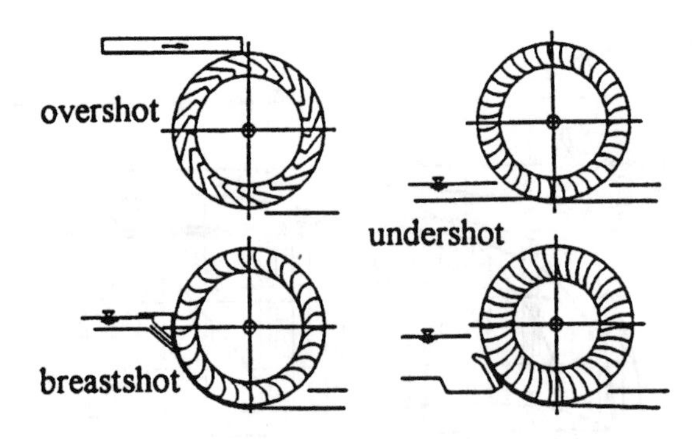


**Figure 2.76. Types of vertical water wheels.**

***Mechanical and Electrical Components***

In open-water canals, gates available in various forms are installed to control the flow of water or seal it off completely for revision of canals and plants.

Water pipes and penstocks are equipped with appropriate gate, butterfly, or special valves, depending on the pressure and the hydrodynamic conditions. They are considered as part of penstock design.

Intakes to pipelines or plants are equipped with trash racks to prevent all floating matter from entering the turbine. Hand-cleaned screens are increasingly being replaced by cleaning machines when cleaning is labor intensive because of leaves, weeds, and floating matter in the water.

For electric-power production, the turbine shaft is connected to a generator or alternator, preferably by a direct speed connection or by a gear or a belt transmission.

Depending on whether the generator is operating with an existing grid or in isolated operation, the type of generator and its regulation have to be selected. Asynchronous generators of the simple squirrel cage induction type will not require voltage and speed regulation when operating in parallel with the grid. Synchronous generators, however, need their own exitation and control. In addition, they require a synchronizing system when operating in parallel with the grid.

The synchronous generator may be regulated either by a governor that controls the flow to keep the speed constant or by load control to regulate the load.

For small water turbines working in isolated systems, electronic load controllers gain importance over the traditional mechanical and hydraulic governors.

For switch gear and protection equipment, each country's electricity supply regulations will apply and must be referred to. Even stations operating in isolated areas or by private ownership should follow the prevailing national standards for these plants.

Special reference is given here to the international standard IEC 1116 that describes the requirement for the electromechanical equipment [11].

## 2.5.4 Decentralized Water-Power Development

***Role of Water Power in Development***

In industrialized countries, water power has been the most important source of energy in many regions during times of industrial growth. Although the small power drives have lost their importance with the advent of electrification, 20% of thousands of small hydropower plants still use the direct mechanical drive [3].

In the past decades, infrastructure development has aimed at rural electrification by means of the grids. An evaluation of 40 projects financed by international donors in developing countries has revealed that most projects were not economic and were considered unsuccessful. It resulted in restrictions toward financing grid-based rural electrification in these countries [12].

The key for adaptation of technologies to the place of use lies in the thorough evaluation of local conditions and has to involve the local resources and national production. The development of small hydropower sources in regions with considerable hydraulic potential will lead to economic and sustainable energy infrastructure. It will foster self-sufficiency and improve local economic activities. A micro hydroscheme benefits village people especially. The most important application, the mill, particularly reduced women's work load.

The technique for the locally required water-power plant components may be adapted to the local resources. These comprise the skills, material, and facilities of investors. In community projects the readiness for self-help plays a major role.

Whenever such water power is established, more independence is created, as the power source can be utilized for a flour mill or other postharvest processing, or pump water for irrigation or water supply and other craft development; it could, however, generate electricity as well (Fig. 2.77). The main use during the day serves a high demand of power needs to make the plant economic by continuous use. Economic performance is ensured especially in multipurpose plants that serve agricultural processing, small-scale

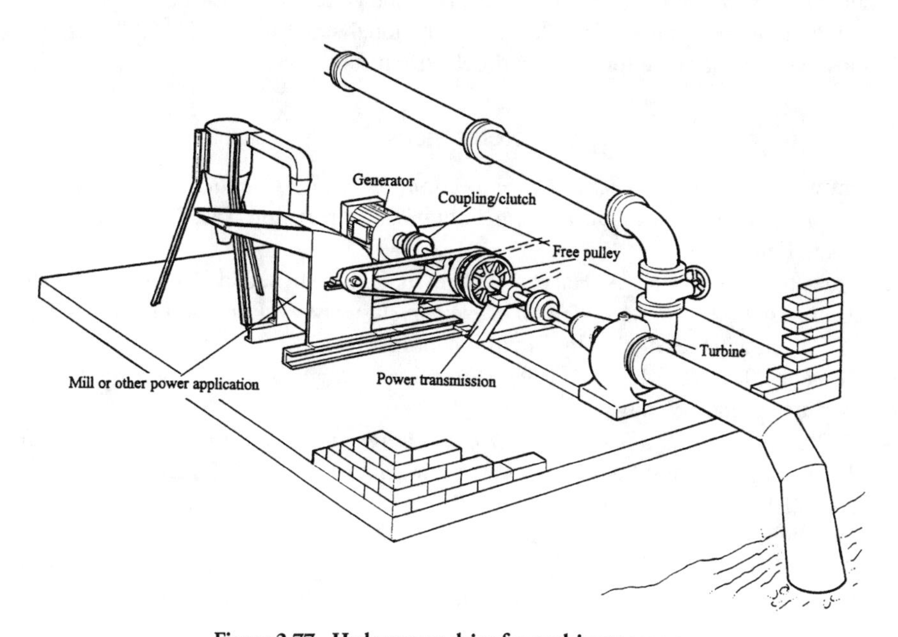


**Figure 2.77. Hydropower drive for multipurpose use.**

industry, and domestic lighting. It paves the way to electrification on the basis of economic performance.

Once several individual power sources are available, interlinkage and grid connection may prove feasible in a further step of decentralized development.

***Guide to Plant Establishment***

Apart from the complex technical requirement as outlined above, economic, legal, and administrative aspects are to be considered.

At an early stage, after the basic technical feasibility has been established with the hydraulic data, demand output, and figures and a project outline has been drafted, the following aspects have to be verified before the overall feasibility of the project can be established:

- permission for power-generation conditions and tariffs,
- water-rights licensing and environmental restrictions,
- land use, transmission rights,
- ecological impact, influence on farming, fishing, forestry, landscape, and protected areas,
- planning and construction permission,
- financing facilities, subsidies, and government support,
- economic evaluation.

As government regulations vary widely in different states the procedural issue has to be examined locally.

In the ensuing detailed planning, a close contact with suppliers is recommended, as this dialogue contributes much to optimal solution finding. Depending on the plant ownership, the normal procedures for tendering and contracting may be followed.

Special reference is made here to the international standard IEC 1116, "Electromechanical equipment guide for small hydroelectric installations."

## References

1. Inversin, A. R. 1986. *Micro-Hydropower Sourcebook, A Practical Guide to Design and Implementation in Developing Countries.* Washington, DC: NRECA International Foundation.
2. Fraenkel, P., O. Paish, A. Harvey, A. Brown, R. Edwards, and V. Bokalders. 1991. *Micro-Hydro Power, A Guide for Development workers.* London: IT Publications, IT Power, and the Stockholm Environment Institute.
3. Giesecke, J. and E. Mosonyi. 1997. *Wasserkraftanlagen, Planung, Bau und Betrieb.* Berlin: Springer-Verlag.
4. Radler, S. 1981. *Symposium on Project Design and Installation of Small Hydro Power Plants.* Wien: Institut für Wasserwirtschaft Universität für Bodenkultur.
5. FAO. 1985. Irrigation and Drainage Paper Design and optimization of irrigation distribution network. Rome: FAO.
6. Scheuerlein, H. 1984. Die Wasserentnahme aus geschiebeführenden Flüssen. Berlin: Ernst & Sohn Verlag für Architektur und technische Wissenschaften.

7. Moog, B. 1994. *The Horizontal Watermill History and Technique of the First Prime Mover*. Sprang Capelle: The International Molinological Society.
8. Bachmann, A. and A. M. Nakarmi. 1983. *New Himalayan Water Wheels*. Katmandu: Sahayogi.
9. Müller, W. 1939. *Die Wasserräder, Berechnung, Konstruktion und Wirkungsgrad*. Detmold: Verlag Moritz Schäfer.
10. Schnitzer, V. and W. Gross. 1993. *Do Water Wheels have a Future*?: Hydroenergia 93 Forum für Zukunftsenergien, Bonn.
11. International Electrotechnical Commission 1992–10. *International Standard Electromechanical Equipment Guide for Small Hydroelectric Installations*. Geneva: Bureau Central de la Commission Electrotechnique Internationale.
12. Biermann, E. R. K., F. Corvinus, T. C. Herberg, and H. Höfling. 1996. *Basic Electrification for Rural Households*. Eschborn: Deutsche Gesellschaft für Technische Zusammenarbeit GmbH.

# 3 Biomass Engineering

## 3.1 Biomass Liquid Fuels

### 3.1.1 Ethanol and Methanol

*T. Saiki, I. Karaki and K. Roy*

***General Properties of Ethanol***

Ethanol or ethyl alcohol [1, 2], which is commonly known as alcohol, is a colorless transparent liquid with a pleasant odor and burning taste. The molecular formula for ethanol is $C_2H_5OH$ and the molecular weight is 46.07. It has some important physiological characteristics; for example, it is intoxicating and can be used as a sterilizing and disinfecting agent. By nature, ethanol is volatile and combustible. It mixes well with water and many other organic solvents and dissolves a number of organic chemical compounds. Because of these properties, ethanol is extensively used in drinks and foods, medicines, cosmetics and in chemical industries. Moreover, in recent years, its use as a clean substitute for petroleum fuel has been popular because it does not cause any environmental pollution and destruction. The physical properties of ethanol are shown in Table 3.1 [2].

***Properties of Ethanol as a Fuel***

Table 3.2 [2] presents a comparison between gasoline and ethanol-added gasoline (gasohol). For gasoline, the heating value per 1 kg of fuel is 44.37 MJ while that of ethanol is 26.79 MJ, which indicates that the heating value of ethanol per unit weight is 60% of that of gasoline. The molecular formula shows that the weight percentages of C, H, and O for ethanol are 52%, 13% and 35%, respectively, while for gasoline ($C_nH_{2n+2}$) those of C and H are ~84% and 16%, respectively. Therefore, unlike gasoline, ethanol contains 35% oxygen, which reduces the heating value per unit weight and activates the combustion efficiency well. In addition, the presence of oxygen gives ethanol a lower combustion temperature than gasoline, resulting in the production of no soot and less $NO_x$.

The ratio of theoretical air required per unit weight of fuel combustion is called the theoretical air–fuel ratio. For example, the required air volume per 1 kg of gasoline is 1.51 kg, which gives gasoline a theoretical air–fuel ratio of 15.1. In the case of ethanol, the theoretical air–fuel ratio is 9.0. If the heating values of the air–fuel mixture per 1 $m^3$ for gasoline and for ethanol are expressed in terms of their theoretical air–fuel ratios, these two liquids show values almost identical to each other: 3.55 MJ for gasoline and

Table 3.1. Physical properties of ethanol

| | |
|---|---|
| Appearance | Colorless, transparent liquid |
| Smell | Special sweet smell |
| Taste | Burning taste |
| Specific gravity | $d_4^{15}$0.79360 $d_4^{20}$ 0.78934 |
| Boiling point | 78.325°C (101 kPa) |
| Melting point | −114.15°C |
| Refractive index | $n_D^{20}$ 1.3614 $n_D^{25}$ 1.35941 |
| Critical temperature | 243.1°C |
| Critical pressure | 6.38 MPa |
| Critical density | 0.276 g/cm$^3$ |
| Flash point | 14.0°C (closed condition) 16.0°C (open condition) |
| Ignition point | 390–430°C |
| Explosion limit | minimum 3.3 vol. % maximum 19.0 vol. % (in air) |
| Vapor pressure | 1.6 kPa (0°C) 5.8 kPa (20°C) 46.8 kPa (60°C) |
| Heat of combustion | 1.37 MJ/mol |
| Heat of vaporization | 43.27 kJ (20°C)<br>38.95 kJ (bp) |
| Heat of dissolution | −11.18 kJ (water 200 mol, 18°C) |
| Coefficient of volume expansion | 0.001101/$K$ (0–30°C) ($K$, absolute temperature) |
| Adiabatic compression efficiency | 92.9 × 10$^{-11}$/Pa, 20°C<br>99.8 × 10$^{-11}$/Pa, 30°C |
| Surface tension | 24.05 × 10$^{-5}$–8.32 × 10$^{-8}t$ N/cm ($t$, temperature in °C) |
| Standard enthalpy of formation | 2771.6 kJ/mol |
| Standard Gibbs energy of formation | 1741.4 kJ/mol |

*Source:* Ref. [1].

3.53 MJ for ethanol. Moreover, if gasoline and ethanol are compared from the standpoint of heating values per 1 m$^3$ due to 1-kg suction of air adjusted to the theoretical air–fuel ratio, the values are also almost identical: 2.94 MJ for gasoline and 2.98 MJ for ethanol. The latent heat of vaporization of gasoline is 272 kJ/kg, while that of ethanol is 862 kJ/kg. As ethanol is hard to vaporize at low temperatures, the ignition of the engine is inferior. The octane value of ethanol (106) is higher than that of gasoline (100). Therefore, if ethanol is mixed with regular gasoline, the octane value can be increased.

***Production and Uses of Ethanol in Different Countries***

Of an estimated amount of total ethanol production [1, 2] of ~24.2×10$^6$ kl, ~93%, i.e., ~22.5 × 10$^6$ kl, is produced by the fermentation method. The remaining ~1.7 × 10$^6$ kl, ~7% of the total production, is currently produced by the synthetic method [2]. Table 3.3 [2] presents estimates of the world production of ethanol (with a concentration over 90%) in 1993. It shows that Brazil was the greatest producer of fermentation ethanol in the world, with almost 52% of the total. Following Brazil, the USA and the former USSR had 18% and 8% of world ethanol production, respectively.

**Table 3.2. Properties of gasoline, ethanol, and their mixed materials**

| Parameter | Gasoline (Iso-octane) | Ethanol | Ethanol Mixed Gasoline: Ethanol 10% Gasoline 90% | Ethanol Mixed Gasoline: Ethanol 20% Gasoline 80% |
|---|---|---|---|---|
| Molecular formula | $C_8H_{18}$ | $C_2H_5OH$ | — | — |
| Molecular weight | 114 | 46 | — | — |
| Specific gravity (20°C) | 0.7 | 0.79 | 0.72–0.76 | 0.73–0.77 |
| Composition (wt. %) | | | | |
| Carbon | 84 | 52 | 81–84 | 78–81 |
| Hydrogen | 16 | 13 | 12–15 | 12–15 |
| Oxygen | 0 | 35 | 4 | 7 |
| Theoretical air–fuel ratio (air kg/fuel kg) | 15.1 | 9 | 14.2 | 13.6 |
| Net heating value (MJ/kg) | 44.37 | 26.79 | 42.28 | 40.19 |
| Heating value in theoretical air–fuel ratio (MJ/m$^3$ mix) | 3.55 | 3.53 | — | — |
| (MJ/kg air) | 2.94 | 2.98 | — | — |
| Latent heat of vaporization (kJ/kg) | 272 | 862 | 381sp. gr.0.783 | 435sp. gr.0.738 |
| Research octane number | 100 | 106 | — | — |
| Motor octane number | 100 | 89 | — | — |
| Cetane value | 12 | 8 | — | — |
| Ignition point (°C) | 275 | 420 | — | — |
| Flash point (°C) | −43 (mix. of $C_4 \sim C_{12}$) | 14 | — | — |
| Boiling point (°C) | 99.4 | 78.3 | — | — |

*Source:* [1] and Technology Survey Index No. 15, "Resource-economical background of ethanol and its applicability to internal combustion engines" (Japan Automobile Research Institute, Inc., 1977).

The production of ethanol and its uses in some countries are described below.

*USA*

In 1995, the USA produced ~5.5 × 10$^6$ kl of ethanol in total, of which ~90% was produced by the fermentation method, with corn as a feedstock, and the remaining 10% by the synthetic method, with ethylene. 90% of the fermentation ethanol was consumed for automotive purposes, i.e., as fuel in different states of the corn-belt zone of the Midwest. It is assumed that to minimize air pollution and global warming and to promote agriculture, the production and the use of ethanol will continue at an increased rate. Apart from its use as a fuel, ethanol is commonly consumed in the production of paints, detergents, cosmetics, medicines, vinegar, chemicals (e.g., ethyl acrylate), and is used domestically as rubbing alcohol.

*UK*

According to 1989 statistics, total ethanol production in the UK was 3.4 × 10$^5$ kl, all of which was produced by the BP Chemicals Ltd. by the synthetic method. Approximately 1.9 × 10$^5$ kl of the total amount was consumed within the country, and the remainder exported. Most internal consumption was in the chemical and the industrial fields and

Table 3.3. **Estimates of ethanol production in different countries of the recent world**

| | Production (Unit, 1000 kl) | |
|---|---|---|
| Country | Fermentation Ethanol | Synthetic Ethanol |
| USA | 4000 | 600 |
| UK | 0 | 340 |
| France | 585 | 110 |
| Germany | 190 | 140 |
| Brazil | 11,680 | 0 |
| Argentina | 120 | 0 |
| Australia | 60 | 0 |
| Thailand | 200 | 0 |
| Indonesia | 42 | 0 |
| Philippines | 90 | 0 |
| New Zealand | 10 | 0 |
| India | 800 | 0 |
| Pakistan | 60 | 0 |
| South Korea | 200 | 27 |
| China | 900 | no information |
| Taiwan | 35 | 0 |
| Mexico | 100 | 0 |
| Poland | 280 | 0 |
| Hungary | 100 | 0 |
| Italy | 400 | 0 |
| Spain | 220 | 0 |
| Switzerland | 8 | 0 |
| Canada | 64 | 0 |
| Sweden | 30 | 0 |
| Denmark | 14 | 0 |
| Norway | 22 | 0 |
| Holland | 100 | 0 |
| Saudi Arabia | 0 | 350 |
| Cuba | 300 | 0 |
| USSR (former) | 1500 | no information |
| Jamaica | 70 | 0 |
| Japan | 346 | 98 |
| Total | 22,526 | 1665 |

This table is prepared based on data available in Ref. [1] and in the reports of JETRO and others; it indicates the total amount of alcohol used for industrial purposes and as fuel and drinks (liquors).

for use in medicines and cosmetics. Moreover, in 1989, $0.25 \times 10^5$ kl of fermentation ethanol was imported for use in the production of vinegar, etc.

*France*

In 1991, France had a total production of $6.3 \times 10^5$ kl of ethanol: $5.2 \times 10^5$ kl by the fermentation method and $1.1 \times 10^5$ kl by the synthetic method. 70% of the fermentation ethanol was made from beet and a major part of the rest was made from grapes and other agricultural products. The main areas of ethanol consumption were 40% of total production for chemical and industrial uses; 25% for medicines and cosmetics; 20% for inks; and 15% for paints, detergents, and others. Ethanol was also used in perfumes and for some domestic uses, such as cooking fuel and cleaning agents.

*Germany*

The total amount of ethanol production in 1991 in Germany was $3.4 \times 10^5$ kl, with $1.9 \times 10^5$ kl produced by the fermentation method and $1.4 \times 10^5$ kl by the synthetic method. Potato, cereal, and beet molasses were predominantly used to make fermentation ethanol, of which potato and cereal supplied 30% and 40% of raw materials, respectively. The main raw material used to produce synthetic ethanol was ethylene. However, sulfite pulp waste liquor (SWL) was also used in the production of some synthetic ethanol. The major uses of fermentation ethanol were 57% of total production for foods and drinks, 28% for medicines and cosmetics, 3% for vinegar, and 12% for common industrial uses. On the other hand, synthetic ethanol was used in only general industrial fields.

*Brazil*

The production of ethanol in Brazil was $0.6 \times 10^6$ kl in 1975, which jumped up to $12–13 \times 10^6$ kl at the beginning of the 1990's because of the introduction of the country's national fuel alcohol policy. The increase in production was therefore almost 20 times in 20 years. The high dependency on imported oil and bad price conditions for sugar at that time forced the country to increase the production of ethanol as a petrol substitute for automotive use. This also served the purposes of saving foreign currencies, creating jobs in rural areas, and preventing global warming. Approximately 90% of the total production has been used as fuel. For fermentation ethanol, 90% of the feedstock was sugar cane juice and the remaining 10% was sugar cane molasses and others. As ethanol is more expensive than gasoline, the situation in Brazil is now less favorable for subsidies and other financial assistance. However, the country has clearly been successful at reducing its foreign exchange burden from imported liquid fuels. By the year 2015, the production of fuel ethanol is estimated to be $\sim 16.5–19.4 \times 10^6$ kl for an estimated demand of $22.4 \times 10^6$ kl [3].

***Raw Materials for Ethanol***

The common agricultural products used as raw materials for ethanol fermentation are starchy and saccharine materials. Apart from these, SWL is a special type of raw material used to make fermentation ethanol. On the other hand, the main raw material for synthetic ethanol is ethylene.

*Saccharine Materials*

Sugar cane is a crop cultivated mostly in tropical regions. Sugar cane juice contains 11%–17% sugar (a mixture of sucrose, glucose, and fructose). Following the steps of clarification, concentration, and crystallization, sucrose is obtained from sugar cane juice. Sugar cane juice is also used as a feedstock for ethanol fermentation. Brazil is the predominant user of sugar cane juice as the fermentation feedstock for ethanol. The yearly production of sugar cane in Brazil is 260 Mt (where Mt is millions of metric tons), of which ~60% is used to make ethanol, and ~40% to make of sucrose.

The main countries that cultivate sugar beet are the former USSR, the USA, Poland, France, Greece, and China. In the case of sugar beet juice, the sugar density is 14%–20%, which is higher than that of sugar cane juice. The use of sugar beet as a fermentation feedstock for ethanol in France reached 53% of the total beet production in 1990–1991.

There are two types of molasses: sugar cane molasses and sugar beet molasses. Sucrose is manufactured through the steps of clarification and concentration of juice, crystallization, and centrifugation. The residue fluid is obtained from the mother liquor after sucrose separation. This residue is called molasses. Molasses contains some sugars, as it is not economically profitable to extract much more sucrose. Altogether, ~55% of molasses consists of sugars, including sucrose, glucose, and fructose. Molasses is a very viscous liquid, containing lime and other ingredients that are added for the clarification of the juice. Molasses is frequently used as a feedstock for ethanol fermentation and as an animal feed, a raw material for sodium glutamate, bread yeast, and other products.

India, Brazil, Thailand, Cuba, Indonesia, Pakistan, Australia, and some other countries produce sugar cane molasses. Of these, the productions of the first four countries were 5.08, 4.25, 1.9, and 2.5 Mt in 1991, respectively. On the other hand, among the main countries producing sugar beet molasses—the former USSR, Germany, Poland, Hungary, and France—the 1991 production levels of the former USSR and Germany were 3.18 and 1.18 Mt, respectively. The production of molasses in the USA is 1.91 Mt and that in China is 2.17 Mt.

Milk whey is known as a special type of saccharine material. In New Zealand, a large amount of milk whey is produced as a by-product ($2.8 \times 10^5$ kl in 1985), containing ~4% lactose. To reduce the biological oxygen demand of drainage and to recover value-added substances, this milk whey is used as the raw material for ethanol fermentation.

Juice squeezed out from the peel of oranges after the fresh juice is extracted contains ~8% sugar. It is then concentrated to a density of over 40% sugar. This is called citrus molasses and is used as a fermentation feedstock for ethanol. A small amount of ethanol is produced from citrus molasses in Japan, where oranges are available in abundance.

*Starchy Materials*

Starch is a polymer of glucose. Among starchy materials, cereals (like corn and wheat), sweet potatoes, potatoes, and others are extensively used as fermentation feedstocks. Table 3.4 shows the production of the major agricultural products usable as ethanol feedstocks in different countries.

The main corn-producing countries are the USA, China, Brazil, the former USSR, Mexico, South America, and India. Of a total world production of 475 Mt in 1990, the production of corn in the USA was ~200 Mt. Of this amount, 8–9 Mt, i.e., ~4% was used as a raw material for ethanol. Following the USA, China and Canada also make extensive use of corn as a raw material for ethanol.

China, Indonesia, Japan, India, Brazil, and the USA are the major cultivators of sweet potatoes. China is the leading producer, with 112 Mt in 1990, amounting to almost 80% of the total world production of 132 Mt in that year. At present, China is the main country that produces ethanol from sweet potatoes.

**Table 3.4. Production of the major agricultural products usable as ethanol feedstocks in different countries**

| | Source (unit, 1000 tons) | | | | | |
|---|---|---|---|---|---|---|
| | Sweet potato | Potato | Corn | Tapioka | Molasses | Sugar |
| | Year | | | | | |
| Country | 1990 | 1990 | 1990 | 1990 | 1990 | 1989 |
| World | 131,707 | 269,561 | 475,429 | 157,656 | 39,411 | 107,864 |
| USA | 591 | 17,866 | 201,509 | | 1849 | 6193 |
| Brazil | 760 | | 21,298 | 24,611 | 4245 | 7326 |
| China | 112,220 | 33,050 | 87,345 | 3222 | 1748 | 5400 |
| India | 1300 | 15,000 | 9500 | 4500 | 4880 | 9912 |
| Indonesia | 2180 | | 6741 | 17,064 | 1171 | |
| South Korea | 432 | | 4400 | | | |
| Japan | 1450 | 3500 | 1 | | 75 | 998 |
| France | | 6000 | 8996 | | | |
| East Germany (former) | | 8900 | | | | |
| West Germany (former) | | 7716 | | | 743 | |
| Poland | | 36,313 | | | | 1939 |
| UK | | 6504 | | | | |
| USSR (former) | | 63,700 | 16,000 | | 3260 | 9600 |
| Italy | | | 5864 | | | |
| Romania | | | 6810 | | | |
| Canada | | | 7033 | | | |
| Mexico | | | 14,762 | | | 3570 |
| Argentina | | | 5049 | | | |
| Paraguay | | | 1139 | 4100 | | |
| Philippines | | | 4854 | | 845 | 1878 |
| Thailand | | | 3675 | 20,701 | 1719 | |
| South Africa | | | 9442 | | 795 | |
| Nigeria | | | | 26,000 | | |
| Zaire (former) | | | | 17,500 | | |
| Tanzania | | | | 5500 | | |
| Pakistan | | | | | 995 | |
| Australia | | | | | 747 | 3844 |
| Cuba | | | | | 2635 | 7579 |

*Source: Pocket Statistical Yearbook of Agriculture and Fisheries, 1992.* Tokyo, Association of Agriculture and Forestry Statistics.

Potatoes are produced in large quantities in the former USSR, Poland, China, and in some other countries. The total production of potatoes in the world was 270 Mt in 1990. In Poland and Germany, potatoes are used extensively in the production of ethanol, amounting to 64% and 48%, respectively, of the total raw material used for that purpose in the two countries.

In addition to the above, cassava is cultivated for starch (tapioca) in the tropical parts of the world, with a total production of 158 Mt in 1990. Although most of this is presently consumed as food or livestock feed, tapioca could be an important source of ethanol in the future.

*Lignocellulosic Materials*

SWL, a liquid waste of the pulp industry, contains ~4% sugar of pentoses and hexoses. In Germany, Canada, and the USA, as well as in the three Scandinavian countries, it is used for ethanol production.

*Ethylene*

Ethylene, the raw material of synthetic ethanol, is obtained by the processing of crude petroleum or by the processing of the ethane present in natural gases into ethylene.

***Principle of Ethanol Fermentation***

*Ethanol Fermentation of Saccharine Materials*

In most saccharine materials, the main components are mixtures of glucose, fructose, and sucrose. Metabolized by the yeast *Saccharomyces cerevisiae* under anaerobic conditions, these sugars generate ethanol. In this process, the yeast acquires energy for its growth. One molecule of glucose generates two ethanol and two $CO_2$ molecules:

$$\begin{array}{ccc} C_6H_{12}O_6 \rightarrow & 2C_2H_5OH + & 2CO_2 \\ 181\ g & 92.1\ g & 88.0\ g \end{array}$$

Therefore 51.4 g of ethanol and 48.8 g of $CO_2$ can be theoretically generated from 100 g of glucose. Fructose is metabolized in the same way. In the above equation, the fermentation ratio is 100%. In the case of sucrose, it must first be decomposed into glucose and fructose by the enzyme called sucrase in yeast cells before it can be metabolized. In most cases, however, saccharine materials contain some percentages of nonfermentable sugar. Moreover, a part of the fermentable sugar is used to grow the yeast biomass. Therefore, in practice, a 100% fermentation ratio cannot be achieved in ethanol fermentation.

Cane juice and beet juice are seasonal raw materials, which cannot be preserved and thus are impossible to transport over long distances. The juice squeezed out from the harvested crop is immediately fermented into ethanol by yeast. Approximately 15 h later, mash containing 7–8 vol. % of ethanol is obtained. Assuming the sugar content, the fermentation ratio, and the distillation recovery of sugar cane to be 12%, 85%, and 99%, respectively, to produce 1 kl of 95 vol. % of ethanol, the required amount of sugarcane will be ~14.6 t.

When cane molasses or beet molasses is used as a fermentation feedstock, the sugar concentration is usually diluted to 20%–24%. Because of its high sugar concentration (45%–55%) as well as its high salt concentration, molasses can be used as a feedstock throughout the year and so is exportable. Assuming the sugar content, the fermentation ratio (for total sugar), and the distillation recovery to be 55%, 82%, and 99%, respectively, ~3.3 t of molasses will be required for producing 1 kl of 95 vol. % of ethanol.

In ethanol factories in Brazil, Europe, and America, the use of the bread yeast *S. cerevisiae* is common. On the other hand, in some tropical countries, yeast strains with thermotolerant properties are used.

*Ethanol Fermentation of Starchy Materials*

The fermentation method [1] used in Japan, with raw sweet potatoes as a feedstock, is described. Raw sweet potatoes are washed, crushed into 5–10-mm pieces by a hammer crusher and then cooked for 30 min at 90–100°C by a steam cooker. Mash is prepared with a total sugar concentration of ~15%, then cooled to 55°C. Two types of amylases, $\alpha$-amylase and glucoamylase, are then added to the mash, and liquefaction and saccharification are carried out for ~2 h. After the mash is cooled to 34°C, inoculated with yeast, and fermented at 30–33°C for 4 days, a matured mash with ~8-vol. % ethanol is obtained.

Assuming the starch value to be 24.3% (glucose content 27%), the fermentation ratio to be 92%, and the distillation recovery to be 98.5%, the production of 1 kl of 95-vol. % ethanol will require ~6.03 t of raw sweet potato.

A typical method used in America, with corn as a fermentation feedstock, is as follows. First the corn is divided into germ, fiber, and gluten by steeping in a dilute $SO_2$ solution and then mashing by a wet mill. The remaining starch is steam cooked, liquefied with amylases, and converted into glucose. After cooling, yeast is added for fermentation, just as it is for the saccharine materials described above. In this method, no waste remains as oil is collected from the germ, and fiber and gluten are used as animal feed. Assuming the starch value to be 63%, the glucose content 70%, the fermentation ratio 90%, and the distillation recovery 98.5%, to produce 1 kl of 95-vol. % of ethanol will require ~2.4 t of corn.

*Ethanol Fermentation of Special Materials*

Ethanol is made from the lactose of milk whey in New Zealand. As the yeast *S. cerevisiae* cannot ferment lactose, some special types of yeast, such as *Kluyveromyces fragilis* and *K. lactis*, are used instead. When SWL is used as the fermentation feedstock, sulfite-adaptable yeast cultured in SWL is used.

### ***Process of Ethanol Fermentation***

*Batchwise Fermentation*

Batchwise fermentation is the simplest fermentation process. In this process, the mash is put into the fermentation tank and inoculated with yeast. Fermentation is done at a constant temperature (30–33°C) until the sugar is completely consumed and a matured mash obtained. In tropical regions, the fermentation temperature sometimes rises to ~40°C. In this method, yeast, once it is used, is not collected for reuse. This simple process is also used in many other fermentation industries.

*Fed-batch Fermentation Process*

In the case of fermentation at a high sugar concentration in the batchwise process, alcohol fermentation by yeast and yeast growth is inhibited, resulting in a low fermentation ratio and delayed fermentation. In the case of molasses, a high salt density also has a bad effect. Inhibition due to a high-density substrate can be avoided if the density of the substrate is kept comparatively low at the initial stage of fermentation, and then new amounts are added at regular intervals once the previous amounts have been completely consumed. In this way, high-density fermentation products can be generated.

In the factories of New Energy and Industrial Technology Development Organization (NEDO) in Japan, a fed-batch process called the "soe-gake" process is used in the fermentation of molasses [1]. In this process, a seeding mash is added to a starting mash of 17% total sugar concentration, and the mash is fermented at 32°C for ~20 h.

At the point of sufficient yeast growth, additional mash (soe-mash) with 42% sugar concentration is added continuously for 24 h, bringing the final total sugar concentration of the mash to ~24%. Within 4 to 5 days after inoculation with the yeast, a matured mash containing ~13-vol. % ethanol is obtained. This process is time consuming, but results in higher-concentration ethanol and less waste liquid.

*Melle–Boinot Semicontinuous Fermentation Process*

This is a semicontinuous fermentation method, patented in 1933 by the Melle Co. of France, that enables the reuse of yeast in the same process. When the fermented mash is passed through a centrifuge, the yeast-cell paste is first separated. Then, when dilute sulfuric acid is added in a yeast processor, the yeast cells are processed at pH 3. This kills various contaminating bacteria and makes the yeast cells able to be reused in the next fermentation. This process is used in most of the ethanol factories in Brazil.

*Continuous Fermentation Process*

The continuous fermentation process is a method in which concentrations of yeast and ethanol are kept constant at a steady state so that the withdrawal of mash and the supply of the sugar solution can be performed continuously at a constant rate. For example, in the case of fermenting 100-kl matured mash inside a fermentation tank, if 10-kl mash/h is supplied continuously and 10-kl matured mash withdrawn simultaneously from the fermentation tank, the mash inside the tank will be totally replaced in 10 h. In this case, the retention time is 10 h, while its reciprocal number, 0.1 $h^{-1}$, is the dilution rate. There are many variations of the continuous fermentation process. The number of fermentation tanks varies from one to several, and although the yeast is recycled in some cases, in some cases it is not.

The productivity of the continuous fermentation process (kilograms times ethanol per kiloliter of fermentation tank per hour) is higher than that in the batchwise process, enabling the use of a smaller fermentation tank. On the other hand, a disadvantage of this process is that the concentration and the yield (fermentation ratio) of the fermentation product ethanol are rather lower than those of the batchwise fermentation process, resulting in an increase in the amount of waste liquid. In addition, the possibility of bacterial contamination is higher because of the long operation time. Some preventive measures, however, are possible. For example, after collection of the yeast, it can be washed by acid and inoculated in high density.

***a. Multivessel Continuous Fermentation Process [4, 5].*** The continuous fermentation process that uses 4–7 fermentation vessels from Vogelbusch Gmbh is one of the most common continuous processes used for fuel ethanol production. In this process, the fermentation vessels are connected in series and yeast is collected at the end of the vessels by centrifugation, and then recycled. When corn is used as a feedstock, following the stages of pretreatment by wet mill, cooking the starch, and saccharification, mash with 22% glucose can be derived. In a retention time of 20 h, the final concentration of ethanol becomes 10.5 vol. % with a fermentation ratio of 92.5%. In this method, bread yeast is used.

***b. Biostil Process [6, 7].*** The fermentation equipment used in the Biostil process is a combination of a continuous fermentation vessel and distillation towers. Molasses, cane juice, hydrolysate liquid of starch, etc., are used as fermentation feedstocks. In this process, yeast is recycled by centrifugation to a fermentation vessel. Beer (mash) is passed through a heat exchanger and is then supplied to the beer column (mash column) of the distillation equipment. From the top of the column, 40–50 vol. % of ethanol is recovered and sent to a rectification column. The greater part of stillage coming out from the bottom of the beer column is reused as charge water. One characteristic of the Biostil process is that fermentation and distillation proceed in parallel, and therefore the concentration of ethanol inside the fermentation vessel is kept as low as ~5 vol. %, which causes no inhibition in fermentation. Moreover, as the process uses recycled yeast, the yeast density can be kept high. Depending on the type of feedstock, the production rate (in grams times ethanol per hour times liters) of ethanol may be 80 times that of the batchwise process. The substrate is supplied at high density (40%–50% in the case of molasses), sugar is fermented quickly, and the residual sugar inside the fermentation vessel is kept at 0.2%. With a high sugar concentration of feed mash and the use of recycled distilled waste liquid, in the case of molasses, final waste liquid reduces to 1/3–1/4 of that in the batchwise process. In this process, the salt density of beer (mash) increases because of the use of recycled waste liquid. Therefore a distinguishing characteristic of this process is the use of the salt-resistant yeast *Schizosaccharomyces pombe*. In Australia, Indonesia, and some other countries, industrial plants based on this process are now in operation. Chematur Engineering AB (Sweden) at present holds the patent for this process.

***Fermentation Technology Undergoing Research Development***

Several promising fermentation techniques that are either yet to be used practically or not yet in wide-scale industrial use are described below.

*Continuous Fermentation Process with Immobilized Yeast Cells*

The maximum yeast concentration in batchwise fermentation is $\sim 1 \times 10^8$/ml. However, if the yeast is immobilized in the form of beads of 1–2 mm in diameter, the yeast concentration inside the fermenter or reactor can be significantly increased. Thus the productivity of batchwise fermentation—typically 1–2 g ethanol/l h—can be raised several times or more by the use of an immobilized yeast reactor because of the high concentration of the yeast. Therefore, with respect to productivity, the immobilized yeast-cell process is an excellent fermentation technique. Nevertheless, the stability of the structure of the immobilized yeast depends on the types and the strengths of the immobilizing agents, which, in most cases, cannot stand continuous operation for a long time. Moreover, the immobilizing process is relatively expensive, and the preservation of the immobilized yeast during the pause in operation time is also problematic. All these factors appear as barriers to the use of the process on an industrial scale.

Ghose and Bandyopadhyay [8] have carried out a continuous fermentation process by immobilizing the yeast cells with calcium alginate, filling up the column with immobilized yeast, and adding a 10% glucose solution. For a retention time of 10 h, the obtained ethanol was 40 g/l and the fermentation ratio was 90%. Wada *et al.* [9, 10] have developed another continuous ethanol fermentation process by immobilizing the yeast cells with $\kappa$-carrageenan. In this case, the yeast concentration was $4–6 \times 10^9$ cells/ml while the ethanol productivity was 50 g/l h for 10% glucose and 25–30 g/l h for 25% glucose.

Some yeast-immobilizing techniques have also been developed that use nonnatural gelling agents that are artificial polymers, such as photo-cross-linkable resins, of which the main chains are polyethyleneglycol and polypropyleneglycol [11]. In this method, yeast can be immobilized in either bead form or sheat form. However, the resin is strengthened up to 30–50 times more than $\kappa$-carrageenan gel. Using the 2 types of immobilized yeast (in bead form and sheat form), Matsui et al. have developed the Flash alcohol fermentation process as described below [12].

*Flocculating Yeast Process*

By its flocculating property, flocculating yeast forms floc ~1–2 mm in diameter. When a column-type reactor is filled with yeast cells in floc instead of immobilized yeast, continuous fermentation can be achieved. When a conventional fermenter is used, repeated batch fermentation is possible. In the latter case, at the end of fermentation, once agitation is stopped, the yeast cells subside. Then the supernatant mash is withdrawn from the distillation process. If a sugar solution is added to the yeast sediment, the next fermentation can be started immediately. Therefore, by use of flocculating yeast in continuous fermentation, the problem of immobilization and associated expenses can be avoided. If a repeated batch fermentation process is used, the energy and the equipment required for centrifugation are unnecessary.

Using molasses as the substrate, Kuriyama *et al.* [13, 14] have carried out continuous fermentation and repeated batch fermentation at 30°C on a laboratory scale. In continuous fermentation, the ethanol concentration of mash was 69–70 g/l and the productivity was 25–31 g/l h, whereas in repeated batch fermentation, for a 24-h fermentation period, the results were 95–108 g/l and ~ 4 g/l h, respectively. On the other hand, Dorsemagen [15] has carried out continuous fermentation for a fixed period by using flocculating yeast in a demonstration plant with a scale of ethanol production capacity of 15 kl/day. The yeast was maintained at a high concentration of 50 g/l. The productivity obtained in this process (15 g/l h) was comparatively high, while the ethanol concentration of the mash was 8.0–8.3 vol. %.

*Miscellaneous Processes*

In addition to the above, a number of further types of fermentation processes have been developed to date. Some of them are given below (see references for details).

1. Flash alcohol fermentation process. This is a combination of the continuous fermentation process that uses immobilized yeast cells and flash vaporization. It is an energy-saving process with less waste liquid [12].
2. Energy-saving fermentation process that uses raw starch saccharifying amylases [16, 17].
3. Alcohol production from starch with genetically improved yeast strains that have amylase activity [18, 19].
4. Extractive fermentation process [20].
5. Ethanol fermentation process that uses *Zymomonas mobilis* [21, 22].
6. Ethanol production from lignocellulosic materials [23–26].

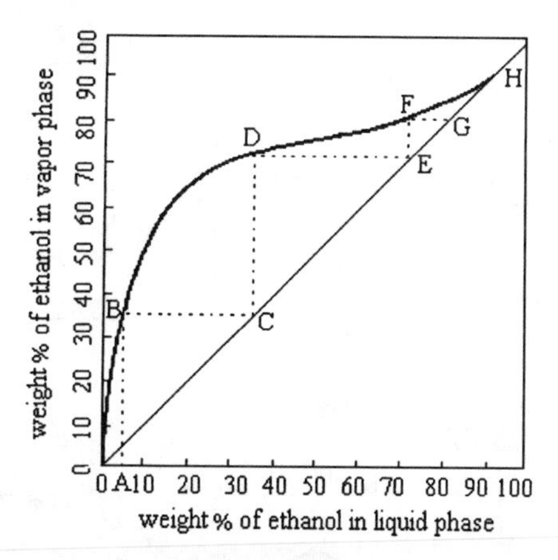


**Figure 3.1. Equilibrium curve of ethanol–water solution.**

## *Ethanol Distillation Technology*

### *Principle of Distillation*

Distillation technology [2] is based on the different boiling points of various liquid components. Liquid containing several components is first evaporated by heat; then the vapor is condensed by cooling water. During the process, each of the components can be separated, concentrated, and purified according to their boiling points and relative volatilities.

Figure 3.1 illustrates the concentration of ethanol in an ethanol–water solution. This figure is the equilibrium curve of the ethanol–water mixed solution and represents the relation between the ethanol concentration of the liquid and that of the vapor during evaporation. As shown in the figure, if a 5% (by weight) ethanol solution is heated, as shown by point A, the ethanol concentration of evaporated vapor becomes 36% (by weight), as shown by point B. The liquid obtained after the vapor is condensed at point B by cooling is then boiled at point C. The ethanol concentration of the evaporated vapor, as shown by point D, becomes 73% (by weight). If the above operations of boiling, separation, and condensation (cooling) are repeated, the ethanol concentration of the condensed liquid gradually increases.

However, if the ethanol concentration of the liquid reaches 95.57% (by weight), as shown at point H, the liquid boils at 78.15°C. At this temperature, the ethanol concentration of the liquid and that of the vapor to be evaporated will be the same. It is impossible to obtain higher ethanol concentrations in the usual distillation process. This boiling point (78.15°C) is called the azeotropic point of the ethanol–water solution and the mixture at the azeotropic point is called the azeotropic mixture. The ethanol obtained in this way is usually called hydrated alcohol.

The azeotropic distillation with a solvent is used to make dehydrated (anhydrous) ethanol, which usually is called absolute alcohol. A solvent like cyclohexene, n-pentane, or benzene is used to break the azeotrope that forms between ethanol and water.

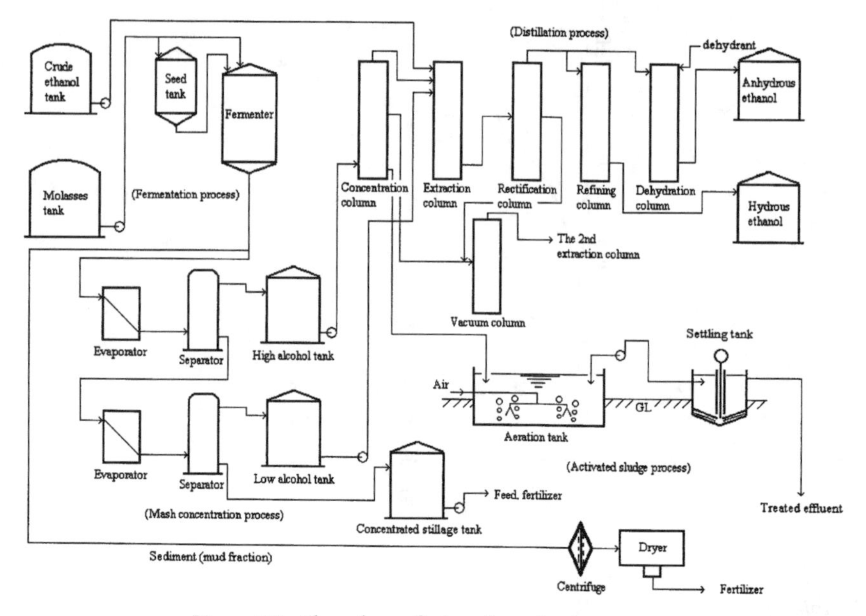


**Figure 3.2. Flow chart of ethanol production process.**

*Distillation Apparatus*

There are two types of distillation apparatus: complex continuous distillation apparatus and the relatively simple pot still. The former is composed of several kinds of distillation columns, heaters and condensers. This type of apparatus is highly effective in concentrating ethanol from fermented mash that contains 7%–14% ethanol and rectifying and producing hydrous (~95%) or anhydrous ethanol of high quality in continuous operation. Figure 3.2 shows an outline of the continuous distillation apparatus.

The pot still is accompanied by a heater and a condenser and is used mainly in the liquor industries for making whiskey, brandy, and other spirits. Matured mash is boiled in the still batchwise, and ethanol vapor with useful aromatic ingredients is recovered through the condenser at ~50% ethanol (by weight).

The typical continuous distillation apparatus for the production of high-quality hydrous ethanol is, for example, composed of a mash column, a concentration column, an extraction column, a rectification column, a refining column, an impurity column, etc. In addition, a dehydration column, a dehydrant recovery column, and a dehydrant separator are necessary for the production of anhydrous ethanol.

The distillation apparatus necessary for the production of anhydrous ethanol for fuel use includes only a mash column, a rectification column, and a dehydration unit, because high quality is not required for fuel ethanol.

*Role and Functions of Distillation Columns*

***a. Mash Column.*** The mash column has conventionally been used to separate the ethanol part of the matured mash and the distillation waste liquid. In recent years, however, the mash concentration apparatus has become popular in place of the mash column. This apparatus is a type of multieffect evaporator that separates the ethanol part of the matured mash into two ethanol condensates (one is 5 vol. % and the other is ~35 vol. %). Simultaneously, it concentrates the solid part of distillation waste to ~45% by weight. The condensates are treated through the continuous distillation apparatus to produce hydrous ethanol.

***b. Extraction Column.*** Many kinds of ingredients (mainly aldehydes) other than methanol can be separated from the top of the extraction column. In general, volatile components with higher boiling points have lower volatility than those with lower boiling points. For example, methanol with a boiling point of 64.7°C is more volatile than ethanol with a boiling point of 78.3°C. However, the relative volatility of ethanol and methanol is reversed at low ethanol concentration, and the relative volatility of ethanol becomes higher than that of methanol at a water concentration of more than 70 mol. %. The relative volatility of higher alcohols (components of fusel oil) becomes higher than that of ethanol at low ethanol concentration. Since the extraction column is usually operated at a low ethanol concentration (10%–25%), ethanol together with methanol, withdrawn from the bottom of the extraction column, is sent to the rectification column, and components of fusel oil can partly be separated from the top of the extraction column.

***c. Rectification Column.*** Diluted ethanol from the extraction column is concentrated to ~95% at the top of the rectification column and then sent to the refining column. By-product fusel oil is separated mostly from the middle part of the rectification column. The composition of fusel oil is isoamylalcohol 45%, isobutylalcohol 10%, active amylalcohol 5%, and *n*-propylalcohol 1%–2%. From 0.1 to 0.7 kg of oily yellowish fusel oil per 1 kl of product ethanol is obtained. Most of the fusel oil components are metabolic products from amino acids. Fusel oil from synthetic ethanol, different from fermentation ethanol, contains diethylether, acetaldehyde, secondary butylalcohol, tertiary butylalcohol, and ketones. This fusel oil can also be separated in the same process.

***d. Refining Column.*** Methanol can be concentrated and separated at a high ethanol concentration at the top of the refining column because the relative volatility of methanol is higher than that of ethanol at a high ethanol concentration. Product hydrous ethanol is obtained from the bottom of the column.

***e. Dehydration Column.*** To make anhydrous ethanol, there are three dehydration processes that use solid dehydrant, liquid dehydrant, and azeotropic distillation. The latter is now the most common process.

An azeotropic mixture is formed when an ethanol–water solution is mixed with a third component, such as cyclohexane, that is insoluble (or slightly soluble) in ethanol–water or that dissolves in either ethanol or water. The boiling point of the mixture as a whole (62.1°C) is lower than that of each component; ethanol, water, and cyclohexane having boiling points of 78.3, 100, and 80.7°C, respectively. An azeotropic mixture containing cyclohexane 76%, ethanol 17%, and water 7% is obtained from the top of the dehydration column, whereas anhydrous ethanol is recovered from the bottom of the column. Cyclohexane in the mixture is recovered and recycled for reuse as a dehydrant (Fig. 3.2).

***f. Vacuum Column.*** The boiling point and volatility of *n*-propylalcohol and isopropylalcohol are very close to those of ethanol under atmospheric pressure. However, in a vacuum, the differences between the boiling point and the volatility of the higher alcohols and those of ethanol become larger and consequently their separation becomes easier. Impure ethanol, containing higher alcohols from the middle part of the rectification column, is sent to the vacuum column, which is operated at ~9.33 kPa. Ethanol is then recovered from the top of the column, and the higher alcohols are separated effectively from the middle part of the column.

*New Technologies for Concentrating and Refining Ethanol*

There are several promising new technologies for separating and purifying ethanol from a matured mash that contains ethanol or an ethanol–water mixture. These new technologies include pressure-vacuum multieffect distillation, membrane separation technology, and supercritical fluid extraction of ethanol.

***a. Pressure-Vacuum Multieffect Distillation.*** Multieffect distillation has been widely used on a commercial scale in the petrochemical industry. In this process, the latent heat of the top spirit vapor from one column under pressure is reused as a heat source for other columns under atmospheric pressure or under vacuum. The Ishioka Alcohol Plant of New Energy and Industrial Technology Development Organization (NEDO) built in 1986 (Ishioka City, Japan) is a typical example of the multipressure multiple-effect process, in which the columns are operated individually at different pressure levels to optimize energy saving, steam, and cooling water. In addition, the heat of the top spirit vapor of the rectification column is recovered by a heat pump apparatus [27].

***b. Dehydration of Ethanol with Cornmeal.*** Ladisch *et al.* [28–30] have invented a unique ethanol dehydration process that can substitute for conventional azeotropic distillation. In this process, aqueous ethanol mixtures in the vapor state are exposed to cornmeal, which operates as the dehydration agent. Most of the energy consumption in distillation occurs when the ethanol concentration is over 85%. Therefore dehydration with cornmeal is a much more energy-efficient process when the aqueous ethanol mixture has a high ethanol concentration. In 1986, the Archer Daniels Midland Co. industrialized this process in an alcohol plant in the USA.

***c. Membrane Separation Technology.*** Separation of ethanol and water by membrane has become possible because of the recent progress of research in this field. There are two types of membranes, one for selective ethanol permeation and the other for selective water permeation (dehydration). In a membrane separation process called the pervaporation process, one side of the membrane is kept in atmospheric pressure and the reverse side in vacuum pressure. When hydrous ethanol is passed through the atmospheric-pressure side, water is emitted as vapor from the vacuum-pressure side. Hydrous ethanol becomes anhydrous after the removal of moisture, which is condensed into water by cooling and then discharged. Improvement of the membrane efficiency, modulation of the membrane, and systematization of the whole distillation apparatus are substantial technological problems that remain to be solved.

Ethanol dehydration by pervaporation membrane was industrialized in an ethanol plant in France in 1988 by use of a membrane produced by GFT GmbH (Germany) [31].

***d. Supercritical Fluid Extraction of Ethanol.*** The principle behind this process is that when an supercritical fluid is brought into contact with an ethanol–water solution, ethanol can easily be extracted into the supercritical fluid. $CO_2$ is commonly used as the extraction agent under conditions, for example, of 31°C and 7.4 MPa. Under these conditions, $CO_2$ exhibits properties intermediate between liquid and gas. Therefore, with a slight change of temperature and pressure, the solubility of ethanol may change substantially. When the differences of solubility are used, ethanol and impurities can be separated from each other.

Selective extraction of ethanol from an ethanol–water mixture by supercritical fluids of $CO_2$, $C_2H_4$, or $C_3H_8$ is evidently possible [32, 33].

***Production Process of Synthetic Ethanol***

*History of Synthetic Ethanol*

Until the petrochemical industry was developed in the USA in the 1920's, only fermentation ethanol was produced. In 1929–1930, synthetic ethanol [2] was first produced and industrialized by the Union Carbide Chemicals Corp. of the USA by the sulfate method.

In Japan, the production and the use of synthetic ethanol commenced after World War II. The modern Japanese petrochemical industry began with the introduction of foreign technologies after 1950. In 1954, oil refineries were modernized, and in 1958, an ethylene plant that used naphtha as raw material was started. With the increasing utility of different chemical goods derived from ethylene, two alcohol factories in Japan, the Japan Synthetic Alcohol Co. Ltd. Kawasaki Factory and the Japan Ethanol Co. Ltd. Yokkaichi Factory, started to produce ethanol by the Shell process in 1965 and 1972, respectively. Both plants gave priority to the use of inexpensive raw materials. The total productive capacity of the two companies in 1975 was $1.0 \times 10^5$ kl, and at present has reached $1.44 \times 10^5$ kl.

Even though ethanol has been known for a long time, it has been produced synthetically for only ~70 years in the USA and for just over 30 years in Japan.

*Raw Material of Synthetic Ethanol*

In Japan, the raw material used to produce synthetic ethanol is ethylene, which is obtained from naphtha, which in turn is derived from crude petroleum. However, in some oil-producing countries, ethylene obtained from oil-accompanied gas is also used as the ethanol feedstock.

World production of ethylene was 54.2 Mt in 1989, while that in Japan was 7.13 Mt, of which ~47 kt is used as feedstock for synthetic ethanol.

*Production Method of Synthetic Ethanol*

Worldwide, there are two main methods for producing synthetic ethanol from ethylene. One is the direct hydration method, with phosphoric acid as a catalyst, and the other is the sulfate method.

***a. Direct-Hydration Method.*** There are two main types of the direct-hydration method. One is the Shell process that was developed by the Shell Chemical Co. in 1947 and industrialized in the USA; and the other is the VEBA process, which was developed by the Hibernia Chemie Co. of the former West Germany and then taken over and improved on by the VEBA · AG Co. (recently, Hülls Company). The USA, the former USSR, and some other East European countries also use their own individual technologies. But essentially these are the same as the Shell process, which involves a reaction between ethylene and pure water, with a catalyst soaked into a carrier (e.g., silica gel) of a solid.

The reaction occurs at 240–260°C temperature and 6.86-MPa pressure. The reaction formula is

$$\underset{\text{ethylene}}{C_2H_4} + \underset{\text{water}}{H_2O} \xrightarrow[\text{(phosphoric acid catalyst)}]{H_3PO_4} \underset{\text{ethanol}}{C_2H_5OH} .$$

This is an exothermic reaction in which diethyl ether is produced by the side reaction, as well as a small amount of aldehydes, hydrocarbons with high boiling points, higher alcohols, and ketones.

As mentioned above, synthetic ethanol in Japan is produced in the two factories, which both use the Shell process.

***b. Sulfate Method.*** The Union Carbide Chemicals Corp. of the USA first introduced the method of producing ethanol from ethylene and sulfuric acid in 1930. This method is less economically efficient than the Shell method (direct-hydration method), and the apparatus is susceptible to rust because of the use of sulfuric acid. Because of these constraints, its use at present is apparently limited to only one company, namely Societe D. Ethanol de Synthese (SODES) of France.

The reaction in the sulfate method is divided into two steps. In the first step, ethylene is absorbed into concentrated sulfuric acid and produces ethyl hydrogen sulfate. In the second step, diethylsulfate is hydrolyzed to ethanol. The absorptive reaction occurs at 55–85°C temperature and 1.67–3.43 MPa pressure, while the hydrolysis takes place at 60–80°C. The temperature and the pressure required in this method are less than these in the direct-hydration method. The reaction formula is as follows.

Absorption of ethylene by concentrated sulfuric acid:

$$\underset{\text{ethylene}}{C_2H_4} + \underset{\text{sulfuric acid}}{H_2SO_4} \rightarrow \underset{\text{ethylhydrogen sulfate}}{C_2H_5HSO_4}$$

$$C_2H_4 + C_2H_5HSO_4 \rightarrow \underset{\text{diethyl sulfate}}{(C_2H_5)_2SO_4} .$$

As in the above reaction, ethylene and concentrated sulfuric acid react and form ethylhydrogen sulfate and diethyl sulfate. There is equilibrium between ethylhydrogen sulfate and diethyl sulfate:

$$2C_2H_5HSO_4 \rightarrow (C_2H_5)_2SO_4 + H_2SO_4$$

The Hydrolysis reaction of the sulfate ester is

$$C_2H_5HSO_4 + H_2O \rightarrow C_2H_5OH + H_2SO_4,$$
$$(C_2H_5)_2SO_4 + H_2O \rightarrow C_2H_5OH + C_2H_5HSO_4.$$

In the above equation, ethylhydrogen sulfate and diethyl sulfate are hydrolyzed to ethanol. In this sulfate method, ethanol is synthesized through a two-step reaction.

*Outline of the Production Method of Synthetic Ethanol in Japan*

The production process of the direct hydration method adopted in Japan is presented in Fig. 3.3.

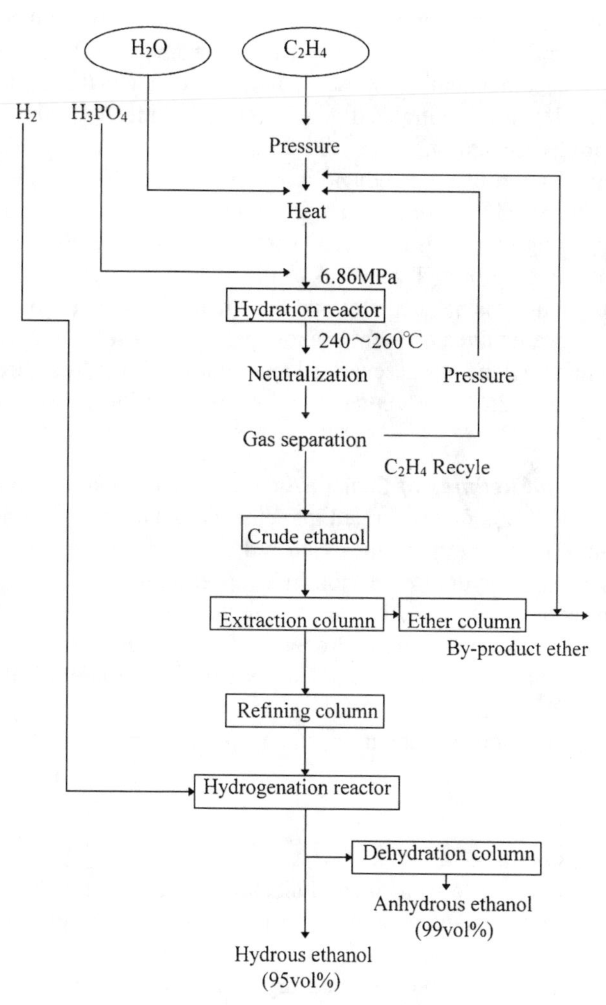


**Figure 3.3. Production process of synthetic ethanol (direct-hydration method) [2].**

***a. Hydration Reaction.*** Ethylene supplied by the petrochemical ethylene center is compressed to 6.86 MPa by a compressor, then heated and sent to the hydration reactor. In the hydration reactor, silica gel soaked with phosphoric acid is filled up as a catalyst. The pure water of the auxiliary material is mixed with nonreacted circulation ethylene gas at a pressure of 6.86 MPa, then heated to 240–260°C and sent to the hydration reactor in a vapor condition.

The hydration reactor is filled with silica gel soaked with the phosphoric acid catalyst, in which ethylene and pure water react with each other (the hydration reaction) and form ethanol. In one reaction, ~5% of ethylene is converted to ethanol and the nonreacted ethylene is recycled to the hydration reactor.

***b. Separation of Crude Ethanol.*** From the bottom of the hydration reactor, the mixture of nonreacted ethylene, pure water, and ethanol can be taken out. The mixture of ethylene, water, and ethanol becomes acidic because it passes through with the layer of phosphoric acid in the reactor. It is then neutralized by sodium hydroxide. Next, the mixture is sent to a high-pressure gas separator, where it is separated into a gas containing nonreacted ethylene gas and a liquid of crude ethanol containing water. The nonreacted ethylene gas is cooled to 70–80°C and added to a circulation gas washing column in which the ethanol accompanying the gas is removed. The gas is compressed to ~76.86 MPa by a circulation gas compressor and sent back to the hydration reactor. The crude ethanol separated by the high-pressure separator and the ethanol collected from the circulation gas washing column are mixed and sent to a low-pressure gas separator at 294–588 kPa. The accompanying ethylene gas is separated and returned to the hydration reactor. The crude ethanol contains 25%–30% ethanol and some impurities as by-products of the reaction.

***c. Concentration and Refining of Crude Ethanol.*** Crude ethanol is concentrated and refined by distillation. It is first separated into ethanol and by-product impurities in the extraction column. From the impurities gathered at the top of the column, ethers are collected and sent to the hydration reactor. The concentration of ethanol released from the bottom of the column is ~12%, which is sent to the refining column and concentrated up to 95%. It is then passed through a hydrogenation reactor, in which the minute amount of accompanying acetaldehyde is converted into ethanol by the addition of hydrogen and hydrous ethanol is thus refined.

The production method of anhydrous ethanol is the same as that of fermentation ethanol.

### *Methanol*

#### *General Properties of Methanol*

Methanol (commonly known as methyl alcohol) [2], a single-carbon compound, is the simplest of the monovalent alcohols. The International Union of Pure and Applied Chemistry officially defines this structure as methanol.

Methanol was first discovered in wood dry distillation products(also called wood spirits) in 1661. The main component of one of the derivatives, pyroligneous acid, is methanol.

Methanol was produced by the dry distillation of wood until 1923. Recently, however, hydrogen and carbon monoxide obtained from natural gas, liquid petroleum gas, naphtha, etc., are used as raw materials to produce methanol. The molecular formula for methanol is $CH_3OH$ and the molecular weight is 32.04. Its other characteristics include specific gravity 0.7928 (20/4°C), melting point −96°C, boiling point 64.6°C, and flash point 11°C. It also dissolves in water, alcohols, and ethers. It is colorless, combustible, volatile, and poisonous. Methanol is primarily used as a basic raw material in many chemical industries and fields. For example, it is used as a feedstock of organic compounds, and as a solvent, detergent, and perfume.

During the chaotic period after World War II, there were some accidents such as loss of eyesight and life due to drinking methanol. Recently, it has become popular as the raw material of methyl tertiary butyl ether (MTBE), an additive (agent to raise the octane value) of premium gasoline. The demand for MTBE is now increasing in the USA and other countries. Therefore, as one of the raw materials of MTBE, the demand for methanol is also increasing. Japan started to sell MTBE-added gasoline in December, 1991.

Methanol can also be used as automotive fuel, either as a single substance or as a gasoline additive, and is already used in this way in parts of America and Europe. In Japan, research is currently being carried out on the use of methanol for the reduction of $NO_x$ and hydrocarbons in exhaust gas from petroleum fuel.

*Uses of Methanol*

Methanol is used mainly as a basic feedstock in the chemical industry and as a petroleum substitute. Its largest use in the chemical industry is as a basic material in the production of formalin. If formalin is brought into contact with a mixture of methanol and air (in gaseous form) in the presence of a catalyst and the generated gas is absorbed into water, a formalin–water solution with ~40% formaldehyde concentration will be formed. Formalin is used as a feedstock for melamine synthetic resin and polyacetal resin, as a disinfectant and antiseptic, and as folmalin adhesives.

Methanol is now gaining attention as a possible gasoline substitute. The reasons for this include the existence of larger reserves of its raw material—natural gas—than the diminishing petroleum reserves, and its low manufacturing cost relative to that of the biomass resource feedstock, ethanol.

*Methanol in Fuel Use*

The use of methanol as a substitute fuel in gasoline- and diesel-operated engines can be taken as representative of its use as an energy source. At present, there are two methods of using methanol in gasoline engines. One method uses methanol as a complete substitute for gasoline, while the other uses it as a partial substitute or enhancer of the octane value of gasoline. Research on the use of methanol in diesel engines is currently in progress, focusing on the effectiveness of its use with vegetable oil (mainly, rape seed oil) after esterification.

In Brazil, fluctuations in the demand and supply of fuel–ethanol have led to shortages of ethanol, resulting in the use of methanol as an emergency fuel.

In the USA, methanol is used as an oxygenated fuel in the form of MTBE.

Research on methanol as a diesel substitute fuel is also underway in America, with the esterification of mainly soybean oil and methanol or ethanol.

In Germany, before the unification of East and West Germany, an experimental project to use methanol derived from coal was started in 1974. Since then, 300 passenger cars and 200 trucks were tested over a 3-year period. The fuels used in the test were 15% methanol added gasoline (M15) and 100% methanol (M100). Another running test with 300 cars and M100 was also carried out. In addition, a large-scale running test with M85 (a mixture of 85% methanol and 15% gasoline) was conducted. In Germany, bioalcohol (fermentation ethanol) derived from a biomass resource is also used as a fuel along with the main fuel alcohol methanol.

There are also some projects on the use of vegetable oil with methanol or ethanol as a diesel substitute.

In France, a research plan, known as the M10 plan, was formulated in 1981. The first stage of the plan involved the use of a fuel mixed with no more than 10% methanol. The second stage involved increasing the mixture ratio and establishing pilot plants. The plan also envisaged a $1.5 \times 10^7$ km running test by 1177 cars, establishing a satisfactory methanol–gasoline ratio and legalizing the sale of methanol-based fuel.

Research and development of fuel–methanol in France has centered on the M10 plan. Nevertheless, research is also being carried out on methanol as a diesel substitute through the combination of small percentages of alcohols with vegetable oil (mainly rape seed oil).

*Methanol as Oxygenated Fuel*

Fuel that contains oxygen is called oxygenated fuel or an oxygenated chemical compound. The most familiar oxygenated fuels are alcohols like ethanol, methanol, tertiary butyl alcohol, etc., and ether chemical compounds like MTBE, ETBE, etc. The main characteristics of oxygenated fuel are that it emits no soot during combustion, produces less $NO_x$ and CO, and has a higher octane value than gasoline. As mentioned above, the use of methanol as MTBE has recently become popular.

As an octane value improver, MTBE was first developed and industrialized by an Italian company named Ente Nazionale Idrocarburi in 1970 and then by Arco in the USA. In Japan, MTBE is mainly used as an octane value improver whereas in the USA, it is also used as an oxygenated fuel for the purpose of preventing air pollution.

MTBE is an ether produced from methanol and isobutylene. Isobutylene is obtained from the C4 fraction of naphtha cracked gas and is used as a raw material for isobutylene-isoprene rubber (butyl gum). MTBE is derived from the following reaction:

$$\underset{\text{isobutylene}}{CH_3{-}\overset{\overset{\displaystyle CH_3}{|}}{C}{=}CH_2} + \underset{\text{methanol}}{CH_3OH} \rightarrow \underset{\text{MTBE}}{CH_3{-}\overset{\overset{\displaystyle CH_3}{|}}{\underset{\underset{\displaystyle CH_3}{|}}{C}}{-}O{-}CH_3}.$$

To produce 1 l of MTBE, 0.76 l of isobutylene and 0.34 l of methanol are necessary. The molecular formula of MTBE is $C_5H_{12}O$ and the molecular weight is 88. It is a stimulus, volatile, and colorless transparent liquid with a boiling point of 55.2°C, a specific gravity of 0.7455, and an octane value of 110 (average of research octane number and motor octane number). It dissolves freely into most of the organic solvents by different optional rates.

The world demand for MTBE was 7.76 Mt (5.13 Mt in the USA and 2.33 Mt in Europe, 300 kt in other countries) in 1990, which was estimated to be 12.25 Mt in 1995 and to be 17.8 Mt in 2000. The production capacity of MTBE in 1989 was 7.66 Mt and the increase in 1994 was estimated to be 19.9 Mt.

*Production Process of Methanol*

About 70% of the raw materials of methanol come from natural gas [34]. Besides this, methanol is also synthesized from CO and $H_2$ derived from liquid petroleum gas, naphtha.

The BASF Co. of Germany developed the synthesis process of producing methanol in 1923. In Japan, Mitsubishi Gas Chemicals Co. Ltd. started to produce methanol by using natural gas produced in the Niigata Prefecture in 1952. After that, active pursuit of technological developments led to its peak in the production of methanol in 1973. However, because of the first oil crisis, domestic production of methanol reached zero by 1995, forcing the country to depend on imported methanol (~2.0 Mt in 1995).

The world production of methanol is reported to be over 20 Mt. It is produced mainly in the natural-gas-producing countries such as the USA, Canada, Germany, the USSR (former), New Zealand, Malaysia, Saudi Arabia, Chile, Norway and Venezuela.

***a. Production Process of Feedstock Gas (Hydrogen and Carbon Monoxide).*** There are two methods of using hydrogen and carbon monoxide gas to produce synthetic methanol from natural gas: the catalytic steam reforming method and the partial oxidation method. At present, the catalytic steam reforming method is mainly used. In this method, the main components of the natural gas methane react as shown by the following equations:

$$CH_4 + H_2O \rightarrow CO + 3H_2,$$
$$CH_4 + 2H_2O \rightarrow CO_2 + 4H_2,$$
$$CH_4 + CO_2 \rightarrow 2CO + 2H_2.$$

***b. Synthetic Process.*** Methanol synthesis occurs at high temperatures (250–400°C) and high pressures (5.07–30 MPa) with the use of a catalyst of the Cu–Zn or Zn–Cr–Cu group:

$$CO + 2H_2 \rightleftharpoons CH_3OH.$$

***c. Distillation Process.*** Crude methanol from a synthetic process is rectified in the prepurifying column. Ingredients with low boiling points, such as dimethyl ether and methylformate, are removed from the top of the column and an aqueous methanol solution is obtained from the bottom of the column.

Next, the methanol–water solution is distilled in the rectifying column, and ingredients with higher boiling points, like higher alcohols, paraffin, ethanol, etc., and water come out from the bottom of the column. At this point, refined methanol is obtained from the top of the column.

In recent years, some researchers have been developing a process that circumvents the step of deriving hydrogen and carbon monoxide from natural gas. This process can be outlined as follows:

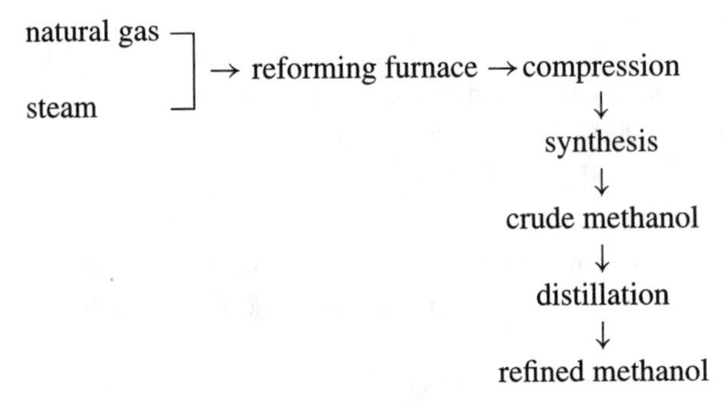


## References


1. 1997. *Alcohol Handbook*, 9th ed. Tokyo: Gihodo Shuppan Co. Ltd.
2. Karaki, I. 1994. *Story of Alcohol*. Tokyo: Research Institute of International Trade and Industry.
3. Fernandes, E. L., *et al.* 1995. Will the Brazilian alcohol accomplish vehicles future demand? *Proc. Second Biomass Conference of Americas*: 1713–1721. Golden, Colorado: National Renewable Energy Laboratory.
4. Brodl, G. B. 1988. Continuous fermentation technology. *Proc. Fuel Ethanol Workshop*.
5. Brodl, G. B. 1987. Technology of fuel alcohol production. *Proc. Alcohol Fuels US and Caribbean*.
6. Alfa Laval AG (Sweden). 1986. A process for the production of ethanol. International Patent 8603514.
7. Saiki, T. 1987. New technology of ethanol fermentation. Biseibutsu 3:555–564.
8. Ghose, T. K. and K. K. Bandyopadhyay. 1980. Rapid ethanol fermentation in immobilized yeast cell reactor. *Biotechnol. Bioeng.* 22:1489–1496.
9. Wada, M., J. Kato, and I. Chibata. 1980. Continuous production of ethanol in high concentration using immobilized growing yeast cells. *Eur. J. Appl. Microbiol. Biotechnol.* 10:275.
10. Wada, M., J. Kato, and I. Chibata. 1981. Continuous production of ethanol using immobilized growing yeast cells. *Eur. J. Appl. Microbiol. Biotechnol.* 11:67.
11. Sakamoto, M., T. Iida, H. Izumida, and K. Takiguchi. 1988. Yeast cell immobilization by photo-crosslinkable resins and its application to ethanol production. *Proc. 8th International Symposium on Alcohol Fuels*: 15–20.
12. Kuriyama, H., T. Murakami, H. Kobayashi, K. Kida, and Y. Sonoda. 1988. Ethanol production with continuous and repeated-batch fermentation using newly isolated flocculating yeast. *Proc. 8th International Symposium on Alcohol Fuels*: 3–8.
13. Kuriyama, H., Y. Seiko, T. Murakami, H. Kobayashi, and Y. Sonoda. 1985. Continuous ethanol fermentation with cell recycling using flocculating yeast. *J. Ferment. Technol.* 63:159–165.

14. Dorsemagen, B. 1985. Practical experience with continuous fermentation for the production of ethanol. US DOE Rep. ANL-CNSV-TH-157, pp. 25–34. Washington, DC: US Department of Energy.
15. Matsui, S., S. Sejima, and H. Izumida. 1988. Development of a flash alcohol fermentation process. *Proc. of 8th International Symposium on Alcohol Fuels*: 107–112.
16. Ueda, S. and Y. Koba. 1980. Alcoholic fermentation of raw starch without cooking by using black-koji amylase. *J. Ferment. Technol.* 58:237.
17. Matsumoto, N., O. Fukushi, M. Miyanaga, K. Kakihara, E. Nakajima, and H. Yoshizumi. 1982. Industrialization of a non-cooking system for alcoholic fermentation from grains. *Agric. Biol. Chem.* 46:1549.
18. T. Ashikari, N. Matsumoto, and H. Yoshizumi. 1988. Alcohol production with a genetically improved yeast which carries a modified Rhizopus glucoamylase gene. *Proc. 8th International Symposium on Alcohol Fuels*: 9–14.
19. Tubb, R. S. 1986. Amylolytic yeasts for commercial applications. *Trends Biotechnol.* 4:98–104.
20. Kollerup, F. and A. J. Daugulis. 1986. Ethanol production by extractive fermentation-solvent identification and prototype development. *Can. J. Chem. Eng.* 64:598–606.
21. Bringer, S., H. Sahm, and W. Swyzen. 1984. Ethanol production by *Zymomonas mobilis* and its application on an industrial scale. *Biotechnol. Bioeng. Symp.* No. 14, pp. 311–319. New York: Wiley.
22. Lawford, H. C. 1988. *Zymomonas*—A superior process organism for motor fuel alcohol production. *Proc. 8th International Symposium on Alcohol Fuels*: 21–27.
23. Lynd, L. R. 1989. Production of ethanol from lignocellulosic materials using thermophilic bacteria: critical evaluation of potential and review. *Adv. Biochem. Eng.* Biotechnol. 38:1–52.
24. Sato, K., M. Tomita, S. Yonemura, S. Goto, K. Sekine, E. Okuma, Y. Takagi, K. Hon-Nami, and T. Saiki. 1993. Characterization of and ethanol hyper-production by *Clostridium thermocellum* I-1-B. *Biosci. Biotech. Biochem.* 57:2116–2121.
25. Etoh, H., H. Michiki, H. Ishibashi, Y. Ado, Y. Murata, T. Ogoshi, M. Furuichi, Y. Morikawa, and T. Hirotani. 1988. Fuel ethanol production from cellulosic biomass. *Proc. 8th International Symposium on Alcohol Fuels*: 155–160.
26. Michiki, H., H. Miyakawa, H. Ishibashi, Y. Ado, Y. Murata, T. Ogoshi, H. Yokouchi, K. Sanse, Y. Morikawa, and T. Hirotani. 1988. Integrated bench-scale plant for ethanol production from cellulosic biomass. *Proc. 8th International Symposium on Alcohol Fuels*: 61–64.
27. Watanabe, H., M. Takahashi, S. Konishi, T. Saiki, and I. Karaki. 1987. Development of energy-saving distillation process based on multi-pressure multiple effect distillation. *Hakko to Kogyo* 45:17–29.
28. Ladisch, M. R. and G. T. Tsao. 1982. Vapor phase dehydration of aqueous alcohol mixtures. US Patent 4,345,973, 24 August 1982.
29. Ladisch, M. R., M. Voloch, J. Hong, P. Blenkowsky, and G. T. Tsao. 1984. Cornmeal absorber for dehydrating ethanol vapors. *Ind. Eng. Chem. Process Des. Dev.* 23:437–443.
30. Neuman, R., H. Voloch, P. Blenkowsky, and M. R. Ladisch. 1986. Water sorption properties of a polysaccharide adsorbent. *Ind. Eng. Chem. Fundam.* 25:422–425.

31. Bruischke, H. E. A., G. F. Tusel, and R. Rautenbach. 1985. *ACS Symp. Ser.* 281:467–478.
32. Horizoe, H., T. Tanimoto, T. Haruki, Y. Nishimoto, I. Yamamoto, K. Ogawa, M. Maki, and S. Sato. 1988. Supercritical fluid extraction of ethanol from water using hydrocarbon solvent. *Proc. 8th Int. Symp. on Alcohol Fuels*: 113–117.
33. Furuta, S., N. Ikawa, and R. Fukuzato. 1990. Extraction of ethanol from aqueous solution using compressed carbon dioxide. *Proc. of 2nd International Symposium on High Pressure Chemical Engineering*: 345.
34. The Chemical Daily Co. Ltd. 1997. *13197 Chemical Products* (in Japanese).

### 3.1.2 Vegetable Oils and Their Esters (biodiesel)

#### *Overview*

*G. Riva and F. Sissot*

Vegetable oils are used for food, as industrial raw materials, and for the generation of energy. In 1995, world oil production exceeded 100 Mt (millions of metric tons) [1], nearly all of which was destined for use as food. An increase in production (to 120 Mt by 2005) is expected because of the current market trends of two major buyers, the Community of Independent States and China, and the population and income patterns of developing countries. As a result, interest in the food end uses of oils is expected to grow, with a consequent rise in product price. However, there are also certain regions that are characterized by surplus oil production (e.g., the US, for soybean), the capability of producing nonfood crops (e.g., the EU), the availability of marginal land or indigenous oleaginous crops suitable for an extensive cropping (e.g., certain developing countries). Moreover, the need to control pollution processes and to explore more sustainable production chains is fostering interest in renewable raw materials. This means that, in almost all countries, vegetable oils have excellent potential for use in nonfood applications. Among these, energy production is one of the most feasible applications for two main reasons: relatively low-quality requirements and the possibility of using large quantities of product.

In general terms there are more than 4000 plants from which oils can be extracted (primarily from the seeds) [2]: Table 3.5 summarizes the characteristics of the most important ones. Yields can vary substantially, as can cropping techniques and agricultural inputs. Mass and energy balances for rape and sunflower, the most common crops in industrialized countries, are shown in Table 3.6. In addition to the common terrestrial plants, microalgae with a fatty content of 60% by weight also merit consideration [3].

For energy production, it is possible to use oils that are crude or refined to varying degrees (simple filtering, extraction of waxes and gums, esterification, etc.). It is also possible to process used oils (e.g., fried oils) [3].

#### *Chemistry of Vegetable Oils*

Vegetable oils are a blend of free fatty acids (FFA's); monoglycerides, diglycerides, and triglycerides; phosphatides, lipoproteins, and glycolipids; waxes; terpenes, gums, and other less important compounds.

**Table 3.5. Seeds and oil yields [2]**

| Common Name | Scientific Name | Indicative Yield of Seeds (t/ha) | Oil Content in the Seeds (%) |
|---|---|---|---|
| Almond | *Prunus dulcis* | 3.0 | 25–50 |
| Bean, African locust | *Parkia filicoidea* | 5.0 | 16–18 |
| Bean, broad | *Vicia faba* | 6.6 | 1–2 |
| Bean, lablab | *Dolichos lablab* | 1.4 | |
| Bigseed falseflax | *Camelina sativa* | 1.0 | 40 |
| Canicha | *Sesbania bispinosa* | 1.0 | |
| Cashew | *Anacardium occidentale* | 1.0 | 38–46 |
| Castorbean | *Ricinus communis* | 5.0 | 35–55 |
| Chickpea | *Cicer arietinum* | 2.0 | 5–6 |
| Cocoa | *Theobroma cacao* | 3.3 | 50 |
| Coconut | *Cocos nucifera* | 6000[a] | 0.63[b] |
| Coloynth | *Citrullus colocynthis* | 6.7 | 19–20 |
| Cotton | *Gossypium spp.* | 1.5 | 20 |
| Crambe | *Crambe abyssinica* | 5.0 | 36 |
| Croton, purging | *Croton tiglium* | 0.9 | 50–60 |
| Crownvetch | *Coronilla varia* | 0.5 | |
| Fenugreek | *Trigonella foenum-graecum* | 3.0 | 6–7 |
| Flax | *Linum usitatissimum* | 1.3 | 38 |
| Gourd, buffalo | *Cucurbita foetidissima* | 3.0 | 24–30 |
| Groundnut, Bambarra | *Voandzeia subterranea* | 4.2 | |
| Guar | *Cyamopsis tetragonoloba* | 2.0 | |
| Jatropha | *Jatropha curcas* | 8.0 | 50 |
| Jojoba | *Simmondsia chinensis* | 2.2 | 48–52 |
| Lablab | *Dolichos lablab* | 1.4 | |
| Lesquerella | *Lesquerella spp.* | 1.3 | 30 |
| Lupine, European blue | *Lupinus angustifolisus* | 1.0 | 5–6 |
| Lupine, European yellow | *Lupinus luteus* | 1.0 | 5–6 |
| Lupine, white | *Lupinus albus* | 1.0 | 6–9 |
| Marijuana | *Cannabis sativa* | 1.5 | 30–35, 23–26[c] |
| Meadowfoam, Baker's | *Limnanthes bakeri* | 0.4 | 24–30 |
| Meadowfoam, Douglas's | *Limnanthes douglasii* | 1.9 | 24–30 |
| Mu-oil tree | *Aleurites montana* | 5.5 | 7–8 |
| Mustard, blanck | *Brassica nigra* | 1.1 | 27–35 |
| Mustard, white | *Sinapis alba* | 8.0 | 30–35 |
| Niger seed | *Guizotia abyssinica* | 0.6 | 31–41, 30–35[c] |
| Nut, macadamia | *Macadamia spp.* | 1.0[d] | 65–75 |
| Oilvine, Zanzibar | *Telfairia pedata* | 2.0 | 35–38 |
| Palm, African oil | *Elaeis guineensis* | | 2.2[c] |
| Pea, cow | *Vigna unguiculata* | 2.5 | |
| Peanut | *Arachis hypogaea* | 5.0 | 36–50 |
| Perilla | *Perilla frutescens* | 1.5 | 49–51, 26–33[c] |
| Poppy, opium | *Papaver somniferum* | 0.9 | 40–50 |
| Rape | *Brassica napus* | 3.0 | 33–40 |
| Safflower | *Carthamus tinctorius* | 4.5 | 25–37, 17–25[c] |
| Sesame | *Sesamun indicum* | 0.5 | 50 |
| Soybean | *Glycine max* | 3.1 | 17–26 |
| Stylo, townsville | *Stylosanthes humilis* | 1.2 | |

(Continues)

Table 3.5. (Continued)

| Common Name | Scientific Name | Indicative Yield of Seeds (t/ha) | Oil Content in the Seeds (%) |
|---|---|---|---|
| Sunflower | *Helianthus annus* | 3.7 | 35–40 |
| Tallow tree | *Sapium sebiferum* | 14.0 | 55 |
| Velvetbean | Mucuna deeringiana | 2.0 | |
| Walnut, black | Juglans nigra | 7.5 | 60 |
| Walnut, Persian | Juglans regia | 7.5 | 60 |

[a] Nuts.
[b] Oil (t/ha).
[c] Mechanical extraction.
[d] Kernels.

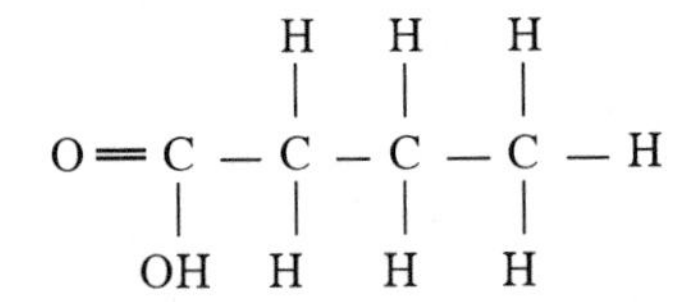

**Figure 3.4. Fatty acid (propanoic acid).**

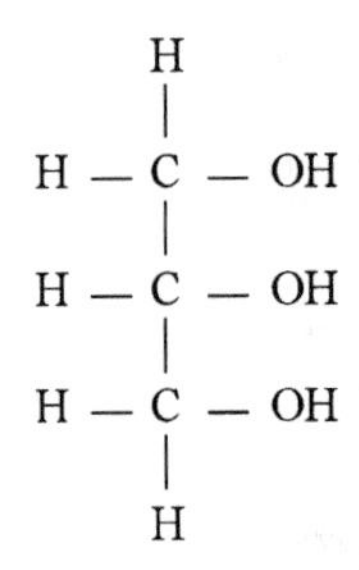

**Figure 3.5. Glycerol.**

*Fatty Acids*

Fatty acids (FA's) consist of a hydrocarbon chain—a sequence of C atoms bonded by single (saturated FA) or double/triple bonds (unsaturated FA). The first C atom is bonded to a carboxyl group (COOH). H atoms are bonded to the free valence of the C chain (Fig. 3.4). The number of C atoms and double bonds is defined by the formula $n$:$n$ (e.g., 18:1 oleic acid is composed of 18 C atoms and one double bond). Some of the principal characteristics of FA's are given in Table 3.7 [4].

*Glycerol*

This trihydric alcohol (Fig. 3.5) is slightly viscous, sweet smelling at room temperature, fully soluble in water and alcohol, slightly soluble in diethyl ether, and fully insoluble in hydrocarbons. The term glycerol refers specifically to the pure compound, whereas glycerine refers to commercial products that contain at least 95% glycerol. The physical properties of this alcohol are given in Table 3.8.

**Table 3.6. Mass and energy balance for rape, sunflower, and soybean (average values for Europe)**

| Parameter | Rape | | | Sunflower | | | Soybean | | |
|---|---|---|---|---|---|---|---|---|---|
| | Yield (t/ha) | Conversion Factor (MJ/t) | Energy (MJ/ha) | Yield (t/ha) | Conversion Factor (MJ/t) | Energy (MJ/ha) | Yield (t/ha) | Conversion Factor (MJ/t) | Energy (MJ/ha) |
| Inputs | | | | | | | | | |
| Direct consumption | | | 6792 | | | 11,838 | | | 7786 |
| Fuel | | | 6491 | | | 11,487 | | | 7577 |
| Lubricant | | | 301 | | | 351 | | | 209 |
| Indirect consumption | | | 16,056 | | | 13,894 | | | 13,495 |
| Machinery | | | 1112 | | | 1573 | | | 2298 |
| Fertilizers | | | 14,600 | | | 11,516 | | | 8733 |
| Others[a] | | | 344 | | | 805 | | | 2464 |
| Total | | | 22,848 | | | 25,732 | | | 21,281 |
| Outputs | | | | | | | | | |
| Seeds | 3.00 | | | 3.50 | | | 3.50 | | |
| Oil | 1.20 | 37,600 | 45,120 | 1.40 | 36,800 | 51,520 | 0.80 | 36,800 | 29,624 |
| Cake | 1.80 | 15,000 | 27,000 | 2.10 | 15,000 | 31,500 | 2.70 | 15,000 | 40,425 |
| Crop residues | 4.80 | 12,500 | 60,000 | 9.21 | 12,500 | 115,063 | 4.48 | 12,500 | 56,000 |
| Total | | | 132,120 | | | 198,083 | | | 126,049 |
| Balance (oil) | | | +22,272 | | | +25,788 | | | +8343 |
| Balance (total) | | | +132,120 | | | +172,351 | | | +104,768 |

[a] Seeds, pesticides, buildings, etc.

**Table 3.7. Some FA's and their physical properties [4]**

| Chain Length | Systematic Name | Common Name | Melting Point (°C) | Boiling Point (°C)[a] | Specific Gravity (kg/m$^3$)[b] | Molecular Weight |
|---|---|---|---|---|---|---|
| 4:0 | Butanoic | Butyric | −5.3 | 164 | 958.3 | 88.10 |
| 6:0 | Hexanoic | Caproic | −3.2 | 206 | 927.6 | 116.16 |
| 8:0 | Octanoic | Caprylic | 16.5 | 240 | 910.5 | 144.21 |
| 10:0 | Decanoic | Capric | 31.6 | 271 | 885.8 | 172.26 |
| 12:0 | Dodecanoic | Lauric | 44.8 | 130[a] | 847.7 | 200.31 |
| 14:0 | Tetradecanoic | Myristic | 54.4 | 149[a] | 843.9 | 228.36 |
| 16:0 | Hexadecanoic | Palmitic | 62.9 | 167[a] | 841.4 | 256.42 |
| 18:0 | Octadecanoic | Stearic | 70.1 | 156[a] | 839.0 | 284.47 |
| 20:0 | Eicosanoic | Arachidic | 76.1 | 204[a] | — | 312.52 |
| 22:0 | Docosanoic | Behenic | 80.0 | — | — | 340.57 |
| 24:0 | Tetracosanoic | Lignoceric | 84.2 | — | — | 368.62 |
| 14:1 | Tetradec-9-enoic (*cis*) | Myristoleic | −4.0 | — | — | 226.34 |
| 14:1 | Tetradec-9-enoic (*trans*) | — | 18.5 | — | — | 226.34 |
| 16:1 | Hexdec-9-enoic (*cis*) | Palmitoleic | 0.5 | — | — | 254.40 |
| 16:1 | Hexdec-9-enoic (*trans*) | — | 32.0 | — | — | 254.40 |
| 18:1 | Octadec-9-enoic (*cis*) | Oleic | 16.0 | — | — | 282.45 |
| 18:1 | Octadec-9-enoic (*trans*) | — | 45.0 | — | — | 282.45 |
| 22:1 | Docos-13-enoic (*cis*) | Erucic | 34.0 | — | — | 338.56 |
| 22:1 | Docos-13-enoic (*trans*) | — | 60.0 | — | — | 338.56 |
| 18:2 | Octadec-9, 12-dienoic | Linoleic | −5.0 | — | — | 280.44 |
| 18:3 | Octadec-6, 9, 15-trienoic | α-Linolenic | −11.0 | — | — | 278.42 |
| 20:4 | Eicosa-5, 8, 11, 14-tetraenoic | Arachidonic | −49.0 | — | — | 304.46 |

[a] Boiling point at 133.322 Pa pressure instead of 101 324.72 Pa.
[b] Measured at 20°C.

**Table 3.8. Physical properties of glycerol**

| Characteristic | Unit | Value |
|---|---|---|
| Melting point | °C | 18.7 |
| Boiling point at 101.3 kPa | °C | 290 |
| Specific gravity | kg/m$^3$ | 1.47 |
| Vapor tension | | |
| At 50°C | Pa | 0.33 |
| At 100°C | Pa | 26 |
| At 150°C | Pa | 573 |
| At 200°C | Pa | 6100 |
| Viscosity at 20°C | cP | 1499 |
| Flash point | °C | 199 |

*Monoglycerides, Diglycerides, and Triglycerides*

In vegetable oils, free FA's (known as FFA's) account for just a small proportion (e.g., 2.5%–3.0% in crude rape oil). Most of them are esterified with a glycerol molecule to form monoglycerides (single FA), diglycerides, and triglycerides (two and three FA's; Fig. 3.6).

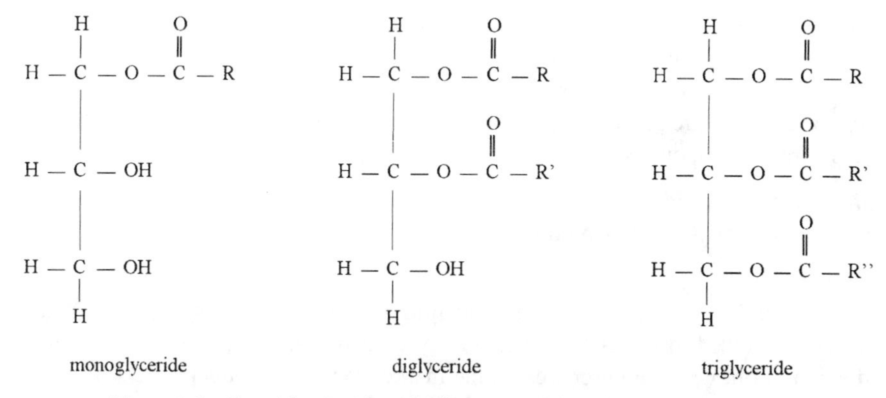


**Figure 3.6. Glycerides (R, R′, and R″, represent three hydrocarbon chains).**

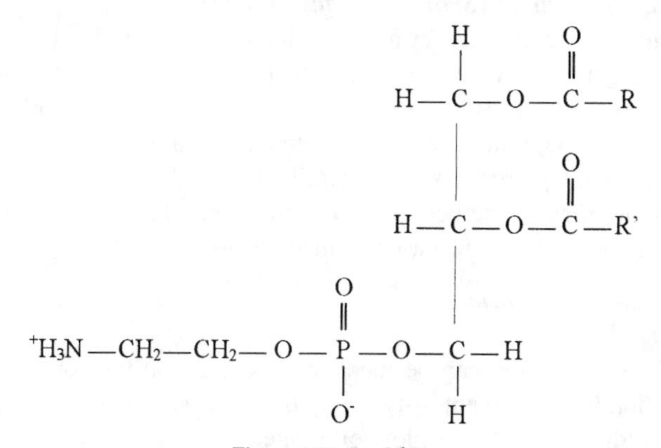

**Figure 3.7. Lecithin.**

*Phosphatides, Glycolipids, and Lipoproteins*

In phosphatide molecules, two of glycerol's hydroxyl (-OH) groups are esterified with FA's, while the third is esterified with a phosphoric acid molecule. This compound, called phosphatidic acid, bonds with a choline molecule to form lecithin (Fig. 3.7), or with a diglyceride to form a glycerophosphatide, which, within lecithin, is the most important portion of phosphatides. Glycolipids are mainly monogalactosyl and digalactosyl diglycerides (present only in chloroplasts). Lipoproteins consist of a lipid combined with a protein.

*Waxes*

These are esters of FA with a long-chain alcohol called fatty alcohol. Waxes precipitate at low temperatures, forming deposits in tanks, manifolds, and filters.

*Terpenes, Gums, and Other Compounds*

Terpenes are chains of isoprene units. Their combustion is responsible for the bad smell of exhausts. Gums, which are made up of terpene chains, can occlude manifolds and filters. Other compounds formed by terpene units are tocopherol (which gives good stability against oxidation) and carotene.

Vegetable oils also contain trace amounts of certain hydrocarbons: paraffins and olefins. The former are composed of saturated carbon chains ($C_nH_{2n+2}$), while the latter are unsaturated compounds ($C_nH_{2n}$).

***Production Chains for Energy Applications***

We can identify the following different cases.

*Large Networks (Large Cycle)*

Vegetable oils are processed in large quantities to produce esters, which can either replace or be added to diesel fuel. The energy industry does not usually consider the production of oil, which it purchases on the marketplace. By-products could be refined to compensate for the high costs of processing.

*On-Site or Local Applications (Short or Medium Cycle)*

On-site applications refer to on-farm production and use of vegetable oils. The quantities involved are generally very small (less than 5000 t/year) and the process should be very simple. Local applications refer to vegetable oil produced and refined locally for the use of a small community. A higher degree of refining or a simple esterification process could be used. By-products are generally disposed of as waste.

In this section we discuss methods for producing and refining oil for energy applications within the context of the two cases defined above.

***Raw Oil Production from Seeds***

*Industrial Plants*

The methods of extraction can be mechanical extraction (hot or cold pressing) or chemical extraction (using solvents—typically hexane). The process depends on the oil content of the seeds: for seeds with low oil content, only chemical extraction is used, whereas for seeds with high oil content, chemical extraction may follow mechanical extraction.

When only mechanical extraction is used, the end products are raw oil and oil cake (expeller), whereas for chemical extraction the final products are raw oil and oil meal. Cakes (10%–15% oil) and meals (0.7%–3.0% oil) are normally used as feed (Table 3.9) [5].

The complete oil-extraction process involves the following steps (Fig. 3.8) [6].

*a.* ***Cleaning.*** The seeds, picked up during harvest, are separated from the impurities (i.e., iron, rubble, stones, leaves, etc.) and stored. Figure 3.9 shows the seed-cleaning phases.

*b.* ***Drying.*** The moisture levels suitable for storage (13% for soybeans, 10% for cottonseed, 8.5% for sunflower, 7% for rape seed) are not necessarily the same as were required for oil extraction. Therefore seeds may be dried one more time before being processed. Gravity flow dryers are generally used.

*c.* ***Cracking.*** The seeds are passed between two corrugated rollers (turning at high speed in opposite directions). This reduces the size of the seeds, thereby facilitating the subsequent hulling (if necessary) and flaking operations.

**Table 3.9. Indicative cake and meal composition [5]**

| Product | Dry Matter (%) | Proteins | | Fats (%) | Nitrogen-Free Extraction (%) | Crude Fiber (%) | Ashes (%) |
|---|---|---|---|---|---|---|---|
| | | Crude (%) | Digestible (%) | | | | |
| Peanut | | | | | | | |
| Cake | 93.0 | 52.3 | 47.6 | 1.6 | 26.3 | 6.9 | 5.9 |
| Meal | 91.8 | 42.7 | 38.0 | 1.9 | 25.4 | 17.0 | 4.8 |
| Rape | | | | | | | |
| Cake | 90.0 | 36.0 | 29.0 | 7.2 | 28.7 | 11.0 | 7.1 |
| Meal | 90.0 | 36.0 | 29.0 | 3.2 | 30.2 | 13.5 | 7.1 |
| Sunflower[a] | | | | | | | |
| Cake | 91.4 | 36.4 | 33.1 | 11.0 | 25.0 | 14.2 | 4.8 |
| Meal | 90.4 | 39.0 | 34.7 | 2.0 | 25.7 | 16.5 | 7.2 |
| Soybean | | | | | | | |
| Cake | 91.0 | 44.0 | 39.0 | 4.9 | 30.0 | 5.9 | 6.2 |
| Meal | 90.3 | 45.7 | 42.0 | 1.3 | 31.4 | 5.8 | 6.1 |
| Cotton | | | | | | | |
| Cake | 91.5 | 38.0 | 31.9 | 7.0 | 30.3 | 10.2 | 6.0 |
| Meal | 89.3 | 42.0 | 35.3 | 0.7 | 30.2 | 9.6 | 6.8 |
| Flax | | | | | | | |
| Cake | 91.0 | 33.6 | 26.8 | 4.5 | 37.5 | 9.0 | 6.4 |
| Meal | 91.0 | 36.0 | 28.8 | 1.0 | 38.9 | 9.3 | 5.8 |
| Palm | | | | | | | |
| Cake | 91.6 | 20.4 | | 8.3 | 56.6 | 9.0 | 5.7 |
| Meal | 90.8 | 18.6 | | 1.7 | 38.2 | 37.0 | 4.5 |

[a] Decorticated seed.

***d. Hulling.*** This operation is used to eliminate the hull (pericarp of certain seeds; e.g., sunflower and cottonseed), which does not contain oil and generally has a low protein content. Hulling makes it possible to reduce press size, abrasion, and the quantity of solvent used. A slight pressure is exerted on the seed to open the pericarp, and then the kernel is separated by means of an air draft. Hulled seeds are often referred to as meats.

***e. Conditioning.*** The temperature and the moisture level of the meats are adjusted to speed up the oil-extraction process, improving the drainage of the proteic matrix and facilitating the subsequent flaking process. In conditioning, a film of water is deposited on the seed surface, improving oil diffusion from the inside to the outside by breaking the cell wall. Conditioning is usually performed with steam-heated rotary dryers.

***f. Flaking.*** The cell structures that contain the fatty matter are broken up when the meats are passed between two smooth cast iron rollers rotating at high speed in opposite directions. This increases extraction efficiency, which is, in fact, related to seed flake thickness (Table 3.10). It is therefore advantageous to crush the seeds as much as possible. However, laminae that are too thin may cause the formation of dusts, resulting in problems in drainage during the extraction phase. The optimum flake thickness generally ranges from 0.25 to 0.40 mm.

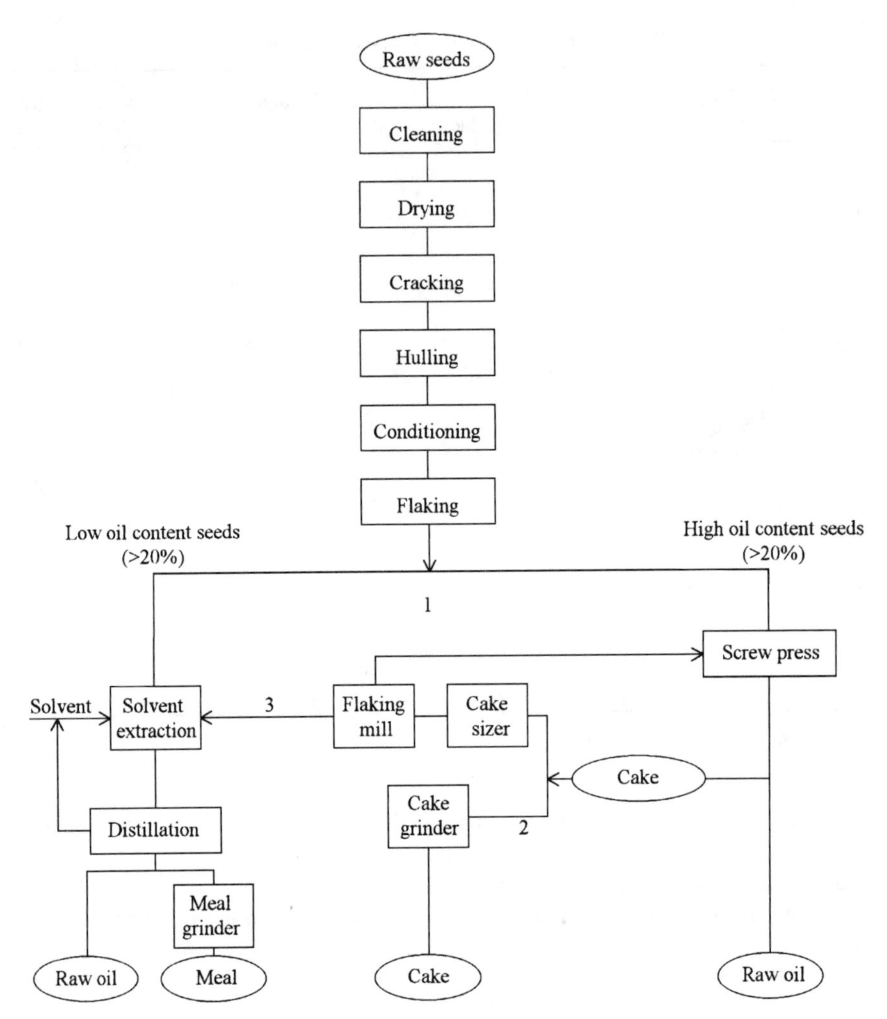


**Figure 3.8. Oil extraction (industrial process: 1, two-stage way; 2, full-press way; 3, prepress way. [6]).**

***g. Pressing.*** Mechanical extraction is carried out with a screw press. For increased throughput, the press can be equipped with two screws. The extracted oil is collected in a screening tank, at the bottom of which the larger solid particles (called foots) settle (Fig. 3.10). The foots are conveyed back to the screw, while the oil is collected in a tank for unfiltered oil. From there, oil is filtered and finally stored in a different tank. The unfiltered oil tank and the filter are cleaned from time to time, and the deposits are conveyed to the screw. A part of the oil in the screening tank is cooled and used for cooling the press in order to avoid damaging the screw or the oil quality and to prevent the cake from fusing into a solid mass. Pressing can be a total (full pressing) or a partial operation (prepressing).

**Table 3.10. Residual oil of hexane-extracted soy flakes [6]**

| Flake Thickness (mm) | Residual Oil (%) |
|---|---|
| 0.20 | 0.50 |
| 0.35 | 0.74 |
| 0.40 | 1.12 |
| 0.55 | 2.00 |

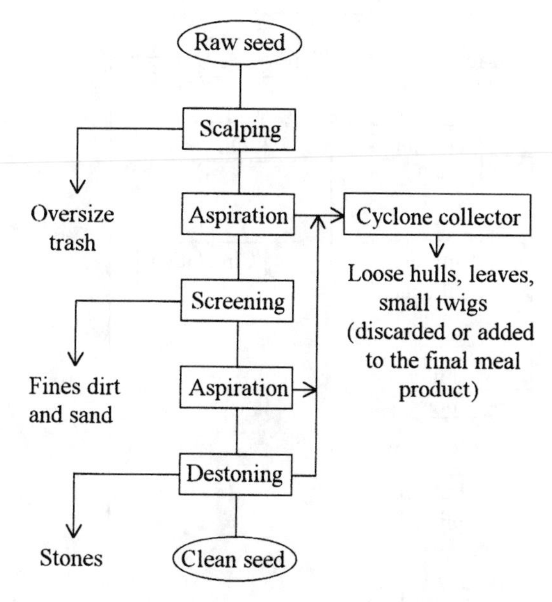


**Figure 3.9. Seed cleaning [6].**

- Full pressing This method is applicable only to seeds with high oil content, i.e., with a fatty content greater than 20% (e.g., rape and sunflower) and is suitable for final yields of ~10%–15%. In full pressing, most of the oil is generally extracted in a single step by continuous presses with a power consumption of 45 kWh/t of seeds, and the resulting expeller has a minimal oily residue of 8%–12%. The cake is usually milled with a cake grinder to obtain a uniformly sized product. It is also possible to perform full pressing in two stages in order to obtain a better oil yield (final oil content of the cake: 5%–6%): the first press extracts only a part of the oil, after which the cake is milled to reduce the size of the pieces and is pressed again. The second-stage cake is also milled.
- Prepressing In this case, a lower quantity of oil is extracted and the cake (oil content: 15%–18%) is milled, dried, and processed by means of chemical extraction.

***h. Solvent Extraction.*** The cake produced by the prepress operation is first milled and then conveyed to the solvent extractor. The countercurrent percolation extractor is

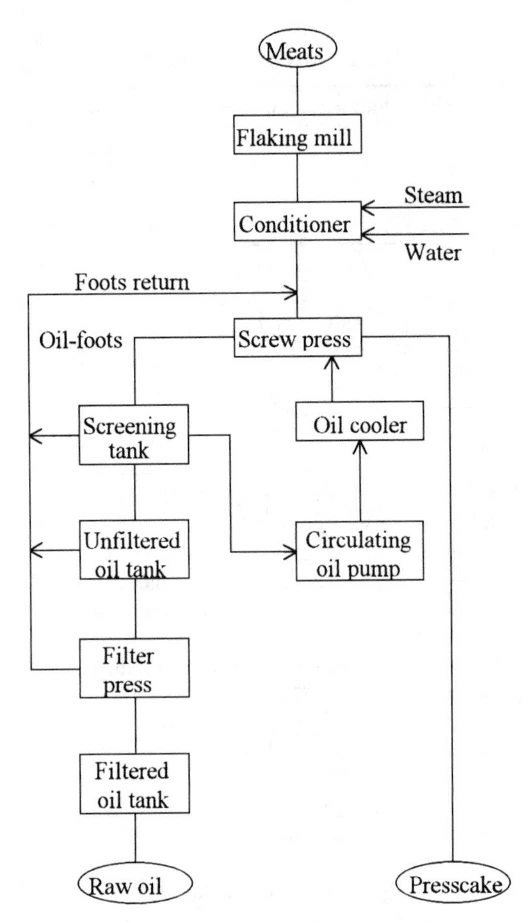


**Figure 3.10. Mechanical extraction [6].**

the most commonly used type (Fig. 3.11). The drained solvent–oil mixture (called the miscella) is spread over the flake bed several times to obtain the highest possible oil content. Finally the high-oil-content miscella leaves at the bottom of the extractor. The extractors differ in the method used for moving the flakes: individual cells, wedge-wire screen, mobile, and perforated screen.

The amount of oil extracted depends on the duration of the process. As a general rule, oil is extracted from broken oleiferous cells by direct dilution in solvents, whereas the oil of whole cells is extracted by diffusion. Therefore the quantity of oil extracted is initially directly proportional to time, after which it follows an asymptotic pattern. Most of the oil, however, is extracted in the first 30 min. After this time it is possible to obtain a meal containing approximately 2.5% fatty residue for sunflower and 1.4% for rape. Achieving residues below 1% requires processing times in excess of 2 h for rape and 1 h for sunflower.

**Table 3.11. Influence of time extraction on rape expeller (starting oil content: 14.6%) and sunflower expeller (starting oil content: 11.9%)**

| Extraction Time (min) | Expeller/Solvent Rate | Final Oil Content in the Meal (%) | |
|---|---|---|---|
| | | Rape | Sunflower |
| 30 | 1:9 | 3.0 | 1.6 |
| 60 | 1:19 | 1.6 | 0.8 |
| 90 | 1:28 | 1.5 | 0.8 |
| 120 | 1:37 | 0.8 | 0.5 |

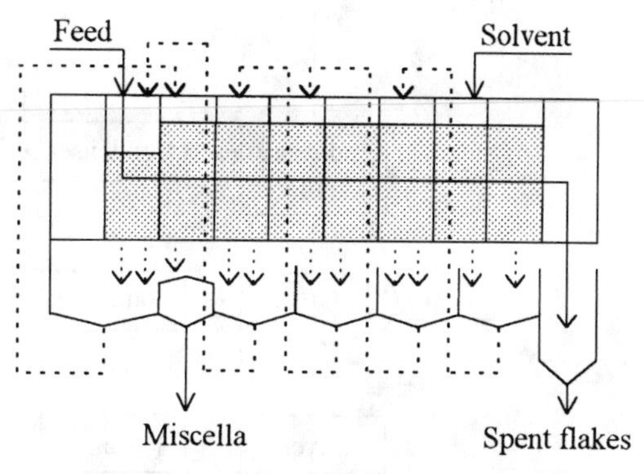


**Figure 3.11. Percolator extractor layout [6].**

Other parameters or technical factors affecting the process are preprocessing (lamination, crushing, conditioning), quantity, temperature, and type of solvent. The expeller–solvent ratio in particular can reach the upper limit of 1:18–1:19. Any further increases will produce smaller increments in yield (Table 3.11). The extraction process also improves with increasing temperature up to 50°C, beyond which efficiency begins to fall off.

The following solvents can be used: hexane, petrol solvent, trichloroethylene, and carbon disulfide. However, the only solvent used commercially today is hexane, which gives good yields, can be separated from water (a step in its recovery process) with simple, low-energy-consumption equipment, and can be easily removed from the meal. Otherwise, if there is a requirement for nonflammable products and if product quality is not a priority, then trichloroethylene can be used.

After extraction, the solvent is recovered from the miscella by means of three-stage evaporation system (Fig. 3.12), while steam evaporation is used to recover the solvent (25%–40% by weight in solvent) from the meal.

*Extraction for On-Site and Local Applications*

The only suitable extraction method for on-site and local applications is full pressing.

The most common process involves seed cleaning, oil extraction, and oil filtering. Screw-type presses can be used for oil extraction (Fig. 3.13). The capacity of the presses ranges from 2 to 5 kg of seeds per hour (hand operated) to over 500 kg of seeds per

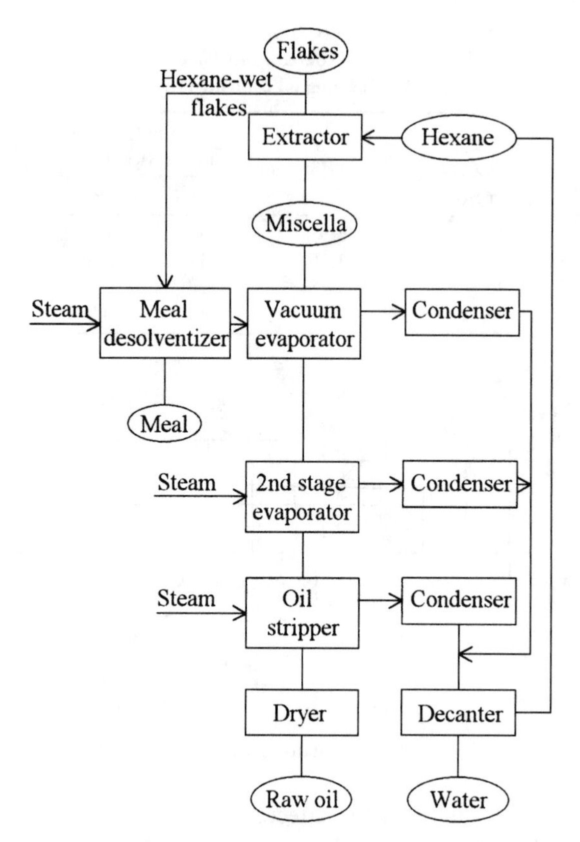


**Figure 3.12. Solvent extraction [6].**

hour (suitable for small groups or cooperatives). They consist of a worm screw rotating within a pressing cylinder or a cage. During extraction, the seeds can be heated to facilitate separation of the oil from the seed fibers.

***Oil Depuration and Refining***

Oil impurities, such as proteins, gums, phosphatides, ketones, and aldehydes, can be present in solution or in suspension. Moreover, all crude oils contain FFA's, which are either naturally present or produced by the action of lipolytic enzymes. In either case, they form a substrate for further oxidation reactions.

*Industrial Processing*

There are basically two methods for refining oils: conventional and physical refining (Fig. 3.14). In the conventional process, FFA's are removed by an alkaline treatment, whereas in physical refining, they are removed by stripping during deodorization.

However, in order to achieve good oil stability it is necessary to remove resins, gums, phosphatides, and other mucilaginous products. This process, which is called degumming or desliming, is always necessary before physical refining can take place.

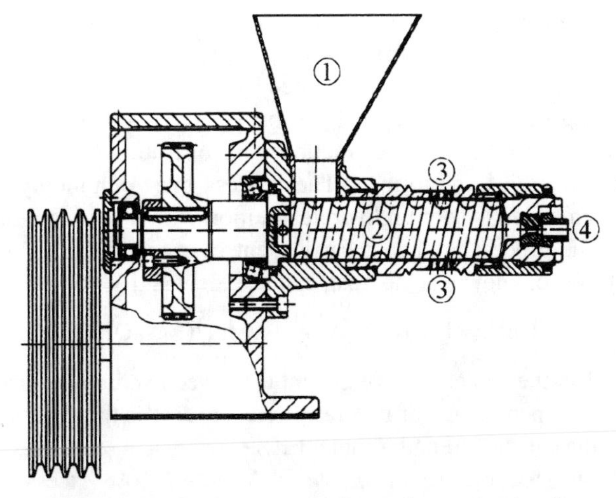


**Figure 3.13. Screw-type press for on-farm extraction (1, charging hopper; 2, screw; 3, oil outflow; 4, cake outlet).**

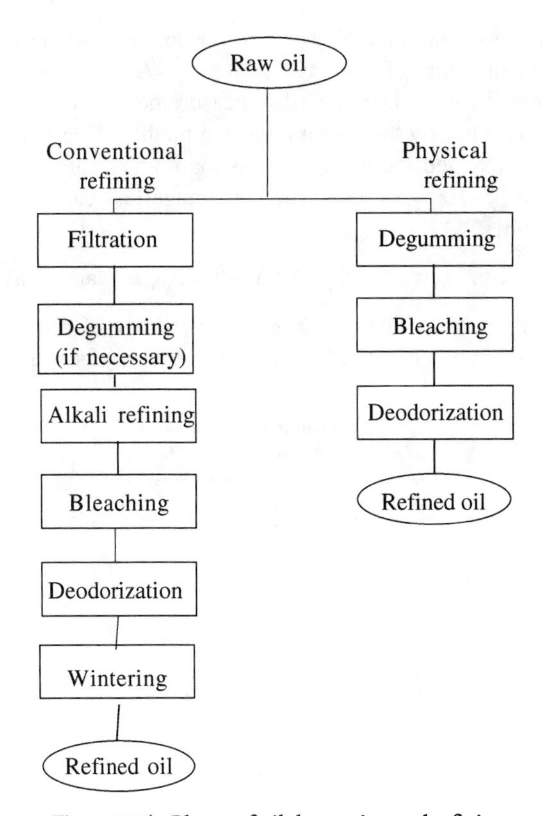


**Figure 3.14. Phases of oil depuration and refining.**

a. ***Conventional Refining.*** This process consists of the following phases:
- Filtration  Filtration is performed with filter presses that retain all the substances in suspension that have not been separated by sedimentation.
- Alkali refining (neutralization)  This process eliminates the FFA's produced by the activity of lipase before oil extraction. This operation is particularly necessary before esterification is performed. The process most commonly used in simple plants uses an aqueous sodium hydrate solution with a concentration between 10 and 30°B. The process is always carried out in excess (0.10%). The quantity of caustic substance, known as the treat, can be calculated as follows [7]:

$$\text{Treat} = [(0.142\text{FFA}) + \text{excess}]/(\%\text{NaOH}/100), \tag{3.1}$$

where 0.142 is the molecular weight ratio between sodium hydroxide and oleic acid, FFA is the percentage of free fatty acids, excess is the excess of caustic, and %NaOH is the caustic strength (Table 3.12).

For example, for an oil with 0.75% FFA, with 0.10% excess caustic at 16°B (strength of 11.06%), the treat is

$$\text{treat} = [(0.142 \times 0.75) + 0.10]/(11.06/100) = 1.86\%.$$

Greater quantities of sodium hydrate determine an increase in saponification losses as well as high quantities of mucilage.

Refining losses can be estimated by measurement of the oil input and output (weight or volume) or by the sodium balance method. The latter method is based on the fact that all added sodium appears as soaps and sodium in the refined oil and requires only a 25–100-g sample. The percentage loss can be estimated with the following formula:

$$\text{loss} = [\text{treat}(\text{Na}_r - \text{Na}_s) - 100\text{Na}_o)/(\text{Na}_s - \text{Na}_o), \tag{3.2}$$

where loss is the refining loss, treat is the percentage of caustic used, $Na_r$ is the percentage of sodium in caustic, $Na_s$ is the percentage of sodium in soapstock, and

**Table 3.12. Sodium hydroxide strength of different degrees Baumé (°B) [7]**

| °B at 15°C | Sodium Hydroxide (%) |
|---|---|
| 10 | 6.57 |
| 12 | 8.00 |
| 14 | 9.50 |
| 16 | 11.06 |
| 18 | 12.68 |
| 20 | 14.36 |
| 22 | 16.09 |
| 24 | 17.87 |
| 26 | 19.70 |
| 28 | 21.58 |
| 30 | 23.50 |

$Na_o$ is the percentage of sodium in the refined oil. If there are large differences between the moisture in raw and refined oil, the formula must be corrected.

Finally, there are metering systems for measuring instantaneous and batch losses.

- Bleaching or decolorization This eliminates most of the pigments (carotenoids and chlorophyll) still present in the oil. Small quantities of activated earth (0.5%–4%) mixed with active coal (10% of the earth) at temperatures of 60–100°C for a period of 15 to 30 min are used. The mass (oil + earth + coal) is filtered with a filter press.
- Deodorization FFA's, volatile compounds (aldehydes and ketones derived from oxidation processes of unsaturated fatty acids), residual carotene, and natural toxic principles are eliminated. The treatment consists of exposing the oil to deaerated superheated steam (200°C) in vacuum containers.
- Wintering This eliminates the triglycerides with higher melting points, the aim being to minimize this parameter. This process is not necessary for the oils (rape and sunflower) commonly used in energy applications. However, with very viscous or semisolid oils (e.g., palm), it is advisable to lower their melting point to room temperature.

***b. Degumming.*** Degumming can be carried out either at the extraction plant (usual method) or at the refinery. The following methods are used [8]:

- Acid conditioning (dry degumming) This is used mainly to process raw oils with low gum content. An acid (citric or phosphoric acid) is added to the oil (the amount of acid is 0.05%–0.2%) at a temperature of 60–70°C. Intensive stirring is necessary in order to achieve a retention time of less than a minute. Static methods or slow-moving mixers require over 20 minutes and greater quantities of acid. It is possible to add the acid in a stainless steel storage tank, but the need for mixing remains. The oil is then sent through a bleaching phase in which the precipitated gums are removed with adsorbents. After degumming, the oil can be treated with alkali: a pH of 7 is recommended to neutralize the acid without reacting with FA's.
- Modified physical refining The alkali is added after water degumming or acid conditioning in order to produce 300–400 parts in $10^6$ (ppm) of soap, which enhances the adsorbing properties of a special type of silica (0.2%–0.5%) used in the subsequent bleaching phase. In this manner, smaller quantities of bleaching earth are necessary.
- Water degumming This is not suitable for physical refining. Water degumming is used primarily when it is necessary to recover the lecithin. Water, in an equal quantity to that of the gums (1%–3%), is mixed with the oil at 60–80°C. The agglomerated and oil-insoluble gums can be removed by centrifugation after 20–30 min.
- Acid degumming This consists of acid conditioning followed by water degumming, and it is used for oils that are easy to degum. After acid conditioning, the temperature should be lowered to 25°C before water degumming and the mixture held for 3 h before heating and separation are performed.

For nonedible oils, sulfuric acid and phosphoric acid can also be used [7]. In the case of using sulfuric acid, the concentration (1%–2% of sulfuric acid at 66°B for fluid oils at 15–20°C) must be such that it does not cause hydrolysis, sulphonation, and treatment with sulfur of the glyceric part. The acid–oil contact time must be very brief and the temperature must be lower than 25–30°C. After this, dilution with water is necessary to stop the action of the concentrated acid. The two phases are separated by centrifugation, followed by washing in hot water to eliminate acid residues, and then followed by drying. Phosphoric acid, being the most effective and easiest to handle, is the most widely used, principally as a pretreatment of alkali refining (the amount is 0.1%–0.4%).

The following commercial degumming processes are used [8]:

- Superdegumming This was developed by Unilever. The steps of this process are heating the oil (70°C) and mixing it with citric acid, a retention time of 5–15 min, cooling to 25°C, mixing with water, formation of liquid crystals from the gums (in 3 h), heating to 65°C, and centrifugation. The final phosphorous content is between 15 and 30 ppm.
- Unidegumming Superdegummed oil is cooled to 25°C, held for another 3 h, heated, and centrifuged. The residual phosphorous can be reduced to 8 ppm.
- Special degumming This process was developed by Alfa-Laval. The oil, heated to 80°C, is mixed with phosphoric or citric acid. Then concentrated lye is added to neutralize the acid. The mixture reacts with the added water for 30 min; then the gums are removed by centrifugation. The residual phosphorous is less than 15–30 ppm, but if an additional washing phase is performed, it can be reduced to less than 50%.
- Total degumming process This was developed by Vandemoortele. In this two-stage process, the heated oil is mixed with acid and reacts for 3–4 min, after which caustic soda is added. The recycled heavy-phase effluent obtained from the second phase is added to the mixture and then centrifugation is performed (the separator is adjusted to minimize oil losses). Next the oil is washed with water and centrifuged. The separator is then adjusted to minimize the residual phosphatides (below 10 ppm) in the final oil. Residual iron is less than 0.2 ppm.
- Alcon process This is marketed by Lurgi. It consists of a flake treatment that increases the quantity of hydratable phosphatides and decreases the quantity of nonhydratable phosphatides. The flake is heated to 100°C and the moisture content is increased to 15%–16%. After 15–20 min, the conditions are restored to those suitable for extraction. The oil can then be processed by means of water degumming. However, the quality of the cake or meal is lower.

***c. Physical Refining.*** In physical refining, the FA neutralization phase typical of industrial refining—the subsection on conventional refining) is replaced by acid conditioning (especially for palm and lauric oils), and the precipitated gums are removed in the bleaching phase or, in some cases, by centrifugation before bleaching. Physical refining aims to remove FFA's and bad-smelling compounds in a single process (steam deodorization). It can be used only for oils whose total acidity exceeds 5% and that are free of FA's with low molecular weight. This method offers the advantage of speed but involves greater complexity and higher plant costs. Moreover, it is difficult to achieve a final acidity of less than 1%.

**Table 3.13. Refining operations and removed compounds [9]**

| Treatment | Reduced or Removed Compounds |
|---|---|
| Alkali refining | All the hydratable phosphatides and most of the nonhydratable phosphatides; chlorophyll and gossypol (by 1/2); waxes (with low-temperature treatment only); metal salts (by 1/3); peroxides, aldehydes, and ketones (by 1/2); FFA's (to 0.05%); tocopherols and sterols (by 10%); soaps (to 50 ppm). |
| Physical refining | All the phosphatides (they must start at 50 ppm); chlorophyll and gossypol (it depends on clay); all metal salts; peroxides, aldehydes, and ketones (by 1/2); tocopherols and sterols (by 10%); all the soaps. |
| Bleaching | Phosphatides (it depends on clay); chlorophyll and gossypol (it depends on clay); all metal salts; peroxides, aldehydes, and ketones (by 1/2); tocopherols and sterols (by 10%); all the soaps. |
| Deodorization | All carotene; peroxides, aldehydes, and ketones (by 1/2); FFA's (to 0.02%); all sulphur; some tocopherols and sterols; some monoglycerides, diglycerides, and triglycerides. |

Table 3.13 [9] summarizes the refining phases and the compounds removed. The product obtained after processing is called refined oil. Energy and mass balances for oil production and processing are shown in Table 3.14.

***Other Refining Methods [8].***

- Miscella refining The miscella is mixed with sodium hydroxide (neutralization), then reacted with phosphatides and subjected to decolorization. The soaps are removed by centrifugation. This method makes it possible to reduce loss and eliminate the water washing phase, but it requires explosion-proof equipment. It has been applied to cottonseed, soybean, sunflower, palm and coconut oil, but is widely used for cottonseed oil.
- Zenith process This consists of degumming with concentrated phosphoric acid, separation of phosphatides by decantation, and neutralization of FFA's with dilute caustic. Water washing is unnecessary.

*Processing for On-Site and Local Applications*

The methods for refining oils in small plants or on the farm must be extremely simple. Oil can be processed first by filtration, but the final product is not very different from crude oil. Therefore specific measures must be taken into account during its use.

It is also possible to degum the oil by water degumming. In this way, a part of the polar impurities can be removed from the oil. After decantation, the oil must be filtered. The extent of this partial refining will depend on the size of the installation in relation to the sustainable investments. It is also possible to perform alkali neutralization in on-site or local applications (refer to treat evaluation). In any case, much care should be taken during the water washing process.

***Oil Esterification***

*Chemical Aspects*

Esterification aims at breaking up monoglyceride, diglyceride, and triglyceride structures in order to simplify molecules, reduce the viscosity and melting point, and achieve

**Table 3.14. Mass and energy balances for oil production and refining (European average values)**

| Item | Rape | | | Sunflower | | | Soybean | | |
|---|---|---|---|---|---|---|---|---|---|
| | Yield (t/ha) | Conversion Factor[a] (MJ/t) | Total Energy (MJ/ha) | Yield (t/ha) | Conversion Factor[a] (MJ/t) | Total Energy (MJ/ha) | Yield (t/ha) | Conversion Factor[a] (MJ/t) | Total Energy (MJ/ha) |
| Inputs | | | | | | | | | |
| Cultivation | 3.00[b] | 7615 | 22,845 | 3.50[b] | 7354 | 25,739 | 3.50[b] | 6080 | 21,280 |
| Extraction | 3.00[b] | 1245 | 3735 | 3.50[b] | 1245 | 4358 | 3.50[b] | 1245 | 4358 |
| Refining | 1.20[c] | 1660 | 1992 | 1.40[c] | 1660 | 2324 | 0.81[c] | 1660 | 1336 |
| Total | | | 28,572 | | | 32,421 | | | 26,974 |
| Outputs | | | | | | | | | |
| Refined oil | 1.08 | 37,200 | 40,176 | 1.34 | 36,430 | 48,962 | 0.72 | 36,800 | 26,662 |
| Phosphatides and by-products | 0.12 | 38,181 | 4582 | 0.06 | 38,181 | 2138 | 0.08 | 38,181 | 3074 |
| Cake | 1.80 | 15,000 | 27,000 | 2.10 | 15,000 | 31,500 | 2.70 | 15,000 | 40,425 |
| Crop residues | 4.80 | 12,500 | 60,000 | 9.21 | 12,500 | 115,063 | 4.48 | 12,500 | 56,000 |
| Total | | | 131,758 | | | 197,663 | | | 126,160 |
| Balance (oil) | | | +11,604 | | | +14,541 | | | −312 |
| Balance (total) | | | +103,186 | | | +165,242 | | | +99,186 |

[a] Specific consumption or energy content.
[b] Seeds.
[c] Extracted oil.

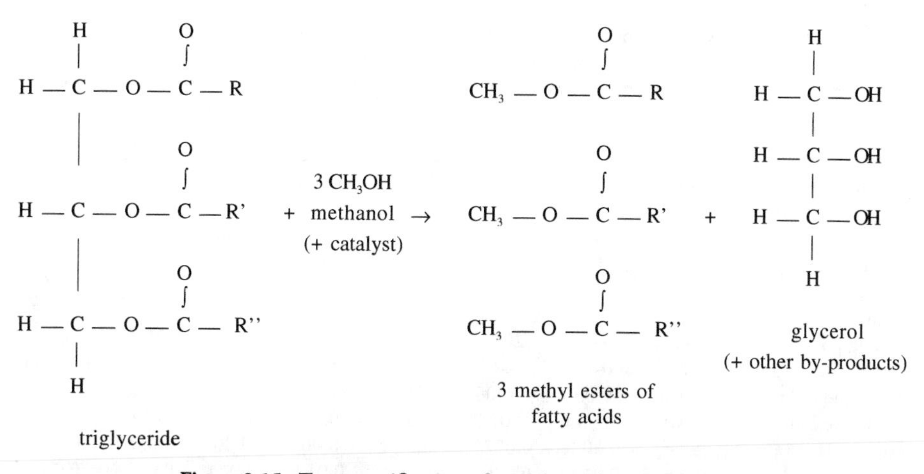


**Figure 3.15. Transesterification of a triglyceride with methanol.**

good combustion. In other words, it is possible to obtain a fuel with physical and chemical characteristics similar to those of diesel fuel.

The process, because of the necessary investments and technology, is suitable only for large networks or local applications.

The chemical aspects of ester production are the following:

***a. Reaction.*** In the esterification (or transesterification) reaction, FA molecules break away from glycerol and esterify an alkyl coming from an alcohol molecule. With methanol used as alcohol, the reaction for a triglycerides is as shown in Fig. 3.15.

***b. Stoichiometric Rate.*** In esterification a triglyceride (or monoglyceride or diglyceride) molecule reacts with three molecules of alcohol (methanol or ethanol) to form a molecule of glycerol and three molecules of FA methyl ester. It is necessary to shift the equilibrium to the right with excess alcohol: to obtain more than a 97% yield of methyl ester, a 1:6 stoichiometric ratio between oil and alcohol is necessary [10].

***c. Alcohol.*** Esterification can occur with either methanol or ethanol. Although the reaction is faster with methanol, after 1 h the yield in esters is the same as that with ethanol. Therefore the choice between ethanol or methanol depends on the availability and the cost of the alcohol. From a comparison of methyl ester and ethyl ester, the aspects that emerge are (Table 3.15) [11] that the heating value is the same for the both and the viscosity of ethyl ester is slightly higher but its clouding point is lower. The emissions are compared in the section on aspects of applications.

***d. Catalysts.*** Without a catalyst, reaction requires either a temperature of at least 250°C or a very long reaction time. As lecithin interferes with catalysts, it is better to use refined oil. It is possible to use either alkaline or acid catalysts. However, for refined oils (with a total acidity lower than 1%), esterification becomes faster with alkaline catalysts [10]. Moreover, any residual traces of the acid catalyst will render the final

**Table 3.15. Physical proprieties of rape seed methyl ester and ethyl ester and their blend with diesel fuel reference [11]**

| Property | Unit | Methyl ester 100% | Methyl ester 50% | Methyl ester 20% | Ethyl ester 100% | Ethyl ester 50% | Ethyl ester 20% | Diesel fuel |
|---|---|---|---|---|---|---|---|---|
| Specific gravity | kg/dm$^3$ | 0.8802 | 0.8632 | 0.8540 | 0.8760 | 0.8620 | 0.8535 | 0.85 |
| Kinematics viscosity at 40°C | cSt | 5.65 | 3.9 | 3.1 | 6.11 | 4.06 | 3.2 | 2.98 |
| Cloud point | °C | 0 | –7 | –12 | –2 | –9 | –12 | –12 |
| Pour point | °C | –15 | –15 | –15 | –10 | –13 | –15 | –18 |
| Flash point | °C | 179 | 85 | 82 | 170 | 79 | 76 | 74 |
| Boiling point | °C | 347 | 209 | 194 | 273 | 204 | 213 | 191 |
| Distillation range | °C | 300–360 | — | — | 300–360 | — | — | 170–360 |
| Water and sediment | %. vol | <0.005 | <0.005 | <0.005 | <0.005 | <0.005 | <0.005 | <0.005 |
| Carbon residue | %. wt | 0.08 | 0.07 | 0.08 | 0.10 | 0.07 | 0.12 | 0.16 |
| Ash | %. wt | 0.002 | 0.002 | 0.00 | 0.00 | 0.00 | 0.00 | 0.002 |
| Sulfur | %. wt | 0.012 | 0.026 | 0.035 | 0.012 | 0.024 | 0.033 | 0.036 |
| Cetan number | — | 61.8 | 55 | 51.4 | 59.7 | 54.2 | 50.7 | 49.2 |
| Heating value | MJ/kg | | | | | | | |
| Gross | | 40.54 | 42.88 | 44.54 | 40.51 | 42.94 | 44.64 | 45.42 |
| Net | | 37.77 | 40.18 | 41.81 | 37.82 | 40.08 | 41.94 | 42.90 |
| Fisher water | ppm | 535 | 288 | 153 | 757 | 308 | 200 | 38 |
| Particulate matter | mg/dm$^3$ | | | | | | | |
| Total | | 1.0 | 1.1 | 1.0 | 1.0 | 1.4 | 1.1 | 0.9 |
| Noncombustible | | <0.1 | <0.1 | <0.1 | <0.1 | <0.1 | <0.1 | <0.1 |
| Elemental analysis | | | | | | | | |
| Phosphorous | ppm | <10 | — | — | — | — | — | — |
| Nitrogen | ppm | 6 | — | — | 11 | — | — | 0.0 |
| Carbon | % | 76.3 | 82.75 | 84.76 | 78.11 | 82.07 | 84.73 | 86.67 |
| Hydrogen | % | 11.56 | 12.75 | 12.89 | 12.66 | 13.49 | 12.73 | 12.98 |
| Oxygen (by difference) | % | 12.2 | 4.47 | 2.35 | 9.22 | 4.42 | 2.51 | 0.33 |
| Iodine number | | 95.8 | 54.2 | 24.5 | 91.9 | 52.6 | 24.3 | 8.6 |
| Acid value | | 0.128 | — | — | 0.097 | — | — | — |
| Esters content | % | 98.02 | — | — | 94.75 | — | — | — |
| Free glycerine | %. wt | 0.4 | — | — | 0.72 | — | — | — |
| Total glycerine | %. wt | 0.86 | — | — | 0.93 | — | — | — |
| Free fatty acids | %. wt | 0.57 | — | — | 0.58 | — | — | — |
| Monoglycerides | %. wt | 0 | — | — | 0.58 | — | — | — |
| Diglycerides | %. wt | 1.35 | — | — | 1.33 | — | — | — |
| Triglycerides | %. wt | 0.45 | — | — | 2.17 | — | — | — |
| Alcohol content | %. wt | <1 | — | — | <1 | — | — | — |
| Catalyst | ppm | 11 | — | — | 12 | — | — | — |

product corrosive. Best results are obtained with sodium hydroxide, sodium methoxide, and potassium hydroxide. After 1 h, better yields are obtained with sodium methoxide, which prevents any saponification reactions. The amount of catalyst depends on the chemical characteristics of the oil.

***e. Temperature.*** Experience has shown that a temperature increase only speeds up the rate of reaction without altering the yield. In fact, after 1 h, the yields obtained at 45 and

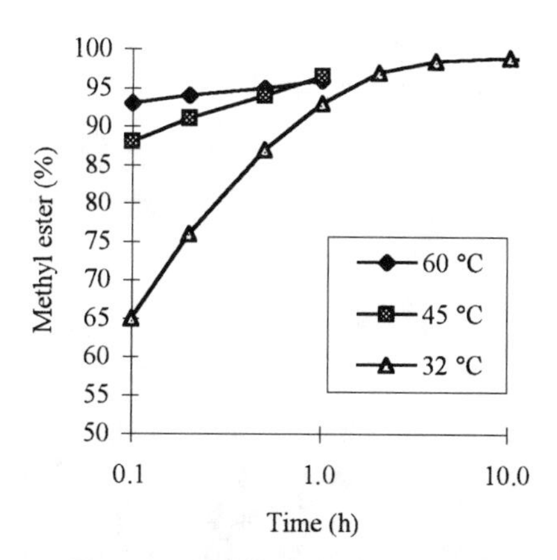


**Figure 3.16. Effect of temperature on methyl ester formation Reaction conditions: soybean oil:methanol (molar ratio) = 1:6, catalyst (sodium hydroxide) = 1% [4].**

60°C are the same, while the yield at 32°C is only slightly lower. In all cases, after 4 h, yield is always 98%–99% (Fig. 3.16). Therefore the key is to strike the right balance between the temperature and the time of reaction.

*Esterification Processes*

The most common methods for producing esters (usually methyl ester) are medium-high temperature process, room-temperature process [12], and continuous and pressurized process [13].

***a. Medium-High Temperature Process.*** The process most commonly used in large-scale plants involves the following steps.

- Pretreatment   Consistent oil quality is a key factor in process optimization. As lecithin and FFA's reduce the obtainable yield, it is preferable to process refined oils that have a total acidity of less than 1%.
- Alcohol–catalyst mixing   First an alkaline catalyst (refer to the above for the quantity) is mixed with methanol. As the reaction between the catalyst and the alcohol releases heat, plants must be sophisticated to guarantee safety.
- Oil–alcohol mixing and reaction   The process can be performed continuously or in the batch mode. The former one takes less time to complete the reaction, but requires a higher level of technology. Consequently it is suitable for production capacities greater than 20,000–25,000 t/year of ester. The temperature and the time required are 70°C and 1 h, respectively (increasing the time makes is possible to lower the temperature). After this time, separation between the hydrophilic (glycerol, excess of methanol, catalyst) and hydrophobic phases (ester) occurs. Separation can be sped up by centrifugation. In any case, to obtain good product quality, the esters must be purified. The process involves highly corrosive reagents. So the plant must be made of steel and suitable plastic material.

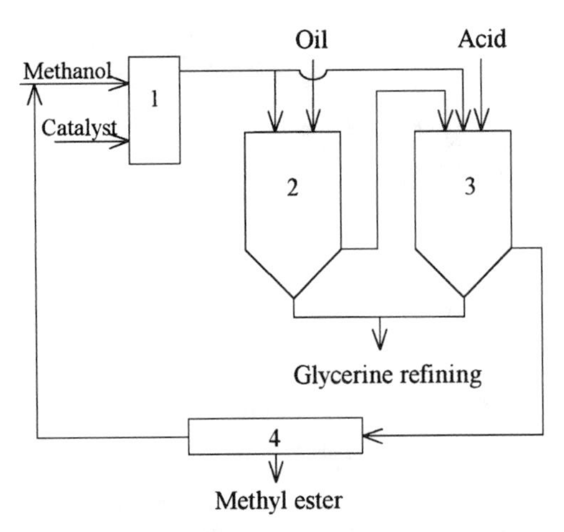


**Figure 3.17. Room-temperature process: 1, alcohol/catalyst mixer: 2, 1st batch, 3, 2nd batch; 4, alcohol recovering [12].**

- Ester purification  Impurities (traces of hydrophilic phase) can be eliminated by washing the ester with water. The purified esters are separated by decantation or by centrifugation.
- Alcohol recovery  The excess alcohol is recovered by evaporation in a vacuum (stripping) of both the hydrophilic phase and the ester and recycled in the process.
- Glycerol refining  If the glycerol is to be used for industrial purposes, it can be purified from water and the catalyst.

***b. Room-Temperature Process.*** This is a simple batch process (Fig. 3.17), suitable for small plants (less than 3000 t/year of esters). Energy input is low, making the process highly economical. Energy and mass balances of a typical process utilizing two batches are shown in Table 3.16. It is possible to esterify unrefined oil or used fried oil. In this case, the quality of oil (in terms of acidity and lecithin content) will obviously be variable. Therefore, to calibrate the quantity of the catalyst, each stock of raw materials must be analyzed before being processed and must never mixed with a different stock.

- Alcohol–catalyst mixing  The most widely used alcohol is methanol, while the catalyst is potassium hydroxide. A fraction of it is consumed for acidity neutralization.
- First batch  Oil and methanol (stoichiometric ratio 1:6) are mixed for 1 h in a tank with a conical base. Then, inside the same tank, decantation occurs over a 7-h period. The hydrophilic phase collected at the bottom of the tank is then removed.
- Second batch  The hydrophobic phase (containing esters) from the first batch is treated in another similar tank. The process is repeated with the same quantity of methanol and catalyst and with the same duration as for the first batch. This second esterification aims at achieving the quality levels required by the current European standards.

**Table 3.16. Mass and energy balances for room-temperature process [12]**

| Refining Degree of Glycerine | Pharmaceutical | 88% | Raw |
|---|---|---|---|
| Inputs | | | |
| Oil processing (kg) | | | |
| Raw oil[a] | 1044.2 | 1044.2 | 1044.2 |
| Methanol | 111.7 | 111.7 | 111.7 |
| KOH[b] | 8.0 | 8.0 | 8.0 |
| Phosphorous content[c] | 20.2 | 20.2 | — |
| Glycerine processing (kg) | | | |
| KOH | 4.1 | 4.1 | — |
| Water content | 9.9 | 9.9 | — |
| Activated carbon | 0.9 | — | — |
| Electric energy (kWh) | 93.5 | 89.6 | 21.0 |
| Thermal energy (–) | — | — | — |
| Outputs (kg) | | | |
| Methyl esters[d] | 1000.0 | 1000.0 | 1000.0 |
| Glycerine | 90.2 | 101.0 | — |
| Water/glycerine blend | — | — | 165.0 |
| Water/oil blend[e] | 8.8 | 8.8 | — |
| Solid residues (fertilizer) | 31.1 | 21.1 | — |
| Water | 15.3 | 15.3 | — |

[a] Phosphorous content: 10 ppm.
[b] FFA's: 0.5%.
[c] Raw oils containing 300 ppm in phosphorous need 0.3 kg of acetic acid and a 30-kg surplus of water.
[d] Content >96%.
[e] For oil burners.

- Catalyst neutralization Potassium hydroxide can be neutralized with phosphoric acid. In this way, potassium phosphate is produced, which can be used as a fertilizer.
- Methanol recovery The excess methanol is recovered from both the esters and the hydrophilic phase by stripping.

*c. **Continuous and Pressurized Process.*** In large plants, esterification can be performed as a continuous process with an acid catalyst. This is possible by an increase in the temperature and the pressure of the reaction. The process therefore offers the following advantages: oils with 4% acidity may be used, the glycerine phase does not require further refining, and corrosion of the building does not occur. Tables 3.17 and 3.18 summarize the mass and energy balances for this process [13]. Before esterification is performed, the oil must be processed as follows: acid degumming to reduce phosphorous content to 25–50 ppm and adsorbing soil filtration to keep the moisture below 0.01% and the phosphorous content at 10–15 ppm (the process cannot take place in conditions of high humidity and phosphorous content). Esterification involves:

**Table 3.17. Electric and thermal needs for 1 t of methyl ester of the De Smet process (steps: 1, seed drying and cleaning; 2, pressing; 3, oil degumming and drying (no neutralization); 4, transesterification; 5, ester drying; 6, ester distillation; 7, glycerine concentration) [13]**

| | Electric Needs (kWh/t) | | | | Thermal and Water Needs | | | |
|---|---|---|---|---|---|---|---|---|
| Step | 50 (t/day) | 50–100 (t/day) | 100–200 (t/day) | >200 (t/day) | Steam (kg) | Cooling Water ($m^3$) | Process Water ($dm^3$) | Thermal Needs (kWh) |
| 1 | Depends on silo capacity | | | | — | — | — | — |
| 2 | ±300 | ±300 | ±300 | ±300 | 525 | — | — | — |
| 3 | 10.1 | 9.8 | 6.8 | 6.8 | 255 | 10 | 30 | — |
| 4 | 28.5 | 24.5 | 22.5 | 18.5 | 1100 | 43 | 100 | 116 |
| 5 | 4 | 4 | 4 | 4 | 170 | 10 | — | — |
| 6 | 4 | 4 | 4 | 4 | 132 | 15 | — | 197 |
| 7 | 1.5 | 1.5 | 1.5 | 1.5 | 125 | 4.5 | — | — |
| Total | 380.6 | 372.3 | 365.3 | 357.3 | 2307 | 82.5 | 130 | 313 |

**Table 3.18. Chemicals needs for 1 t of methyl ester of the De Smet process (for steps see Table 3.17) [13]**

| | Chemicals (kg) | | | | |
|---|---|---|---|---|---|
| Step | Citric Acid | $H_3PO_4$ | Earth[a] | Methanol | Catalyst |
| 1 | — | — | | | |
| 2 | — | — | — | — | — |
| 3 | 0.5 | 1 | 10 | — | — |
| 4 | — | — | — | 135 | 1.5 |
| 5 | — | — | — | — | — |
| 6 | — | — | — | — | — |
| 7 | — | — | — | — | — |
| Total | 0.5 | 1 | 10 | 135 | 1.5 |

[a] Bleaching earth.

- Mixing Methanol (purity of 99.5%; in excess: 135 kg every 1000 kg of ester produced), oil, and catalyst (1.5 kg every 1000 kg of ester produced) are mixed;
- Heating The mixture is heated and continuously fed into the reactor by a high-pressure pump. The conditions during esterification are 5 MPa and 200°C;
- Methanol recovery The excess methanol is recovered by evaporation and stripping;
- Separation Decantation is performed to separate the ester from the glycerine phase;
- Washing The ester is washed with water;

Ester drying The ester is dried;

Distillation The ester is distilled to obtain a purity of 99%. The residual monoglycerides, diglycerides, and triglycerides are reprocessed. The other phases are also distilled to recover methanol, which is reused (purity of 98.5%);

Glycerine concentration Glycerine can be concentrated for industrial uses (purity of 82%–88%) or pharmaceutical applications (purity of 99%).

**Table 3.19. Comparison of physical and chemical characteristics of diesel fuel, raw and refined oil, and methyl ester**

| Characteristic | Unit | Diesel Fuel | Raw Oil | Refined Oil | Methyl Ester |
|---|---|---|---|---|---|
| Specific gravity | $kg/dm^3$ | 0.82–0.85 | 0.915 | 0.910 | 0.86–0.90 |
| Kinematic viscosity at 40°C | cSt | 2.0–3.0 | 32 | 30 | 3.5–5.0 |
| Net heating value | MJ/kg | 42.0–43.0 | 39.4 | 40.2 | 40.0 |
| Filatration limit point | °C | −18 | −6 | −1 | −9–24 |
| Cetane number | | 49.2 | 40 | 40 | 48–49 |
| Distillation range | °C | 180–360 | 359–893[a] | 350–890[a] | 300–360 |
| Flash point | °C | 74 | 300 | 300 | >100 |
| Ashes | % in peso | 0.002 | 0.1 | 0.01 | <0.01 |
| Total acidity | mg KOH/g | — | 2.8 | <1 | <0.5–0.8 |
| Saponification number | mg KOH/g | — | 190.3 | 180–190 | <170 |
| Iodine number | $g\ I_2/100g$ | — | 110–130 | 110–130 | 110–125 |
| Phosphorous content | ppm | — | 180 | <10 | 10–20 |
| Water content | ppm | — | 1000 | <500 | 300–700 |

[a] At 600°C cracking of molecules is beginning.

There are several uses for the process by-products: the cake or meal are used as animal feed; potassium hydroxide, if used as catalyst and neutralized with phosphoric acid, yields a fertilizer salt; and glycerine can be refined for use in the cosmetics and pharmaceutical industries, used for energy purposes (combustion or biogas production), or spread on fields.

***Physical and Chemical Characteristics of Oils and Esters***

*Specific Gravity*

For vegetable oils and their esters, specific gravity related to the origin of the oil. Values range from 0.92 to 0.91 $kg/dm^3$. Methyl esters have a specific gravity of ~0.88 $kg/dm^3$, while that of ethyl esters is slightly lower (Table 3.19).

*Viscosity*

Kinematic viscosity rises with increasing saturated FA content and the length of the FA chains, according to the following equation [14]:

$$KV = -120.05 + 10.07\ CH + 77.8\ UN - 5.22\ CH\ UN, \qquad (3.3)$$

where KV is the kinematic viscosity measured in centistokes, CH are (chain length) times (%FA)/100, and UN are (unsaturated bonds) times (% FA)/100.

For a given oil, viscosity decreases substantially with increasing temperature (Fig. 3.18) [15].

High viscosity affects the atomization of fuel: the spray cone becomes tighter and the jet becomes faster. Esterification makes the viscosity of vegetable products nearly equal to that of diesel fuels (Table 3.19).

*Pour Point*

This is the temperature at which the fuel flow tends to stop because of occlusion of filters and high viscosity. In fact, waxes and saturated FA's (myristic, palmitic, stearic,

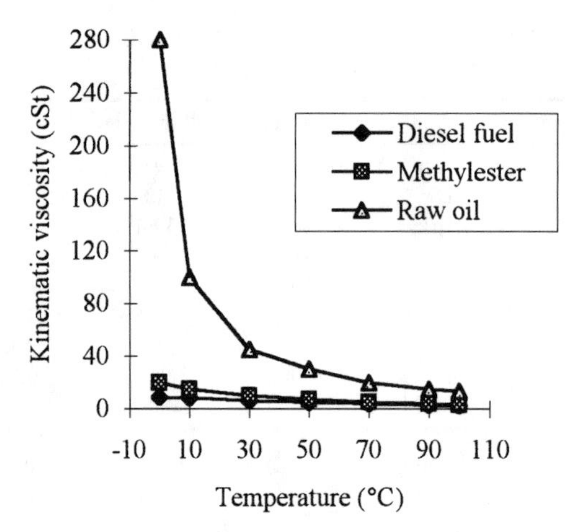


**Figure 3.18. Kinematic visosity of diesel fuel, crude degummed sunflower oil and its methyl ester [15].**

arachidic) reach their point of solidification. Then, in critical conditions, the use of specific additives is opportune.

*Solidification Point*

This is the temperature at which a fuel becomes solid. It depends on the content of unsaturated FA's, which keep down the value of this parameter. Some vegetable oils are already solid at 10–15°C, while others are still liquid at 0°C. In all cases, the point of solidification is reached after the pouring point, when the engine is definitely stopped.

*Cetane Number*

This is an index correlated to the pressure at which the fuel–air mixture burns in the absence of a spark or flame. The cetane number for diesel fuel is 48–51. For vegetable oil it is ~40, and for esters it reaches a value of 49 (Table 3.19), which is related to the unsaturation degree and the length of the carbon chain: long chains with two or more double bonds have a higher bond energy and thus need higher temperatures to vaporize (Table 3.20). The following relation has been proposed [16]:

$$y = 24.48 + 8.431x - 0.1299x^2, \tag{3.4}$$

where $y$ is the cetane number and $x$ is the chain length.

*Heating Value*

The heating value per unit mass of vegetable oils is 15%–20% lower than that of diesel fuel because of the bonded oxygen. FA's and their methyl esters have a heating value that increases with increasing chain length and decreases with increasing numbers of double bonds (Table 3.21) [17]. The heating value affects the performance curve of the engine: specific consumption increases and power decreases.

**Table 3.20. Cetane number of methyl esters of some fatty acids [16]**

| Carbon Number: Double Bonds | Common Name | Purity (%) | Cetane number |
|---|---|---|---|
| 8:0 | Methyl ottanoate | 98.6 | 33.6 |
| 10:0 | Methyl decanoate | 98.1 | 47.2 |
| 12:0 | Methyl laurate | 99.1 | 61.4 |
| 14:0 | Methyl miristate | 96.5 | 66.2 |
| 16:0 | Methyl palmitate | 93.6 | 74.5 |
| 18:0 | Methyl stearate | 92.1 | 86.9 |

**Table 3.21. Gross heating value (GHV) of triglicerides and methyl esters [17]**

| Carbon Number: | GHV of Triglicerides | | GHV of Methyl Esters | |
|---|---|---|---|---|
| Double Bonds | (MJ/mole) | (MJ/kg) | (MJ/mole) | (MJ/kg) |
| 8:0 | 15.2 | 32.3 | 5.5 | 37.7 |
| 10:0 | 19.8 | 35.7 | 6.8 | 39.1 |
| 12:0 | 23.7 | 37.1 | 8.1 | 40.1 |
| 14:0 | 27.6 | 38.2 | 9.4 | 40.9 |
| 16:0 | 31.6 | 39.2 | 10.6 | 41.1 |
| 18:0 | 35.8 | 40.2 | 11.9 | 41.6 |
| 20:0 | 39.5 | 40.6 | 13.2 | 42.0 |
| 22:0 | 43.2 | 40.8 | 14.5 | 42.4 |
| 16:1 | 31.2 | 39.0 | 10.5 | 41.0 |
| 18:1 | 35.1 | 39.7 | 11.8 | 41.5 |
| 18:2 | 34.5 | 39.3 | 11.7 | 41.5 |
| 18:3 | 31.1 | 35.7 | 11.5 | 41.1 |
| 20:1 | 39.0 | 40.3 | 13.2 | 42.3 |
| 22:1 | 42.8 | 40.7 | 14.5 | 42.6 |

*Distillation Curve*

This refers to fuel volatility and provides information concerning the air–fuel mixing system to be used and the regularity of distribution in the combustion chamber. For vegetable oils, it starts at 310–360°C (160–200°C for diesel fuel) and ends at 880–890°C (less than 400°C for diesel fuel). However, cracking of molecules begins just under 600°C, producing an unburned coat on the combustion chamber walls and polycyclic aromatic hydrocarbons in the exhaust. The distillation range for esters is 300–360°C (Table 3.19). Therefore they do not have problems with volatilization.

*Oxidation Stability*

Oxidation increases viscosity by forming gums and waxes in tanks. It is related primarily to the degree of unsaturation. Oil oxidation can occur by various mechanisms: bacteria, hydrolysis, and enzymatic, and autooxidation. The autooxidation is divided into two phases. In the initial phase, oxidation occurs slowly and involves all the compounds (although polyunsaturated compounds are oxidized more quickly than the others). Then, when the critical quantity of oxidation products has been reached, oxidation speed increases rapidly [18]. The oxygen requirements of these reactions depend on oil composition, presence of antioxidizing compounds, the metal content (Fe, Cu), and temperature. Engine tests with stocked products are not available, so the effects on emission are not known.

*Flash Point*

This is the temperature at which fuel vaporizes in the injection phase and provides information concerning the fire hazard during storage. For vegetable oils, it is ~300°C and 74°C for diesel fuel. The difference is due to the longer carbon chains and greater degree of unsaturation of vegetable oils. FFA's can slightly lower the flash point because of their higher volatility. For esters, the flash point depends on the alcohol residue.

*Conradson Index*

This gives information about carbon residue.

*Ashes*

Ashes derive from solid particles of fuels or from water-soluble metallic compounds. Values are shown in Table 3.19.

*Total Acidity*

This is equivalent to the milligrams of potassium hydroxide required for saponification of the FFA's in the sample. Crude oils have an acidity of ~3 mg KOH/g, refined oils 1 mg KOH/g, and for esters it is less than 0.5 mg KOH/g (Table 3.19). FFA's are more prone to oxidation and cause the corrosion of metal. In fact, at high temperatures they form salts with metals, removing them from the matrix and causing damage to engines.

*Saponification Number*

The saponification reaction is a basic hydrolysis of a fat, yielding an alcohol and a salt of FA:

$$\mathrm{RCOOR'} + \mathrm{KOH} \rightarrow \mathrm{RCOO} + \mathrm{K} + \mathrm{R'OH}.$$

It is irreversible and its chemical equilibrium is to the right.

The saponification number is equivalent to the milligrams of potassium needed to neutralize the entire quantity of free and bonded FA's in every gram of fat. It gives information about the FA content and helps to determine the origin of the oil. In fact, there is a relationship between the saponification number and the molecular weight of oil [19]. Values are shown in Table 3.19.

*Iodine Number*

This indicates the number of double bonds (degree of unsaturation). Unsaturation is positive in terms of filterability and pouring point, but negative in terms of oxidation stability (double bonds improve oxidation). What is more, a high iodine number produces a high Conradson index, increasing the formation of solid residue. The iodine number is given by the grams of iodine that can be bonded to 100 g of fat (Table 3.22). Oil refining or esterification does not affect the iodine number. In terms of chain length and degree of saturation, oils can be divided into:

**Table 3.22. Iodine number of some fats. Cocoa, Babassu, and tallow are to be considered saturated**

| Oil | Iodine number (mg $I_2$/100 g) |
|---|---|
| Cocoa | 15.4 |
| Babassu | 15.5 |
| Tallow | 44.5 |
| Palm | 54.5 |
| Peanut | 93.4 |
| Rape | 98.5 |
| Cotton | 105.7 |
| Maize | 122.8 |
| Sunflower | 125.5 |
| Soybean | 130.0 |

1. Lauric oils (babassum, copra): These contain mainly lauric (C12) and myristic (C14) acids. Their iodine number ranges from 5 to 30, so they are basically saturated, and their saponification number is 250 mg KOH/g.
2. Palmitic oils (palm): These are semisolid at room temperature and contain more than 25% palmitic acid (C16). Their iodine number ranges from 45 to 58, and their saponification number is ~200 mg KOH/g. Melting points range from 20 to 40°C.
3. Stearic oils (cocoa): These are solid (saturated C18). Iodine numbers range from 50 to 60, and saponification numbers from 180 to 190 mg KOH/g.
4. Oleic oils (olive, peanut, rape): These are liquid (unsaturated C18). Their iodine number is 80–100. Small quantities of linoleic and linolenic acids are present.
5. Linoleic oils (sunflower, soybean, maize, cotton): Their iodine number is over 110, so they are liquid. The saponification number is ~180–190, while solidification points are very low (−10°C for maize oil, −12/−29°C for soybean).
6. Very unsaturated oils (flax, tobacco; carbon number < 18).

*Phosphorous Content*

Vegetable oils contain phospholipids that can absorb humidity from the air and form insoluble gums in tanks, manifolds, filters, and in the combustion chamber as well (they increase the Conradson index). The quantity of phosphatides varies depending on the oil type and, for any given type, with its degree of refining. Refined oils and esters have a phosphorous content lower than 10 ppm (Table 3.19).

*Bonded Glycerine Content (Only for Methyl Esters)*

This is the quantity of monoglycerides, diglycerides, and triglycerides in the ester (for typical values see Table 3.23). High quantities of these compounds indicate a bad esterification process and increased viscosity. Moreover, oxidation and reaction with water can occur during stocking.

*Free Glycerine Content (Only for Methyl Esters)*

A high glycerine content increases the flash point but causes problems in terms of stability of the ester. During combustion, glycerine loses two water molecules and forms acrolein in the emissions, which is a mutagenic compound. For typical values, see Table 3.23.

*Methanol Content (Only for Methyl Esters)*

The presence of methanol is due to a bad ester purification. The effect of this alcohol is to decrease the flash point and worsen the cetane number. For typical values, see Table 3.23.

*Methyl Esters Content (Only for Methyl Esters)*

This indicates the quality of the product. For typical values, see Table 3.23.

*Lubricity*

Methyl esters have good lubricating properties. Blends with diesel fuel may reduce engine wear and extend the life of fuel-injection systems. Tests with a leading lubricity-measurement system (high-frequency reciprocating rig) show that ester–diesel fuel blends offer better lubricating properties than conventional petroleum diesels. This is important for rotary and distributor-type fuel-injection pumps. In these pumps, the moving parts are lubricated by the diesel fuel itself. When compared with conventional low-sulfur diesel and conventional diesel blended with a lubricity additive, a 20% ester blend offers significant improvements in reducing wear [20]. These results show that the lubricity increase of a 20% ester blend equals the increase offered by a lubricity additive (Table 3.24).

***Energy Uses***

Vegetable oils and their esters can also be used in burners or in engines. Blends with diesel fuel in various proportion can also be used.

*Vegetable Oils*

They are suitable for localized application (on-farm production and use) and are relatively simple to handle as a fuel. Because of their high flash points, no special precautions have to be taken during storage. To prevent oxidation of oils, tanks must be airtight and shielded as much as possible from sunlight. The fuel is biodegradable, so spillage is not dangerous for the environment.

***a. Burners.*** The following aspects are noteworthy.

- Industrial burners Industrial burners can be easily adjusted for vegetable oils. Because of the high viscosity of oil, the geometry of the sprayer nozzle is different from that for fossil fuels (45° and 60°, respectively), and the diameter of transfer pipe between the tanks and the burners should be greater than that used for conventional fuels. Preheating lowers the viscosity of vegetable fuels (the optimal preheating temperature is 65°C). The ignition pressure is 2.2 MPa. Smoke emissions can be controlled: for example, tests conducted on 150–540-kW burners [21] have shown $NO_x$ emissions of 45–55 ppm (60–70 ppm for petroleum fuel) and CO emissions of 7–10 ppm. With continuous running, combustion is the same as for petroleum fuel, whereas frequent starts and stops cause the formation of an unburned deposit on the walls of the combustion chamber. This is due to the high flash point of oil, which ignites 1–1.5 s after atomization. To solve the problem, a 50% blend with

**Table 3.23. Methyl esters standards of vegetable oils**

| | | CUNA NC 635-01 (Italy) | | Önorn C1190 (Austria) | | DIN V 51606 (Germany) | |
|---|---|---|---|---|---|---|---|
| | | Value | | Value | | Value | |
| Properties | Unit | Minimum | Maximum | Minimum | Maximum | Minimum | Maximum |
| Aspect | — | Clear | | — | | Clear | |
| Specific gravity at 15°C | $kg/m^3$ | 860 | 900 | 870 | 890 | 860 | 900 |
| Viscosity at 40°C | $mm^2/s$ | 3.5 | 5.0 | — | — | 3.5 | 5 |
| Viscosity at 20°C | $mm^2/s$ | — | — | 6.5 | 8.0 | — | — |
| Total acidity | mg KOH/g | — | 0.5 | — | 0.8 | — | 0.5 |
| Water | ppm | — | 700 | — | — | — | 0.03[a] |
| Ashes | % m/m | — | 0.01 | — | — | — | 0.01 |
| Sulfur ashes | % m/m | — | — | — | 0.02 | — | — |
| Distillation | | | | | | | |
| Start | °C | 300 | — | — | — | 300 | — |
| 95% in volume | °C | — | 360 | — | — | — | 360 |
| Total glycerine | % m/m | — | — | — | 0.24 | — | 0.25 |
| Bounded glycerine | | | | | | | |
| Monoglycerides | % m/m | — | 0.8 | — | — | — | 0.8 |
| Diglycerides | % m/m | — | 0.2 | — | — | — | 0.2 |
| Triglycerides | % m/m | — | 0.1 | — | — | — | 0.1 |
| Free glycerine | % m/m | — | 0.05 | — | — | — | 0.03 |
| Methanol | % m/m | — | 0.2 | — | 0.2 | — | 0.1 |
| Methyl ester | % m/m | 98.0 | — | — | — | 98.0 | — |
| Iodine number | $g\ I_2/100\ g$ | — | — | — | — | — | 115 |
| Saponification number | mg KOH/g | 170 | — | — | — | 170 | — |
| Phosphorous | ppm | — | 10 | — | 20 | — | 10 |
| Sulfur | % m/m | — | 0.01 | — | 0.02 | — | 0.01 |
| Sodium | ppm | — | — | — | — | — | 2 |
| Suspended matters | $g/m^3$ | — | — | — | — | — | 20 |
| Conradson index | % m/m | — | — | — | 0.05 | — | 0.03 |
| Conradson index (on 10% of residue) | % m/m | — | 0.5 | | | — | 0.3 |
| Cetane number | — | — | — | 48 | — | 49 | — |
| Flash point | °C | 100 | — | 100 | — | 100 | — |
| Pour point | | | | | | | |
| Summer | °C | — | 0 | — | — | — | −9 |
| Winter | °C | — | 0 | — | — | — | −24 |
| Cold filter plugging point | | | | | | | |
| Summer | °C | — | — | — | 0 | — | −5 |
| Winter | °C | — | — | — | −15 | — | −15 |
| Corrosion (3h/50°C) | — | — | — | — | — | — | 1 |
| Oxidation stability | $g/m^3$ | — | — | — | — | — | 25 |

[a] % on weight.

**Table 3.24. Lubricity results with high-frequency reciprocating rig machine [20]**

| Fuel Type | Wear (%) |
|---|---|
| Conventional low-sulfur diesel fuel (reference) | 100 |
| 80% reference/20% methyl ester blend | −60 |
| 70% reference/30% methyl ester blend | −58 |
| 1000-ppm lubricity additive/reference diesel blend | −61 |
| 500-ppm lubricity additive/reference diesel blend | −56 |
| 300-ppm lubricity additive/reference diesel blend | −62 |

fossil fuel can be used (oil pressure must be reduced to 2.0 MPa to compensate for the heating value, viscosity, and specific gravity of diesel fuel, and preheating at 70°C is necessary in any case) without increasing emissions. Alternatively, pure petroleum fuel can be used for the first few seconds, or pure oil can be used if the temperature of the combustion chamber walls is greater than 100–150°C (i.e., steam production).

- Domestic heating burners Small atomizing burners (15–60 kW) are more difficult to adjust than larger units because of the combination of high flash point and high viscosity of vegetable oils. Domestic heating applications need an extension piece to provide sufficient space in the boiler cavity to accommodate a longer flame. However, some firms offer domestic burners suitable for burning a 40/60 blend of oil and diesel fuel.

***b. Engines.*** In terms of engine utilization, vegetable oils can be used pure or blended with diesel fuel in various proportions. However, they are not currently a feasible practical solution. As was mentioned above, on-farm oil processing is not developed apart from filtering and water degumming, so it affects performance and durability of the engines, especially those with direct ignition.

An engine must always be modified in order to use 100% vegetable oil. (An example is the Elsbett engine [22], which is specially designed for use with high-viscosity fuels. However, this engine is not available through conventional commercial channels.)

It is also possible to use normal indirect ignition engines, taking care with the following aspects:

- Because of the high viscosity of the oil, single self-cleaning port-type injectors must be used.
- To reduce viscosity and save filters, fuel preheating is recommended.
- For operation at cold temperatures, engine starting by diesel fuel must be provided.
- The injection timing must be increased slightly to permit good combustion.
- To maintain power and torque comparable with the values for diesel fuel, the fuel flow must be increased.
- High-detergent lubricating oil must be used.
- Because the particular distillation curve of oils, frequent starting and stopping of the engine must be avoided.
- Maintenance and cleaning of the engine must be performed more frequently.

**Table 3.25. Gross heating value (GHV) and kinematic viscosity changing in some sunflower oil (SFO)/diesel fuel (DF) blends [23]**

| Fuel | GHV (MJ/kg) | Kinematic Viscosity (cSt) at 40°C | 60°C | 80°C |
|---|---|---|---|---|
| Diesel fuel | 45.28 | 2.91 | 2.12 | 1.22 |
| SFO[a] | 41.45 | 18.60 | 11.21 | 4.90 |
| SFO[b] | 39.42 | 31.31 | 17.53 | 7.49 |
| SFO/DF: | | | | |
| 20/80 | 44.65 | 5.53 | 3.66 | 1.99 |
| 30/70 | 43.72 | 7.26 | 15.21 | 3.25 |
| 50/50 | 42.47 | 9.43 | 6.17 | 3.19 |
| 60/40 | 41.42 | 11.57 | 7.21 | 3.94 |
| 70/30 | 40.75 | 16.16 | 9.32 | 4.43 |
| 80/20 | 40.10 | 17.84 | 13.09 | 4.89 |
| 90/10 | 39.46 | 23.05 | 14.02 | 5.89 |

[a] Raw oil.
[b] Refined oil.

In direct-ignition engines, the use of vegetable oil blended with diesel fuel alters the characteristics of the fuel-injection process, leading to a longer duration of combustion: increase in injection pressure, longer injection duration, delayed needle closing, and early needle opening. Therefore the engine must be adjusted in accordance with the ratio of vegetable to fossil fuel, which affects the fuel characteristics (Tables 3.15 and 3.25) [23].

There have not been many studies aimed at measuring the emissions of crude oils in engines. What is more, comparisons are difficult because of the higher fuel viscosity, which implies a number of adjustments and the substitution of certain mechanical parts, with marked effects on the quality and quantity of pollutants. Moreover, methods of analysis vary from one analysis labotatory to the other, affecting the comparison of data. In any case, for rapeseed oil, we can say that [24]

- Carbon monoxide tends to increase.
- Nitrogen oxides tend to remain constant or to decrease (−25%, depending on combustion temperature).
- Sulfur oxides are nonexistent because of the absence of sulfur in the fuel.
- Polycyclic aromatic hydrocarbons tend to increase in some cases, especially in direct-ignition engines.
- Hydrocarbons and particulate matters generally increase (more than double).
- Smoke, under the same adjustment of the injection system, decreases considerably.
- Particulate emissions are considerably reduced with indirect-ignition engines (−40%), whereas with direct-ignition engines, they increase as much as 100%.
- The soluble organic fraction increases up to 15%.
- Total emissions of aldehydes and ketones increase by 30%–330%.
- Polycyclic aromatic hydrocarbons increase, especially in direct-ignition engines: anthracene and phenanthrene had the highest concentration, followed by pyrene, chrysene, and fluoranthrene.

**Table 3.26. Comparison between methyl ester and diesel fuel emissions in a boiler of ~1.7 MW [25]**

| Property | Unit | Methyl Ester | Diesel Fuel | Difference (%) |
|---|---|---|---|---|
| $SO_2$ | ppm | 0 | 78 | −100 |
| CO | ppm | <10 | 40 | −75 |
| NO | ppm | 37 | 64 | −42 |
| $NO_2$ | ppm | 1 | 1 | 0 |
| $O_2$ | % | 6 | 6.6 | −9 |
| Total particulate | mg/Nm$^3$ | 0.25 | 5.61 | −96 |
| Benzene | mg/Nm$^3$ | <0.3 | 5.01 | −99.9 |
| Toluene | mg/Nm$^3$ | 0.57 | 2.31 | −99.9 |
| Xylene | mg/Nm$^3$ | 0.73 | 1.57 | −99.9 |
| Ethylbenzene | mg/Nm$^3$ | <0.3 | 0.73 | −59 |

*Esters*

Esters are more suitable for small and large networks because of the investments required for processing the oils.

***a. Burners.*** Esters can also be used in burners, either pure or blended with diesel fuel. Burners need slight modifications, which are related to the lower heating value of esters.

For example, fuel flow must be increased and a lower quantity of oxygen must be used. Manifolds must be constructed in ester-proof material. While the old burners are being converted for the use of esters, the storage tank must be thoroughly cleaned: because of the high cleaning properties of ester, it removes all fossil fuel deposits, causing the occlusion of manifolds. Emission characteristics are shown in Table 3.26 [25].

***b. Engines.*** Many experiments have been conducted on the use of esters in engines. There is no need to modify the engine, and direct-injection engines can be used as well. Only certain manifolds have to be made of specific materials. Transformation kits are available from manufacturers who have approved their products for use with esters. Esters can also be used mixed with diesel fuel in various proportions. As we have seen above, they also improve the lubricity of low-sulfur diesel fuel.

A number of tests have been carried out on emissions. However, because of the differences among collection methods, analysis procedures, and running cycles, a comparison of data is difficult. Another problem is the need for cold-start and hot-start tests. Differences between dynamic conditions and stationary tests have been observed.

Comparing pure esters and diesel fuel and considering the results published in the literature, we can say that [11]

- Hydrocarbon and carbon monoxide emissions are reduced by more than 50% and 45%, respectively, compared with diesel fuel. The best results are obtained with the oxidation catalytic converter [26]. Comparing rape ethyl ester and rape methyl ester, the former reduces HC by 8.7%.
- Nitrogen oxides ($NO_x$) tend to increase in stationary tests, but in dynamic conditions there is no difference between esters and diesel fuel. Other authors [11] have reported

a 10% decrease. It has been observed that a suitable increase in the injection timing can reduce $NO_x$ by over 50% [27]. If an engine is optimized to minimize these emissions, each unit of $NO_x$ reduction will routinely result in a unit increase of particulate matter, and vice versa.

- Sulfur oxides ($SO_x$) are not present.
- Particulate matter is considerably less, while exhaust opacity is half that of diesel fuel and depends on the technology level and operating conditions of the engine.
- Polycyclic aromatic hydrocarbons are roughly the same as those for diesel fuel.

For ester/diesel fuel blends, a significant test has been conducted with an arterial cycle and an Environmental Protection Agency (EPA) cycle [11]. The former has eight repetitions of acceleration to 64 km/h and deceleration to 0 km/h, while the latter is a standard EPA method (Code of Federal Regulations 40, part 86, Appendix I, Cycle D). The results are presented in Table 3.27.

***Other Derived Products***

Vegetable oils can also be used as lubricating and hydraulic oils. The products can be blends of refined oil and additives (to avoid oxidation and the formation of emulsions) or esters of vegetable oils and alcohols with long carbon chains (over 10 carbon atoms). The fields of application are forestry and agricultural machines, outboard motors, wood industry, and metal processing. The advantage is the high biodegradability and, for some products, lower price.

**Table 3.27. Emission percent increase (+) or decrease (−) for blends of rape ethyl ester (REE) and methyl ester (RME) in diesel fuel compared with diesel fuel arterial and EPA cycle [11]**

| | REE | | RME | |
|---|---|---|---|---|
| Emission | Arterial | EPA | Arterial | EPA |
| 20% blend | | | | |
| HC | −19.7[a] | −20.2[a] | −20.2[a] | −18.0[a] |
| CO | −27.6[a] | −34.6[a] | −25.3[a] | −31.3[a] |
| $NO_x$ | −4.1[a] | −6.5[a] | −2.6[a] | −4.1[a] |
| $CO_2$ | +0.3 | +1.5[a] | 0.0 | +1.6 |
| Particulate matter | −5.0 | −0.4 | −10.4 | +7.6 |
| 50% blend | | | | |
| HC | −35.1[a] | −37.2[a] | −37.2[a] | — |
| CO | −43.7[a] | −50.0[a] | −39.1[a] | — |
| $NO_x$ | −8.2[a] | −8.4[a] | −5.3[a] | — |
| $CO_2$ | +0.9 | +0.1[a] | +0.7 | — |
| Particulate matter | +12.1 | +10.8 | +7.6 | — |
| 100% blend | | | | |
| HC | −60.[a] | −55.[a] | −55.[a] | −49.[a] |
| CO | −47.[a] | −53.[a] | −42.[a] | −54.[a] |
| $NO_x$ | −11.[a] | −12.[a] | −9.[a] | −7.[a] |
| $CO_2$ | +1.[a] | +1.[a] | +0.[a] | +0. |
| Particulate matter | +1. | +17. | +6. | +21.[a] |

[a] Significantly different from diesel fuel.

***Standardization***

No studies have been carried on crude oils to define qualitative standards. In fact, for these products, the quality depends on the final use (burners, engines, gas turbines, etc.).

On the other hand, for esters, there are standards that vary from country to country. Germany, France, Austria, Italy, and the USA have defined methods, parameters, and values for ester fuel. At the moment, standards are not related to the final use (burning or engines), but for some characteristics, the differences are substantial (Table 3.23).

## References


1. Mielke, S. 1996. The future trends of the vegetable oils market. Proceedings of the 2nd European Motor Biofuels Forum: 161–164. Graz, A, Joanneum Research.
2. Duke, J. A. and O. Bagby. 1982. Comparison of oilseed yields: a preliminary review. *Proceedings of the International Conference on Plant and Vegetable Oils as Fuels*: 11–23. St. Joseph, MI: ASAE.
3. Mittelbach, M. 1996. The high flexibility of small scale biodiesel plants. Proceedings of the 2nd European Motor Biofuels Forum: 183–188. Graz, A, Joanneum Research.
4. Wan, P. J. 1992. Properties of fats and oil. *Introduction to Fats and Oils Technology*, ed., Wan, P. J., pp. 16–49. New Orleans, LA: SRRC/ARS/USDA.
5. Piccioni, M. 1989. *Animal Feeding Dictionary*. Bologna, Italy: Edizioni Agricole.
6. Tandy, D. 1992. Oilseed extraction. *Introduction to Fats and Oils Technology*, ed. Wan, P. J., pp. 59–84. New Orleans, LA: SRRC/ARS/USDA.
7. Hodgson, A. S. 1996. Refining and bleaching. Bailey's industrial oil and fat products. *Edible Oil and Fat Products: Processing Technology*, vol. 4:157–212. New York: Wiley.
8. Carlson, K. F. 1996. Deodorization. Bailey's industrial oil and fat products. *Edible Oil and Fat Products: Processing Technology*, vol. 4:339–390. New York: Wiley.
9. Duff, H. G. 1992. Refining. *Introduction to Fats and Oils Technology*, ed., Wan, P. J., pp. 85–94. New Orleans, LA: SRRC/ARS/USDA.
10. Freedman, B. and E. H. Pryde. 1982. Fatty esters from vegetable oils for use as a diesel fuel. *Proceedings of the International Conference on Plant and Vegetable Oils as Fuels*: 117–122. St. Joseph, MI: ASAE.
11. Peterson, C. and D. Reece. 1996. Emission characteristics of ethyl and methyl ester of rapeseed oil compared with low sulfur diesel control fuel in a chassis dynamometer test of a pickup truck. *Trans. ASAE* 39:805–816.
12. Vogel & Noot GmbH. 1992. The intelligent alternative. Information paper of Vogel & Noot GmbH. 17 Ruthardweg, Graz, A.
13. De Smet Group 1992. Des huiles vegetales au bio-carburant. Information paper of De Smet Group: T. V. A., No. 408.353.568.
14. Auld, D. L., B. L. Bettis, and C. L. Peterson. 1982. Production and fuel characteristics of vegetable oil from oilseed crops in the Pacific Northwest. *Proceedings of the International Conference on Plant and Vegetable Oils as Fuels*: 92–97. St. Joseph MI: ASAE.

15. Thair, A. R., H. M. Lapp, and L. C. Buchanan. 1982. Sunflower oil as a fuel for compression ignition engines. *Proceedings of the International Conference on Plant and Vegetable Oils as Fuels*: 82–91. St. Joseph, MI: ASAE.
16. Klopfenstein, W. E. 1985. Effect of molecular weights of fatty acid esters on cetane numbers as diesel fuels. *J. Am. Oil Chem. Soc.* 62:1029–1031.
17. Freedman, B. and M. O. Bagby. 1989. Heats of combustion of fatty esters and triglycerides. *J. Am. Oil Chem. Soc.* 66:1601–1605.
18. Romano, S. 1982. Vegetable oils—a new alternative. *Proceedings of the International Conference on Plant and Vegetable Oils as Fuels*: 106–116. St. Joseph, MI: ASAE.
19. Bondioli P., C. Mariani, E. Fedeli, M. Sala, and S. Veronese. 1994. Vegetable oil derivatives as diesel fuel substitutes. Analytical aspects. Note 4: determination of methanol. *Riv. Ital. Sostanze Grasse* LXXI:287–290.
20. National Biodiesel Board. 1996. *Biodiesel Lubricity*. Available at Internet address: http://spectre.ag.uiuc.edu/~nbb/lubricity.html.
21. Morton, R. 1993. Rapeseed oil for combustion—practical trials. Project Report No. OS6. London: Peakdale Engineering Limited.
22. Elsbett, K., L. Elsbett, G. Elsbett, and T. Kaiser. 1985. Vegetable oil utilization—requirements for the engine and experimental results. First technical consultation of the CNRE on biomass conversion for energy (el/9), 14–17 October 1985, Freising, Germany: FAO.
23. Vinyard, S., E. S. Renoll, J. S. Goodling, L. Hawkins, and R. C. Bunt. 1982. Properties and performance testing with blends of biomass alcohols, vegetable oils and diesel fuel. *Proceedings of the International Conference on Plant and Vegetable Oils as Fuels*: 287–293. St. Joseph, MI: ASAE.
24. Hemmerlein, N., K. Volker, and H. Richter. 1991. Performance, exhaust emissions and durability of modern diesel engines running on rapeseed oil. SAE Technical Papers 910848. Warrendale, PA: Society of Automotive Engineers.
25. Pignatelli, V. 1996. Biodiesel for heating purposes in Italy. *The Liquid Biofuels Newsletter*, No. 4: pp. 15–18. Wieselburg, Austria: BLT.
26. Sams, T. and J. Tieber. 1996. Use of rape- and used-frying-oil-methyl ester under real world engine operation. Information paper of the Technical University of Graz (Austria).
27. Syassen, O. 1996. The development potential of diesel engines with biodiesel as fuels. Proceedings of the 2nd European Motor Biofuels Forum: 191–202. Graz, A, Joanneum Research.

## 3.2 Biomass Gas Fuels

### 3.2.1 Methane

*T. Maekawa*

Methane as a biomass gas fuel does not currently play a role as an alternative energy resource. It is now famous for the greenhouse effect because methane is a natural

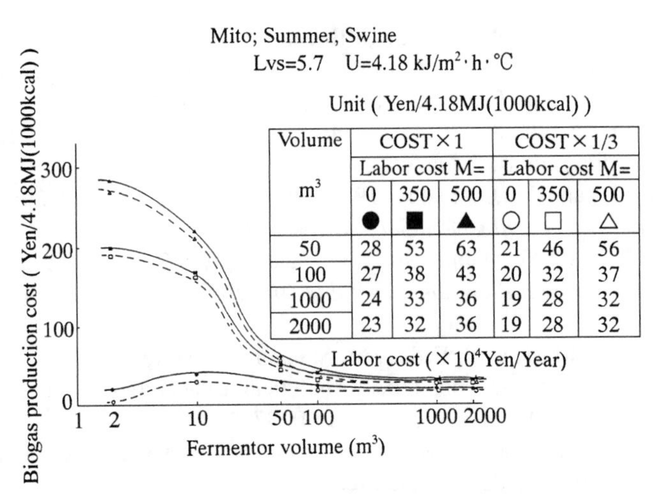


| Volume m³ | COST×1 | | | COST×1/3 | | |
|---|---|---|---|---|---|---|
| | Labor cost M= | | | Labor cost M= | | |
| | 0 ● | 350 ■ | 500 ▲ | 0 ○ | 350 □ | 500 △ |
| 50 | 28 | 53 | 63 | 21 | 46 | 56 |
| 100 | 27 | 38 | 43 | 20 | 32 | 37 |
| 1000 | 24 | 33 | 36 | 19 | 28 | 32 |
| 2000 | 23 | 32 | 36 | 19 | 28 | 32 |

**Figure 3.19. Biogas production cost [Yen/4.18 MJ (1000 kcal)].**

component of the atmosphere, arising from forest fires, swamps, and other wetland. The ghostly will-o'-the wisp that inhabits swamps is nothing but burning methane emitted from the swamp. The methane has caught fire and continues to burn as the gas seeps out of the watery zone. Methane also seeps out of landfills and paddy rice. The gas is also produced in the guts of the ruminant species we use for food such as sheep, steers, cattle, and water buffalo. Fortunately methane molecules survive in the atmosphere only ~10 years, so efforts to control methane emissions would have a near-term pay-off. Although the methane is not a dominant energy resource, methane fermentation is one of the effective methods in biomass conversions and is also effective for wastewater management. It is a special production system for bioenergy production and wastewater treatment in rural areas. The cost analysis concluded by Maekawa [1] indicated that biogas production cost per 4.18 MJ (1000 kcal) generated from a methane fermentor of less than 100 $m^3$ was higher than the cost of fossil oil, as shown in Fig. 3.19. Therefore, since 1985, we have been studying the development of a higher-performance methane fermentation system.

***Outline of Methane Fermentation Technology***

The methane fermentation system was introduced from Southeast Asia and India, where it is in practical use. As most of these methane fermentation systems do not have auxiliary heating systems, there are interruptions of fermentation in the winter when the liquid temperature in the fermentor falls low. Consequently, the methane fermentation systems in advanced countries soon disappeared; soil-fired heating became popular. After the first oil crisis in 1973, some innovative farmers joined those who were concerned about energy problems and began to develop a sophisticated methane fermentation system.

The conventional methane fermentation system has traditionally been used for both acid and methane fermentation in the same bioreactor. In particular, it was estimated that there is little possibility of extra production of methane in cold areas even if the

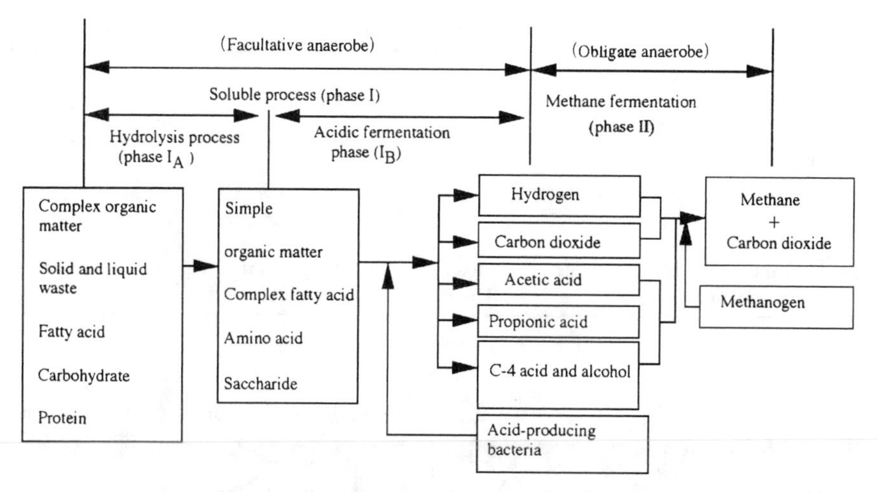


**Figure 3.20. Principles of methane fermentation.**

biogas produced in winter can be used as fuel for heating the bioreactor. The method of separate fermentation of acid and methane, which has been studied by Ghosh *et al.* [2] and Maekawa *et al.* [3], was more effective in increasing the optimum loading of organic matter and in reducing the hydraulic retention time (HRT). This is the so-called two-phase fermentation system, which has been studied for the development of on-site waste management systems.

***Principles of Methane Fermentation***

*Facultative and Obligate Microbes*

A general concept of methane fermentation process is shown in Fig. 3.20 [4]. The organic matter, such as bodies of animals and plants, livestock and human feces, food industry wastes, and garbage waste are degraded from complex organic matter to simple organic matter by facultative anaeorobes. This process is composed of two processes: a hydrolysis process (phase $I_A$) and an acidic fermentation process (phase $I_B$). Low fatty acids such as acetic acids, propionic acid, alcohol, hydrogen, and carbon dioxide are degraded from simple organic matter by acid-producing bacteria in the process of phase $I_B$. Intermediates such as low fatty acids, hydrogen, and carbon dioxide, are converted to digested gas (biogas), which is composed of methane, carbon dioxide, and hydrogen sulfide, by obligate microbes including methanogens. This third process is called methane fermentation (phase II).

*Methane Fermentation as Symbiosis*

In conventional fermentation, the facultative and the obligate anaerobes belonging the methane bacteria that lived symbiotically in the fermentor, participated in fermentation, and decomposed substrates. In this case, as shown in Table 3.28, [4], there was insufficient growth of the anaerobes and a subsequent washout that caused fermentation either to fail or to become unstable when continuous fermentation was performed. This was especially true when it was operated with a high loading of organic matter and with a reduced retention time because of the differences between facultative and

**Table 3.28. Optimum fermentation conditions**

| Phase | Process | Light | Oxygen | Temperature (°C) | Volatile Acid | pH | Oxidation-Reduction Potential (mV) |
|---|---|---|---|---|---|---|---|
| I | Hydrolysis and acid formation | Dark | Facultative | 30–40 | 2%–4% | 4–4.5 | +100~ −100 |
| II | Methane fermentation | Dark | Obligate | Mesophilic: 30–40<br>Thermophilic: 50–55 | less than 3000 mg/L | 6.5–7.5 | −150~ −400 |

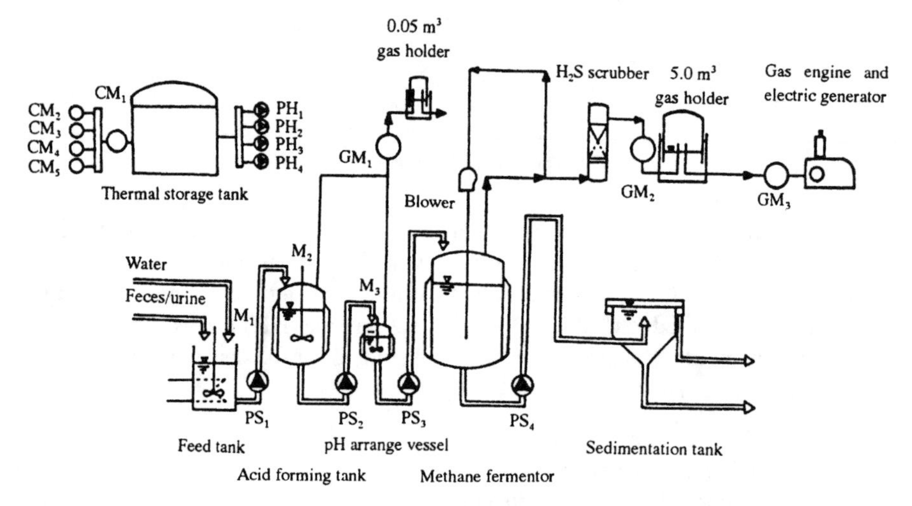


**Figure 3.21. Schematic diagram of two-phase methane fermentation system. CM, calorie meter; PH, hot water pump; PS, slurry pump; GM, gas flow meter.**

obligate anaerobes, the differences of their optimum environments for growth such as content of volatile organic acid, pH value, oxidation-reduction potential (ORP), etc., and because of the difference of minimum generation time between facultative and obligate anaerobes.

*Phase Separation*

A two-phase methane fermentor that connects two different volume reactors in series (Fig. 3.21) separates the process of degradation of organic matter into phase I and phase II by a controlled HRT. The two-phase methane fermentation system was meant to give separate places to facultative and obligate anaerobes by controlling the HRT, and it measures in days [or their reciprocal, which is defined as dilution rate D ($day^{-1}$)] in the operation of continuous fermentation since the minimum generation time of the facultative anaerobes was ~0.5 days and that of the obligate anaerobes was from 2.4 to 4.8 days. In the first small reactor, facultative microbes degraded complex organic matter to simple organic matter. In the second large reactor, obligate microbes predominantly convert the simple organic matter into the mixture gas of methane and carbon dioxide [3].

**Table 3.29. Classified table of methanogens**

| Group of Methanogens (3 Orders, 6 Families, 16 Genera, 44 Species) | | | |
|---|---|---|---|
| *Methanobacteriales* | Gram-positive, pseudomurein | | |
| *Methanobacteriaceae* | $H_2 + CO_2$/formic acid | | |
| *Methanobacterium* | | Long rods | 8 species |
| *Methanobrevibacter* | | Short rods | 3 species (intestinal rumen) |
| *Methanospliaera* | | Coccus | 1 species |
| *Methanothermaceae* | | | |
| *Methanothermus* | | Long rods | 2 species (viable at 90°C) |
| *Methanococcales* | | | |
| *Methanociccaceae* | $H_2 + CO_2$/formic acid | | |
| *Methanococcus* | | Coccus | 6 species (marine species) |
| *Methanomicrobiales* | Gram-negative | | |
| *Methanomicrobiaceae* | $H_2 + CO_2$/formic acid | | |
| *Methanomicrobium* | | Short rods | 2 species |
| *Methanogenium* | | Coccus | 7 species |
| *Methanospirillum* | | Long rods | 1 species |
| *Methanoplanaceae* | | | |
| *Methanoplanus* | | Plate | 2 species |
| *Methanosarcinaceae* | Acetic acid/MeOH | | |
| *Methanosarcina* | | Irregular coccus in packets | 5 species gram-positive |
| *Methanothrix* | | Mold fungi (producing methane mainly from acetic acid) | 1 species |
| *Methanolobus* | | Coccus | 1 species |
| *Methanococcoides* | | Coccus | 1 species |
| *Methanohalophilus* | | | 2 species |
| *Methanocorpasculum* | | | 1 species |
| *Methanohalobium* | | | 1species |

*Lithotrophic Synthesis of Methanogens with $CO_2$ and $H_2$ as Substrates*

Most methanogens synthesize the mixed gas of $CO_2$ and $H_2$ into methane. In the biodegradation of low fatty acids, a few methanogens can convert them into biogas. Classified methanogens are shown in Table 3.29 [5]. It has been found that 44 species of methanogens exist. Most methanogens synthesize methane from the mixed gas of $H_2$ and $CO_2$, as shown in the Table 3.29. Metanogens that can metabolize the low fatty acids are only two genera and 6 species. The two genera *Methanosarcina* and *Methanothrix* contribute to the degradation of organic matter.

***Required Operational Condition for Methane Fermentation***

*Loading of Organic Matter ($L_{VS}$: kg-vs/m$^3$/day, Where vs Is a Volatile Solid)*

The optimum loading of organic matter between a conventional methane fermentor and a two-phase one are quite different, as shown in Fig. 3.22. This figure is a result of mesophilic methane fermentation experiments with pig waste as raw materials compared with the conventional and the two-phase systems [3]. For the conventional system, the optimum loading of organic matter is 3.0 kg-vs/m$^3$/day; the operating condition is limited

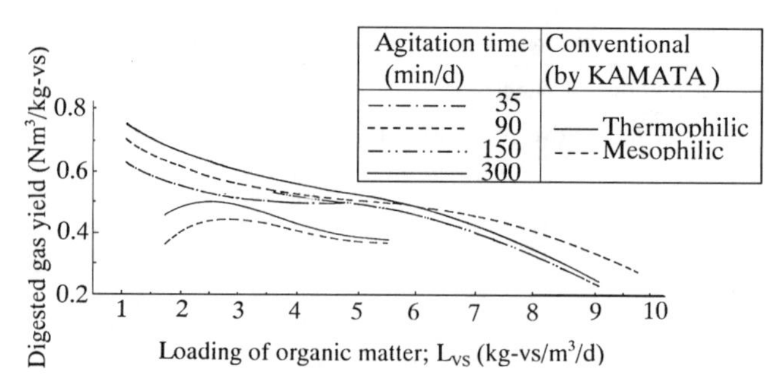


**Figure 3.22. Digested gas yield versus loading of organic matter.**

in the pH range of 6.5–7.5, and the HRT is in the range of 20–30 days. On the other hand, the digested gas yield from the two-phase methane fermentor is 20%–30% higher than that of the conventional, one, as shown in Fig. 3.22. Also, the optimum loading of organic matter turns out to be 6.0 kg-vs/m$^3$/day, which is twice that of the conventional methane fermentor, as shown in Fig. 3.23, which is derived from the results shown in Fig. 3.22 and the data of the methane concentration in the digested gas. The pH's of the acidgenesis (acid former fermentor) and methanogenesis (methane fermentor) are in the ranges of 6.0–6.5 and 7.5–8.0, respectively. The pH of acidgenesis is not so low because the ammonium ion that comes from the biodegradation of protein and urine contained in pig waste undergoes a chemical reaction with carbon dioxide produced from a carbohydrate to make ammonium carbonate.

*Temperature of Methane Fermentation*

The ranges of temperature for methane fermentation are roughly divided into two categories, as shown in Fig. 3.24: mesophilic fermentation for which the peak (optimum) ranged from 35 to 37°C and thermophilic one for which the peak is ranged from 54 to 56°C [6]. In recent studies, psychrophilic fermentation has been tried in the temperature range of 5–20°C from the standpoint of the improvement of heat balance or taking into account the development of a nonheating methane fermentation system [7, 8]. It has been reported that, for the temperature control, a fluctuation rate of fermentor temperature of more than ±2–3°C/h is unacceptable for methanogens because they are so sensitive to temperature change. However, the effect of seasonal changes on methanogens is not apparent to any marked extent from observations of the on-site system of Japanese farmers and the experimental farms [9].

*Nutrient Conditions of the Substrates*

Nutrients are important for methanogens to grow in the fermentor. Acid degradation methanogens, such as the genera of *Methanosarcina* and *Methanothrix*, metabolize the organic carbon sources as their energy for the growth and cellular maintenance of microbes. Lithotrophic synthesis methanogens obtain metabolic energy for their growth and maintenance from hydrogen. In the case of methane fermentation, all methanogens need inorganic nutrients such as nitrogen, phosphorus, and trace metals that are taken into microbe cells.

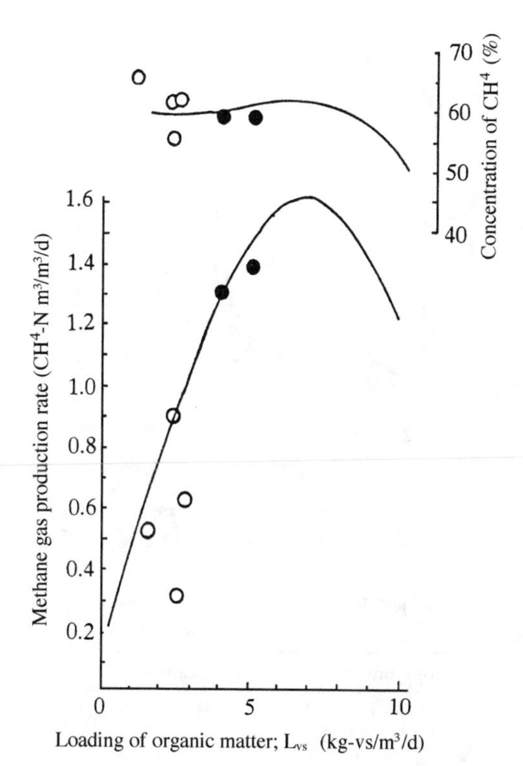


**Figure 3.23. Methane gas production rate and concentration of $CH_4$ versus loading of organic matter (solid curves, swine feces; open circles, swine feces plus swine urine; filled circles, liquid temperatures in feed tank varied by 20 and 30°C).**

***a. Synthesis Wastewater Used for the Laboratory Experiments.*** Actual wastewater has the resources of inorganic nutrients in itself: However, in laboratory work on methane fermentation, these nutrients must be artificially arranged, as shown in Tables 3.30 and Table 3.31 [10,11].

***b. Actual Wastewater Treated by an On-Site Methane Fermentor.*** The operation performance of an on-site methane fermentor is affected by the nutrient composition of treated wastewater. Protein contained in the wastewater is reduced into ammonium nitrogen in the fermentor under anaerobic conditions, and the ammonium gives the growth inhibition of methanogens. The concentration of the upper limit is well known: 3000 mg/L for ammonium [9]. The composition of raw material is quite different, as shown in Table 3.32. The C/N ratio is very low. The optimum ratio for the growth of methanogens is in the range of 10–20. When a C/N ratio lower than 10 is obtained, we should make sense of the concentration of ammonium nitrogen to keep within 3000 mg/L, as given in the following equations:

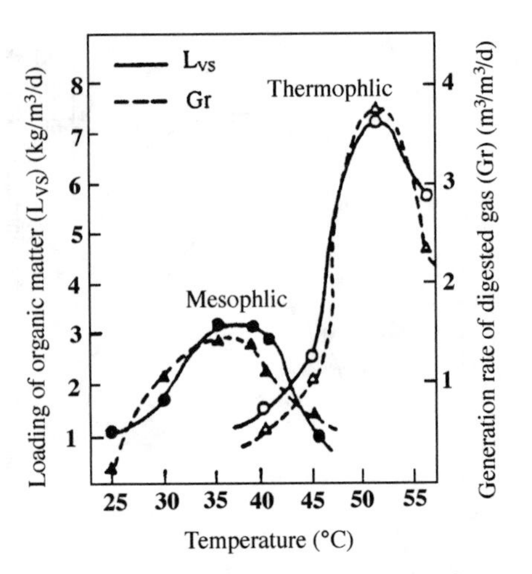


**Figure 3.24. Optimum temperature for methane fermentation [6].**

**Table 3.30. Composition of organic and inorganic nutrition in synthetic wastewater**

| Component | Concentration (mg/L) |
|---|---|
| $(NH_4)_2\ HPO_4$ | 700 |
| $MgSO_4 \cdot 7H_2O$ | 750 |
| $NH_4Cl$ | 850 |
| KCl | 750 |
| $MgCl \cdot 6H_2O$ | 810 |
| $FeCl_2 \cdot 6H_2O$ | 420 |
| $CoCl_3 \cdot 6H_2O$ | 18 |
| $NaHCO_3$ | 6720 |
| $CH_3COOH$ | 3000 |

**Table 3.31. Liquid media composition for methanogens**

| | | | |
|---|---|---|---|
| $KH_2PO_4$ | 50 mM | $NiCl_2 \cdot 6H_2O$ | 5.0 $\mu$M |
| $NH_4Cl$ | 40 mM | $Na_2MoO_4 \cdot 2H_2O$ | 1.0 $\mu$M |
| $(HOCOCH_2)_3N$ | 0.5 mM | Resazurin | 20 $\mu$M |
| $MgCl_2 \cdot H_2O$ | 0.2 mM | $Na_2CO_3$ | 24 mM |
| $FeCl_3$ | 50 $\mu$M | *L*-Cysteine | 1.0 mM |
| $CoCl_2$ | 1.0 $\mu$M | $Na_2S \cdot 9H_2O$ | 1.0 mM |

The ammonium concentration $C_{NW}$, produced in methane fermentor from the substrate, is determined by Eq. (3.5). Usually, as a criterion of the design of the fermentor, we set the maximum concentration of ammonium at the level of 3800 mg/L (~3100 mg/L as $NH_4$-N) to avoid ammonium inhibition [10]:

$$C_{NW} = [C_{TN} - (C_{TOC}/\text{C/N ratio})]18/14R, \tag{3.5}$$

**Table 3.32. Analysis of the screened raw wastewater**

| Parameter | Distillery Wastewater | Pig Wastewater |
|---|---|---|
| pH | 3.7–4.1 | 6.5–7.0 |
| Total solids (%) | 7.6–8.6 | 15.0–16.5 |
| Volatile solids (%) | 7.3–8.4 | 13.2–14.3 |
| Biological oxygen demand (mg/L) | 49,000–83,000 | 55,000–80,000 |
| Chemical oxygen demand(mg/L) | 56,000–97,000 | 92,000–100,000 |
| Total N(mg/L) | 5200–6000 | 41,000–49,000 |
| Total Organic Carbon (mg/L) | 23,000–27,000 | 9800–11,000 |
| C/N ratio | 4.4–4.5 | 4.2–4.5 |

where $C_{TN}$ is the nitrogen content in the wastewater (in milligrams per liter), $C_{TOC}$ is the total carbon content in the wastewater (in milligrams per liter), $O$ and $R$ is the ratio of biodegradation of organic matter.

The ammonium concentration $D_{NM}$, produced from decomposed dead microbes, must be considered in the continuously operated reactor. This is determined by Eq. (3.6) in the steady-state condition:

$$D_{NM} = CR_{NM} K_D 18/14 C_{VSS}/F_R, \tag{3.6}$$

where $CR_{NM}$ is the content ratio of nitrogen to the microbes, and this ratio is based on the composition formula of the microbes, $C_5H_9O_3N$ [12], $K_D$ is the decay coefficient (in inverse days), $C_{VSS}$ is the density of microbes cell (in grams dry matter per liter), and $F_R$ is the flow rate (in inverse days).

When a bioreactor operation condition reaches the steady state, we occasionally meet an ammonium inhibition, especially in the treatment of livestock wastewater. We should monitor the concentration of ammonium in the liquid of bioreactor and reference the calculated $C_{NW}$ from Eq. (3.5) with the actual concentration of ammonium. When the ammonium concentration reaches or is above 3800 mg/L, the effect of ammonium nitrogen caused by the death of microbes occurs in the bioreactor, as shown in Eq. (3.6).

***c. Ammonium Inhibition.*** Inhibition of methane fermentation by ammonium nitrogen has been investigated by Albertson [13] and Lapp *et al.*, [9] who reported that it began to appear at 1500–1800 mg/L as $NH_4$-N and completely inhibited at 3000 mg/L as $NH_4$-N. Maekawa and Kitamura [10] also agree with these descriptions in the above reports, which have same experimental results as those shown in Fig. 3.25. In the figure Pi is an average amount of digested gas evolution in a stabilized condition that is not inhibited by ammonium and P represents digested gas generation inhibited by ammonium. From the progress of P/Pi, the digested gas evolution oscillated in the 2100–3350-mg/L range of ammonium concentration, and the cycle accelerated and the oscillation amplitude increased as the ammonium concentration rose. Also, the complete inhibition of fermentation shows at the concentration of 3800 mg/L as ammonium.

Van Velsen [14] and Van Velsen and Lettinga [15] reported that methane fermentation is better stabilized when methanogens have been sufficiently acclimated in a high ammonium concentration. Maekawa and Kitamura [10] tried long-period acclimation for more than 3 months under the ammonium concentration close to 3800 mg/L as ammonium.

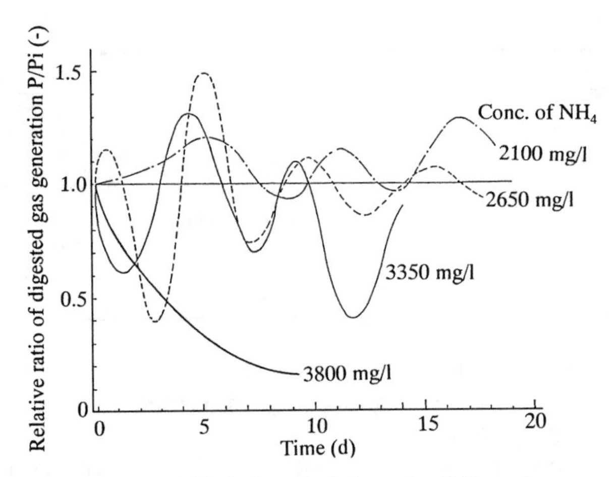


**Figure 3.25. Variation of relative ratio of digested gas generation.**

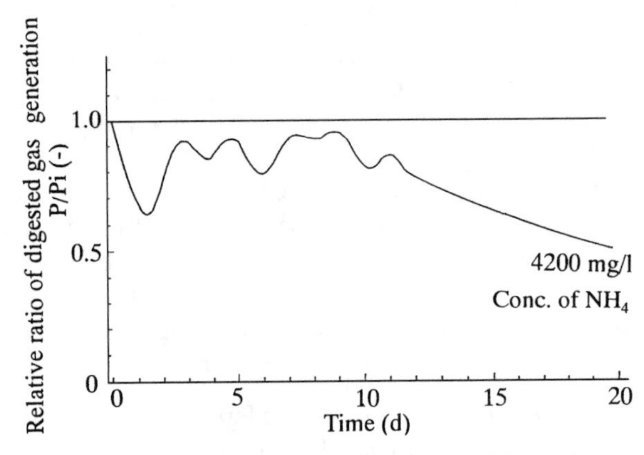


**Figure 3.26. Variation of relative ratio of digested gas generation.**

Figure 3.26 shows that at the 4200 mg/L of $NH_4$, digested gas evolution oscillated for 12 days and then decreased the same as 3800 mg/L of ammonium as shown in Fig. 3.25.

***d. Volatile Fatty Acid (VFA) Inhibition.*** Optimum concentration of VFA in the methanogenesis process is 3000 mg/L, as shown in the Table 3.28. Aguilar *et al.* [16] reported that VFA's, such as propionic, *n*-butyric, and *n*-valeric at concentrations up to 4 g/L caused no inhibition on the acetate degradation by methanogens and that 10g/L were needed to obtain a significant inhibition. Therefore, VFA's would not act as inhibitors of acetate degradations in the conditions normally designed for anaerobic continuous reactors.

*Trace Metals for Methanogens*

Studies on trace metals have not been investigated, but many investigations have been made from the standpoint of physiological toxicants and cell components for methanogens. Peter *et al.* [17] reported that the growth of *Methanobacterium thermoautrophiccum* is very low in the case in which the medium lacked salts of Ni, Co, Mo, and Fe. Diekert *et al.* [18], Graf and There [19], Schere and Sahm [20], and Dixon *et al.* [21] pointed out Ni, which is a component of the coenzyme $F_{430}$ plays the role of methane synthesis at the end of the metabolic pathway in methanogens. Mah *et al.* [22] reported that Mn and Cu are effective for the growth of the acetate-fermenting methanogen *Methanosarcina barkeri* 227. On the other hand, Schere and Sahm [20] reported that Mn is not effective for the growth of *M. barkeri Fusaro*. Although there were some data on the role of Cu, Schere [23] pointed out this metal is important for the growth of *Methanobacterium bryantii* because this methanogen contained 160 paits in $10^6$ of Cu in the dry cells. However, we do not have any information on the physiological role of Cu salts. Recently, Zhang and Maekawa [24] showed that the salts of Fe and Mg are limiting factors for the growth of lithotrophic methanogens (with $H_2$ and $CO_2$ as substrates), and they also showed that the amount of $NH_4^+$ and trace metals, including vitamins added to the medium on the growth of the methanogens are effective. They obtained the highest cell density, a 26-g dry cell/land methane production rate, 21.51/L h under the sufficient mass transfer capacity of the substrate gas ($H_2/CO_2$) from the gas phase to the liquid and fully supplying nitrogen ($NH_4^+$) and sulfur sources.

The improved solution on the trace metals is shown in Table 3.33. Maekawa *et al.* [25] studied the effect of higher concentrations of trace metal solutions than those of Tables 3.33 and 3.34 for acetate fermentation. They found that, at 150 times the trace metal solution that showed the inhibition of the metals only at the start-up, they obtained a result of 3.7 times of methane formation than that of Table 3.35.

**Table 3.33. Composition of trace metal solution**

| Component | Concentration ($\mu$g/L) |
|---|---|
| $MgCl \cdot 6H_2O$ | 410 |
| $MnCl \cdot 4H_2O$ | 50 |
| $FeCl_2 \cdot 4H_2O$ | 50 |
| $NiCl_2 \cdot 6H_2O$ | 12 |
| $ZnSO_4 \cdot 7H_2O$ | 10 |
| $CoCl_2 \cdot 2H_2O$ | 10 |
| $CaCl_2 \cdot 6H_2O$ | 10 |
| $Na_2SeO_3$ | 8 |
| $Na_2MoO_4 \cdot 2H_2O$ | 2.4 |
| $CuSO_4 \cdot 5H_2O$ | 1 |
| $AlK(SO_4)_2$ | 1 |
| $H_3BO_3$ | 1.8 |
| $NaWO_4 \cdot 2H_2O$ | 1 |

Table 3.34. Vitamin solution

| Component | Concentration ($\mu$g/L) |
|---|---|
| Viotin | 20 |
| Folic acid | 20 |
| Pyridoxine hydrochloride | 100 |
| Thiamine hydrochloride | 50 |
| Riboflavin | 50 |
| Ricotinic acid | 50 |
| DL-calcium pantothenate | 50 |
| *p*-Aminobenzoic acid | 50 |
| Lipoic acid | 50 |

Table 3.35. Basic inorganic salts

| Component | Concentration (mg/L) |
|---|---|
| $KH_2PO_4$ | 3400 |
| $K_2HPO_4$ | 3400 |
| $NH_4Cl$ | 2130 |
| $Na_2CO_3$ | 2540 |
| Resazurin | 2 |

*Effect of Methanogen Density on Biogas Production*

From the kinetic study of a continuously stirred tank reactor, the rate of substrate consumption $(\mathrm{d}S/\mathrm{d}t)_C$ is

$$(\mathrm{d}S/\mathrm{d}t)_C = -\mu X/Y_{X/S}, \tag{3.7}$$

where $\mu$ is the specific growth rate of methanogens (in per day), $X$ is the density of methanogens (in gram dry matter per liter), and $Y_{X/S}$ is cellular yield of methanogens (negative).

As shown in Eq. (3.7), when $\mu$ is constant, the density of the methanogens should be kept at a higher level in order to increase the rate of substrate consumption. When $\mu$ is shown by the Monod Eq. (3.8), it is affected by ammonium inhibition and loses the specific growth rate:

$$\mu = \mu_{\max}/[1 + K_S(1 + \mathrm{NH}_4^+/K_N)/\mathrm{VFA}], \tag{3.8}$$

where $\mu_{\max}$ is the maximum specific growth rate (in inverse days), $K_S$ is the constant of substrate saturation (in milligrams per liter), $K_N$ is the inhibition coefficient of free ionized ammonium (in milligrams per liter), $NH_4$ is the concentration of free ionized acid (in milligrams per liter), and VFA is the concentration of the substrate of methanogens (in milligrams per liter).

As $\mu_{\max}$ and $Y_{X/S}$ depend on the species of the methanogens, their values are constant in the operation of the fermentor. When we can control the environment in the fermentor, such as pH and composition of the raw material and fermented liquid, keeping the environment at, e.g., VFA $\gg K_S$ and $K_N \gg \mathrm{NH}_4^+$, we obtain $\mu = \mu_{\max}$ in that operation. For the purpose of accomplishment of a high-performance methane fermentor, we should focus on the cell density of methanogens in the fermentor. From the engineering aspect,

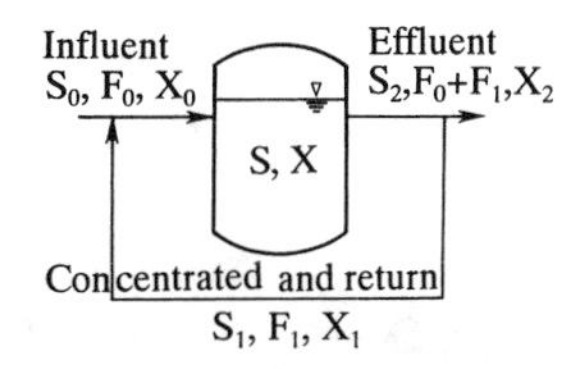


**Figure 3.27. Material flow model for kinetic evaluation.**

it is the sole method for making a high density of methanogens in the fermentor, as we can see in Eq. (3.7). Zhang and Maekawa [24] showed that this hypothesis is possible in their experiments on lithotrophic methanogen fermentation by using mixed gas of $H_2$ and $CO_2$ as the substrate, as mentioned above. There are other ways to obtain a high-density methanogen cell in the bioreactor engineering: UASB (microbial granules) and fluidized-bed and fixed-bed reactors that use carriers with a high density of cells.

***Kinetic Analysis of Methane Fermentation***

*Modeling of Growth Formula*

For the kinetic analysis, we checked the data of a perfectly mixed continuously stirred tank reactor, which is the same as an ideal chemostat. Hill [26] and many investigators showed that their kinetic analyses are based on this chemostat. Kitamura and Maekawa [27] explained a method of kinetic analysis for the fluidized-bed and fixed-bed methane fermentors, which have a high density of methanogens. These fermentors show their density of methanogens as a difference between that of the reactor and the effluent, as shown in Fig. 3.27. The basic equation based on the mass balance of the cell of the methanogens is

$$V(\mathrm{d}X/\mathrm{d}t) = F_0X_0 + F_1X_1 - (F_0 + F_1)X_2 + (\mathrm{d}X/\mathrm{d}t)_G - K_DXV, \qquad (3.9)$$

where $V$ is the working volume of the reactor (in liters), $X$ is the density of the methanogen cell (in milligrams per liter), $t$ is time (in days), $F$ is the flow rate (in liters per day), and $K_D$ is the decay coefficient (in inverse day).

Dividing Eq. (3.9) divided by $F$ and putting $X_0 = 0$, $F_1 = 0$, and $X_1 = 0$, we obtain

$$V(\mathrm{d}X/\mathrm{d}t) = -F_0X_2 + V(\mathrm{d}X/\mathrm{d}t)_G - K_DXV. \qquad (3.10)$$

In the steady state ($\mathrm{d}X/\mathrm{d}t = 0$ in the reactor),

$$-F_0X_2 + V(\mathrm{d}X/\mathrm{d}t)_G - K_DXV = 0. \qquad (3.11)$$

On the other hand,

$$X = [X'V' + X_2(V - V')]/V, \qquad (3.12)$$

where $X'$ and $V'$ express the cell density on the carrier and the volume of the carrier in the fludized-bed and the fixed-bed reactor, respectively.

Equation (3.12) can be expressed as

$$(\mathrm{d}X/\mathrm{d}t)_G = \mu[(X'V' + X_2(V - V')]/V. \qquad (3.13)$$

From Eqs. (3.11)–(3.13),

$$\mu = F_0X_2/[X'V' + X_2(V - V')] + K_D. \tag{3.14}$$

*Mass Balance Based on the Substrate*

$$V(\mathrm{d}S/\mathrm{d}t) = F_0S_0 - F_0S_2 - V(-\mathrm{d}S/\mathrm{d}t)_C, \tag{3.15}$$

where $S$ is the concentration of the substrate (in milligrams per liter) and $c$ is the consumption of the substrate.

In the steady state ($\mathrm{d}S/\mathrm{d}t = 0$),

$$(S_0 - S_2)F_0/V - (-\mathrm{d}S/\mathrm{d}t)_C = 0. \tag{3.16}$$

As the metabolic energy of the cellular maintenance in the case of high-rate methane fermentation is negligible, we can express the coefficient of cell growth yield as

$$Y_{X/S} = (\mathrm{d}X/\mathrm{d}t)_G/(-\mathrm{d}S/\mathrm{d}t)_C. \tag{3.17}$$

From Eqs. (3.13) and (3.17),

$$(-\mathrm{d}S/\mathrm{d}t)_C = \mu/Y_{X/S}[X'V' + X_2(V - V')]/V. \tag{3.18}$$

Then when Eqs. (3.14) and (3.17) are substituted into Eq. (3.18), the following equation can be described:

$$\begin{aligned} &VD(S_0 - S_2)/[X'V' + X_2(V - V')] \\ &= \{F_0X_2/[X'V' + X_2(V - V')]\}(1/Y_{X/S}) + K_D/Y_{X/S}, \end{aligned} \tag{3.19}$$

where $D$ is the dilution rate (in per day), which as defined with $F_0/V$. When $S_0$, $S_2$, $X'$, $X_2$, $V$, and $D$ are known from operation data, we can obtain values of $Y_{X/S}$ and $K_D$. From this $K_D$, we can also obtain the values of $\mu$. From values of $\mu$ that are given in Eq. (3.14) and from Eq. (3.8), we can describe the kinetic equation as

$$1/\mu = 1/\mu_{\max} + [Ks(1 + NH_4^+/K_N)/\mu_{\max}](1/S_2), \tag{3.20}$$

where $S_2$ is the VFA in the effluent from the fermentor. We can obtain kinetic constants, such as $\mu_{\max}$ and $K_S$, from Eq. (3.20) as $\mu$ and $S_2$ are known. The rate of substrate consumption $\nu_{\max}$ can also be calculated from

$$\mu_{\max} = Y_{X/S}\nu_{\max}. \tag{3.21}$$

Kitamura and Maekawa [27] obtained the data of growth and kinetic constants, as compared with the data of the conventional methane reactor reported by Lawrence *et al.* [28] and Chang *et al.* [29], as shown in Table 3.36.

***On-Site Methane Fermentation Technology***

The cost of biogas is more expensive than that of fossil oil and liquid petroleum gas; however, the methane fermentation system is effective in protecting against environmental pollution and also effective for waste management. We must make efforts to develop a high-performance methane fermentation system to obtain get the necessary social acceptance. Possible strategies are

**Table 3.36. Comparison of growth and kinetic constants**

| Parameter | Fluidized Bed | Conventional A[a] | Conventional B[b] |
|---|---|---|---|
| $Y_G$ (mg/mg) | 0.81 | 0.04 | 0.115 |
| $K_d$ (per day) | 0.357 | 0.019 | 0.283 |
| $\mu_{max}$ (per day) | 0.577 | 0.384 | 0.543 |
| $K_S$ (mg/L) | 70 | 166 | 53 |
| $\nu_{max}$ (mg/mg day) | 0.71 | 9.6 | 4.72 |

[a] [28].
[b] [29].

**Table 3.37. Ratio of produced energy to consumed energy**

| Ratio of Organic Loading | Gas Generation ($m^3$/day) | Gas Consumption for Heating ($m^3$/day) | Available Gas ($m^3$/day) | Consumption of Electric Power (kW) | Ratio | |
|---|---|---|---|---|---|---|
| | | | | | 3.6[a] (MJ/kWh) | 10.1[b] (MJ/kWh) |
| full(1) | 609 | 150 | 459 | 190 | 15.4 | 5.5 |
| 3/4 | 476 | 150 | 326 | 180 | 11.6 | 4.1 |
| 1/2 | 315 | 150 | 160 | 175 | 5.8 | 2.1 |
| 1/4 | 182 | 150 | 32 | 170 | 1.20 | 0.43 |

[a] Based on pure energy unit.
[b] Based on electric plant unit.

1. To obtain a high density of cells in the reactor.
2. To control the concentration of ammonium in the reactor.
3. To improve the heat balance under operation of a psychrophilic methane fermentation .

Based on these strategies, the high density of cells has been accomplished by the Upflow Anaerobic Sludge Blanket, and fluidized-bed and fixed-bed bioreactors. The control of ammonium concentration is to set it at 2000 mg/L by dilution with tap water for raw material [30]. We need the study of psychrophilic fermentation to the heat balance [7].

*Development of an On-Site Two-Phase Fluidized-Bed Methane Fermentor*

In the development and the evaluation of the conventional on-site methane fermentation system, there is a good example that verifies the economic performance of centralized biogas plants in Denmark [31]. The 10 centralized large full-scale biogas plants in Denmark receive manure from several farmers and daily treatment from 50–300 tons of manure each: The addition of waste from slaughterhouses and food industries has been described, as well as the aspects of technical achievements, financing and corporate economy, and cost analysis.

Maekawa, T. [32] reported an on-site two-phase fluidized-bed methane fermentor located in south of Okinawa. This biogas plant was financially supported by the Board Resources and Energy of Ministry of Industry, Japan, and it initiated its operation in 1987. This plant, was a centralized biogas plant, the same as that in Denmark, receiving

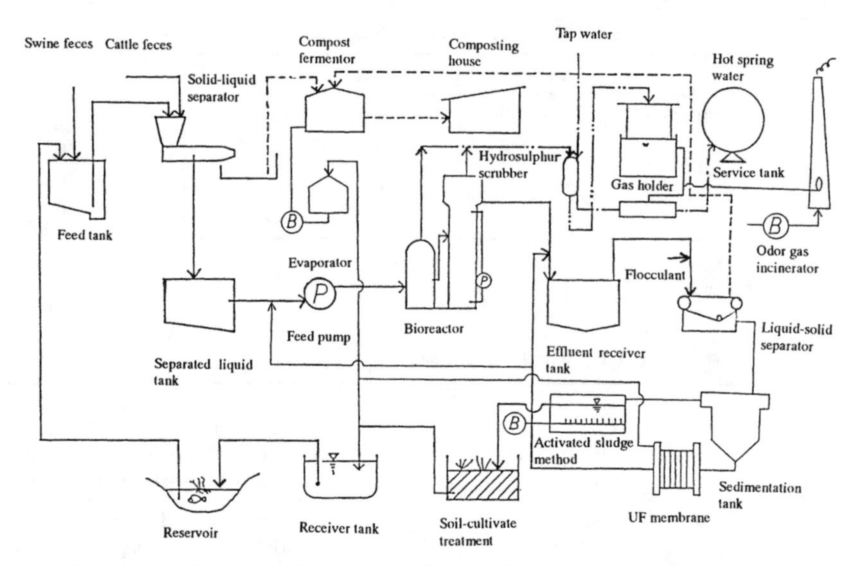


**Figure 3.28. Flow chart of on-site two-phase methane fermentation system located in Okinawa.**

pig manure from five farmers and beef cattle manure from one farmer. Daily treatment of ~12 tons of manure, which was diluted by 8 tons of fresh water and 20 tons/day of raw material, was fed to the plant. The flow chart of this biogas plant is shown in Fig. 3.28. The working volumes of the two reactors of acidogenesis and methanogens are 40 and 160 $m^3$, respectively. The HRT's of their reactors are 2 and 8 days, respectively. The total HRT and loading of organic matter ($L_{VS}$) are designed for 10 days and 6 kg-vs/($m^3$/day) respectively. The designed biogas evolution rate is 600 $m^3$/day at full loading. The maximum volume of biogas consumption for heating is designed for 150 $m^3$/day in winter. The diluted water is designed for reuse by treating the effluent water with the activated sludge method and soil-cultivated treatment incorporated with the Ultra Filter membrane system. The performances of this biogas plant at 80% of full loading are shown in Fig. 3.29. The transient response of the biogas evolution ratio was delayed only 1 day after loading and the rate of biogas evolution was almost 480 $m^3$/day, which accomplished the design values. The biogas was sold out to a hotel as fuel, and 3 ton/day of produced compost was also sold to farmers. The ratio of produced energy to energy consumption in this biogas plant is greater than 1, as shown in Table 3.37. This biogas plant had been operating 7 years with a good cost balance; unfortunately, it was obliged to stop its operation because of the closing of the hotel 3 years ago. As mentioned above, when comparisons are being made among countries, it must be noted that biogas plant dissemination in Japan is favored in respect to energy prices, sales of by-products such as biogas, compost, and liquid fertilizer, financing possibilities, and investment grants.

*Treatment System for Land-Limited Regions*

Land-limited regions such as Japan, Korea, Taiwan, and Malaysia, etc., have water pollution problems from livestock production units. The problems are quite different in

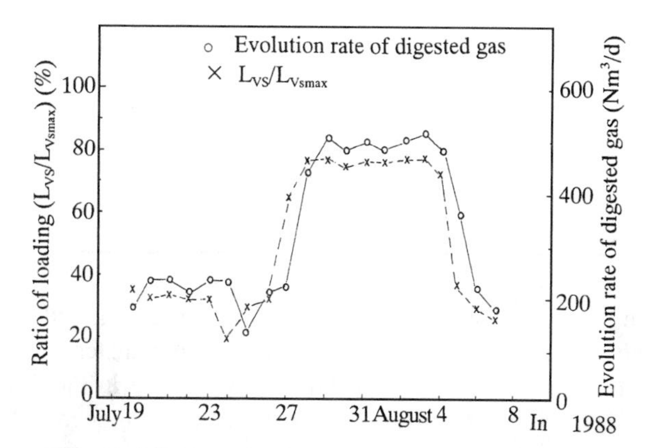


**Figure 3.29. Changes of evolution rate of digested gas on the transient loading ratio ($L_{VS}/L_{VS\ max}$) from 1/3 to 3/4.**

climate, manpower, finance, investment grants, and resources. Consequently, selecting appropriate treatment alternatives for livestock wastewater pollution problems in these countries was difficult for environment and development planners, decision and policy makers, and design engineers [31, 33]. Treatment processes today also have problems such as operation and high running costs. The reason for this is the characteristics of the livestock wastewater itself, its high organic and nutrients contents, as well as an unbalanced carbon:nitrogen:phosphorous ratio. An anaerobic treatment, such as methane fermentation, has a weak point: It is so easy to produce the high concentration of ammonium in the bioreactor that the methane fermentation system needs to install the nitrification and denitrification systems followed by its effluent [34, 35].

*Heat Balance of Methane Fermentation*

Methophilic and thermophilic methane fermentation [1] is generally ongoing. In psychrophilic methane fermentation, to improve obtaining extra usable biogas, the optimum loading condition and the acclimation of methanogenesis to the low-temperature ranges of 10–20°C are studied [7]. Ideal methane fermentation should be processed as nonheating in winter in the temperate and the cold regions.

***a. Calculation of Extra Usable Biogas Ratio.*** It is expected that the loss of heat energy from the methane fermentor is influenced by atmospheric temperature and the values of insulation, as shown Eq. (3.22). Since atmospheric temperatures vary according to locality, three locations were chosen in Japan, as shown in Table 3.38. The overall heat transfer coefficient $U$ from 0 to 3 is used as a parameter. The evolved biogas is supplied by a biogas boiler at 70% efficiency, and the heat energy is supplied by hot water:

$$Q_L = Q_r + Q_h + Q_g, \tag{3.22}$$

where the heating energy of raw material $Q_r = VL_{VS}C_P(t - t_i)/\mathrm{VS}$ (in kilojoules per day), loss of heat transfer $Q_h = UA(t - t_0)24$ (in kilojoules per day), evaporative

**Table 3.38. Atmospheric conditions**

| Location in Japan | Annual Mean Temperature (°C) | Winter Season Temperature (°C) | Summer Season Temperature (°C) | Overall Heat Coefficient (kJ/m$^2$ h °C) |
|---|---|---|---|---|
| Obihiro | 5.9 | −7.3 | 17.5 | $U = 0–13$ |
| Mito | 13.0 | 3.2 | 22.5 | |
| Miyazaki | 16.8 | 7.8 | 25.3 | |

latent heat carried away by evolved gas $Q_g = GVHm$ (in kilojoules per day), $V$ is the volume (in cubic meters), $C_P$ is the specific heat of raw material (in kilojoules per kilogram times degrees Celsius), $t$ is the liquid temperature in the fermentor (in degrees Celsius), $t_i$ is the initial temperature of raw material (in degrees Celsius), VS is the volatile organic matter content, $t_0$ is the atmospheric temperature (in degrees Celsius), $A$ is the surface area of the fermentor ($A = 6V^{2/3}$) (in square meters), $U$ is the overall heat transfer coefficient in kilojoules per square meter times hours times degrees Celsius, $H$ is the latent heat of water on evaporation (in kilojoules per kilogram), and $m$ is the mass of vapor in the digested gas per volume (in kilograms per day).

***b. Calculation of the Production Cost Per 4.18-MJ (1000-kcal) Digested Gas.*** The production cost per 4.18-MJ (1000-kcal) digested gas is calculated as shown in Fig. 3.19. The cost of collecting less than 2 tons/day of raw materials is assumed to be proportional to the total volume of the fermentor; when it exceeds more than 2 tons/day, it is assumed to be proportional to the frequency of being carried by 2-ton dump trucks, and one truck driver will carry four loads daily.

A calculation was made of the total volume of the methane fermentor and the cost per volume from a regression calculation based on the case study data. The construction costs of both conventional and two-phase methane gas fermentation systems that use a heating method is shown as

$$Y = 5.01\mu X^{-0.145}, \tag{3.23}$$

where $\mu$ is the cost coefficient for facilitating the cutting down of construction costs (negative), $X$ is the total volume of the fermentor (in cubic meters), and $Y$ is the construction cost per unit of fermentor volume (in US dollars per cubic meter). In the case of calculation for production cost, the following equation is used:

$$\mathrm{SS} = (\mathrm{SO} + \mathrm{CO} + \mathrm{RI} + \mathrm{Z} + \mathrm{TK})/(\mathrm{SV}8.107), \tag{3.24}$$

where SS is the production cost [US \$/4.18 MJ(1000 kcal)], SO is amortization (construction cost/the life of the plant), RI is the interest of capital (6% compound interest annually), CO are maintenance and management expenses (10% of amortization), Z are personnel expenses (operating and managing the system), Y is the cost of raw material (carrying of raw materials from outside), TK are the expenses of carriage (expenses needed to carry in), and SV is the rate of evolved methane (cubic meters/annually).

$$\mathrm{SSM} = \mathrm{SS} - \mathrm{HH}/(\mathrm{SV}8.107), \tag{3.25}$$

where SSM is the modified production cost and HH is the price of the digested effluent and the sludge as fertilizer.

(1) Estimation as compost:
1 US ¢/kg digested effluent and sludge

(2) Estimation as chemical fertilizer:
Price of digested effluent and sludge as chemical fertilizer

| | | |
|---|---|---|
| Cattle waste | 4 | US ¢ /kg |
| Swine feces | 9.24 | US ¢ /kg |
| Poultry waste | 14 | US ¢ /kg |

Remarks: Calorific value of pure methane, 33.89 MJ/m$^3$ (8107 kcal/m$^3$).

The relationship between the extra usable digested gas ratio $E$ as defined by Eq. (3.25), and the fermentor total volume is shown in Fig. 3.30. As the total volume of the fermentor expands (as shown in Fig. 3.30), $E$ obviously varies until the overall heat transfer coefficient approaches 0. Also, $E$ is naturally larger in higher summer temperatures than in winter, and its effect is more specific when the fermentor total volume $V$ is from 1–10. The relations between the loading of organic matter and the extra usable digested gas are shown in Fig. 3.31, $E$ rises as the scale of the fermentor expands. The amount of extra

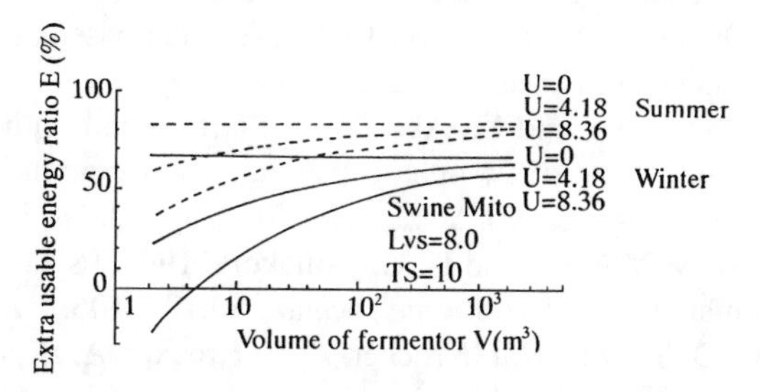


**Figure 3.30. Extra usable energy versus volume of fermentor ($U$: kJ/m$^2$ h °C).**

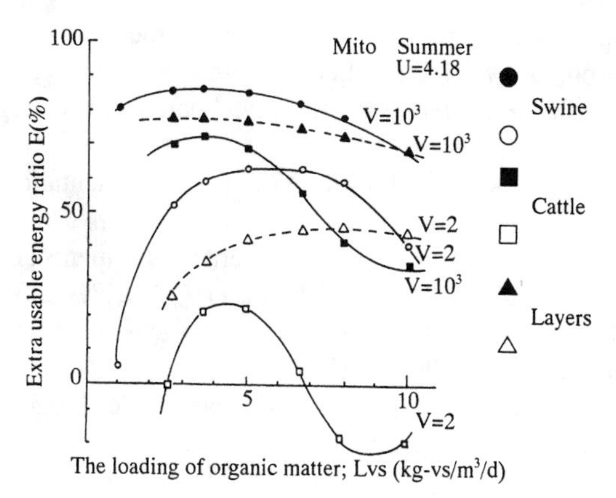


**Figure 3.31. Relation between the loading of organic matter and extra usable energy ($U$: kJ/m$^2$ h °C).**

usable digested gas varies according to the kinds of livestock in the following order: swine feces>layers wastes>cattle wastes. The loading of organic matter at maximum *E* also varies in a similar fashion.

## References

1. Maekawa, T. 1985. Dissemination of methane gas fermentation systems and associated problems in Japan. *Alternative Sources of Energy for Agriculture*. FFTC Book Series 28:163–174.
2. Ghosh, S., J. R. Conrad, and D. L. Klass. 1975. Anaerobic acidogenesis of wastewater sludge. *J. Water Pollut. Control Fed.* 47:30–45.
3. Maekawa, T., S. Yamazawa, S. Yoshikawa, and T. Hanaoka. 1984. On the development of a two-phase methane fermentation system (part 1). *J. Soc. Agric. Struct. Jpn.* 15(1):7–21.
4. Yamauchi, T. and N. Takei. 1980. *Methane Fermentation Engineering and Technology for the Structure and Facilities in Agriculture*, pp. 399–401. Tokyo: Fuji Techno System.
5. Koga, Y. 1988. *Archaebacteria*, part 2, p. 19. Tokyo: University of Tokyo Press.
6. Ono, H. 1962. On the treatment of industrial wastewater based on methane fermentation (part 1). Yousui To Haisui 4(3):233–240.
7. Kurosu, K., T. Maekawa, and M. Abe. 1997. Start-up and optimal organic loading on the plug-flow type of two-phase methane fermentation, using acclimated psychrophilic methanogenesis. *J. Soc. Agric. Struc. Jpn.* 28(1):3–11.
8. Meher, K. K., M. V. Murphy, and K. G. Gollakota. 1994. Psychrophilic anaerobic digestion of human waste. *Bioresource Technol.* 50:103–106.
9. Lapp, H. M., D. D. Schulte, E. J. Kroeker, A. B. Kroeker, A. B. Sparing, and B. H. Topnik. 1975. Start-up pilotscale swine manure digesters for methane production. *Proceesings of the 3rd International Symposium on Livestock Waste*: 234–237. St. Joseph, MI: ASAE.
10. Maekawa, T. and Y. Kitamura. 1989. On the development of anaerobic fluidized bed fermentor incorporated with ultra-filter membrane system. In *Proceedings of 1989 Food Processing Waste Conference*: 415–429. Atlanta, CA: Georgia Tech Research Institute.
11. Zhang, Z. Y. and T. Maekawa. 1993. Kinetic study on fermentation from $CO_2$ and $H_2$ using the acclimated-methanogen in batch culture. *Biomass Bioenergy* 4:439–446.
12. Speece, R. E. and P. L. MacCarty. 1964. Nutrient requirements and biological solids accumulation in anaerobic digestion. *Adv. Water Pollut. Res.* 2:305–322.
13. Albertson, D. E. 1961. Ammonium nitrogen and the anaerobic environment. *J. Water Pollut. Control Fed.* 33:978–995.
14. Van Velsen, A. F. M. 1979. Adaptation of methanogenic sludge to high nitrogen in high rate digestion. *Water Res.* 13:995–999.
15. Van Velsen, A. F. M. and G. Lettinga. 1981. Digestion of animal manure. *Stud. Environ. Sci.* 9:55–64.

16. Aguilar, A., C. Casa, and J. M. Lema. 1995. Degradation volatile fatty acids by differently enriched methanogenic cultures: kinetics and inhibition. *Water Res.* 29:503–509.
17. Peter, S., J. Moll, and R. K. Thauer. 1979. Nickel, cobalt and molybdenum requirements for growth of *Methanobacterium thermoautrophicum. Arch. Microbiol.* 123:105–107.
18. Diekert, D. B., B. Klee, and R. K. Thauer. 1980. Nickel, a component of factor $F_{430}$ from *methanobacterium thermoautrophicum. Arch. Microbiol.* 124:103–106.
19. Graf, E. G. and D. K. There. 1981. Hydrogenase from *Methanobacterium thermoautrophicum*, a nickel-containing enzyme. *Federation of European Microbiological Societies Lett.* 136:165–169.
20. Schere, P. and H. Sahm. 1980. Growth of *Methanosarcina berkeri* on methanol or acetate in a defined medium. *Proceedings of the 1st International Symposium on Anaerobic Digestion*: 45–47. Cardiff: A.D. Scientific.
21. Dixon, N. E., C. Gazzola, R. L. Blakely, and B. Zerner. 1975. Jack bean ureas (EC 3.1.5.5), a metallo-enzeyme. A simple biological role for nickel. *J. Am. Chem. Soc.* 97:4131–4133.
22. Mah, R. A., M. R. Smoth, and L. Baresi. 1978. Studies on an acetate-fermenting strain of *Methanosarcina. Appl. Environ. Microbiol.* 35:1174–1184.
23. Schere, P. 1983. Composition of major elements and trace elements of 10 methanogenic bacteria determined by inductively coupled plasma emission spectrometry. *Biol. Trace Elements Res.* 5:149–163.
24. Zhang, Z. Y. and T. Maekawa. 1995. High productivity of $CH_4$ from $H_2$ and $CO_2$ by optimizing the nutritional conditions of acclimated-methanogens. *J. Soc. Agric. Struct. Jpn.* 25(4):215–222.
25. Maekawa, T., Y. S. Zhang, and Z. Y. Zhang. 1997. Effects of trace metals on the methane fermentation from acetic acid. *J. Soc. Agric. Struct. Jpn.* 28(2):69–75.
26. Hill, D. T. 1983. Simplified monod kinetics of methane fermentation of animal waste. *Agric. Waste* 5:1–16.
27. Kitamura, Y. and T. Maekawa. 1993. Study on the design and development of methane fermentation system (part 3). *J. Soc. Agric. Struct. Jpn.* 24(1):15–20.
28. Lawrence, A. W. and P. L. MacCarty. 1969. Kinetics of methane fermentation in anaerobic treatment. *J. Water Pollut. Control Fed.* 41(2):1–17.
29. Chang, J., T. Noike, and J. Matumoto. 1982. Effect of retention time and feed substrate concentration on methanogenesis in anaerobic digestion. *Dobokugakkai Ronbunhokokushu* 320:67–76.
30. Kitamura, Y., T. Maekawa, A. Tagawa, H. Hayashi, and K. Farell-Poe. 1996. Treatment of strong organic, nitrogenous wastewater by an anaerobic contact process incorporating ultrafiltration. *Appl. Eng. Agric.* 12:709–714.
31. Christensen, J. 1995. *Progress Report on the Economy of Centralized Biogas Plant*, pp. 4–34. Copenhagen: The Biomass Section of the Danish Energy Agency.
32. Maekawa, T. 1993. On the two-phase methane fermentation, pp. 429–440. In *The System of Agricultural Technology-Livestock-*. Tokyo: Nosangyoson Bunka Kyoukai.

33. Maekawa, T. and P. Y. Yang. 1990. Engineering evaluation and cost analysis of appropriate treatment alternatives for a swine waste treatment system in the tropical and temperate regions. Bull. Agrc. Forest. Res. Univ. Tsukuba 2:67–68.
34. Liao, C. M. and T. Maekawa. 1994. Nitrification/denitrification in an intermittent aeration process for swine wastewater. *J. Environ. Sci. Health*. B 28:1053–1078.
35. Maekawa, T., C. M. Liao, and X. D. Feng. 1995. Nitrogen and phosphorous removal for swine wastewater using intermittent aeration batch reactor followed by ammonium crystallization process. *Water Res.* 29:2643–2650.

### 3.2.2 Pyrolysis Gas

*B. M. Jenkins*

***Overview of Gasification Technology***

On heating, biomass fuels spontaneously decompose into a number of gaseous and condensable species, leaving behind a solid carbonaceous residue known as char. This decomposition under heating without the need for added oxygen is referred to as pyrolysis. Pyrolysis is an integral part of solid fuel combustion, and the flame seen when burning wood and other biomass in an open fire is primarily the result of the combustion of volatile compounds emitted during pyrolysis. Pyrolysis can be physically separated from combustion to produce a fuel gas by excluding or limiting oxygen from the reaction. The production of fuel gases from biomass or coal in the presence of oxygen is more appropriately referred to as gasification. The heating of the fuel may be accomplished indirectly by means of heat transfer through the walls of the reactor or some internal heat exchanger or directly by the addition of sufficient oxidant to oxidize the fuel partially with interparticle and intraparticle heat transport resulting from the energy released by the oxidation reactions. The result is a gas rich in carbon monoxide, hydrogen, and hydrocarbons that can be used as fuel or feedstock for furnaces, engines, chemical processing, and possibly advanced power-conversion technologies such as fuel cells.

The technology to gasify solid fuels was developed more than 200 years ago [1] and has been used intermittently since the development of petroleum and natural gas when such fuels were limited in supply. Gasifiers have long been used to convert solid fuels to fuel gases for running internal-combustion engines, both spark ignited and compression ignited (diesel engines). They can also be used for external-combustion devices such as Stirling engines and boilers. Gasifiers were used in many countries for stationary and mobile applications during World War II as replacement for petroleum. Small gasifiers were used to power cars, trucks, and other mobile equipment. Charcoal was the fuel of choice because of the lower production of condensable tars that made use of the gas difficult without substantial precleaning. The energy loss accompanying the loss of volatiles from biomass in the manufacture of charcoal eventually led to the development of improved gasifiers that could use wood with lower tar production rates. Sweden continued a program in gasifier development after the war [2], but the commercial use of small gasifiers was limited by the return of inexpensive petroleum and natural gas supplies. Renewed interest in gasification arose following the oil embargo of 1973–1974, and has remained of interest for biomass fuels as part of an expanding effort in renewable-energy resource development. Research and development in gasification of biomass shares a common basis with coal gasification, although the two materials are sufficiently different in character that reactor design and operating conditions must be separately optimized. Biomass fuels are in general more volatile than coals and produce more reactive chars. The gasifier reactor design must accommodate the higher volatility of biomass and the greater production of tars. Gas cleaning remains the paramount obstacle in the use of biomass gasifiers for small engines in remote and developing country applications and in the use of advanced gasifier technologies for large-scale power systems.

***Chemistry of Gasification***

The decomposition of a solid biomass feedstock by means of pyrolysis and gasification involves a complex reaction system that generates gases, liquids (tars and oils), and solids (chars). Gasification reactors are designed to optimize the production of fuel gases, but liquids and solids are produced as well. In most gasifier applications, especially the use of gasifiers for fueling engines, the production of liquids and carbonaceous solids is undesirable. The partitioning among products is dependent on a number of factors, including the feedstock structure and composition, the reactor design, and the operating conditions, especially temperature, residence time, and heating rate.

Pyrolysis generates primary chars, oils, and gases, which can in turn react in the gasifier to form secondary chars, oils, and gases. The chars can react with oxygen, steam, or pyrolysis products to form reduced gases. The extent to which secondary products form depends in part on the type of gasifier used. Heating rate and final temperature are of critical importance. Slow heating rates and low temperatures favor the formation of chars, whereas rapid heating promotes the formation of liquids. Flash pyrolysis (heating above $10^3$ K $s^{-1}$) is one technique used to generate high yields of liquid pyrolysates intended for use as liquid fuels after processing or refining.

*Feedstocks*

Most solid fuels, including coal and biomass, can be converted by means of gasification to fuel gas or chemical synthesis feedstock. Some liquids can also be converted to fuel or chemical feedstock gas in a similar manner.

Biomass includes all types of terrestrial and aquatic phytomass, along with animal wastes and the organic fractions of industrial and municipal solid wastes, of which a substantial portion is derived from plants. With approximately 0.02% of solar radiation incident at the top of the atmosphere converted to biomass through photosynthesis [3], the total annual global yield of biomass is of the order of $10^{14}$-kg dry matter (that is, plant matter independent of moisture). The global biomass yield has an equivalent heating value (the total heat liberated in complete combustion) of over $10^{21}$ J, or nearly four times the energy content of the world's annual commercial energy consumption [4]. The actual commercial energy to be sustainably generated from biomass is far less than the total photosynthetic yield. World agriculture produces residues in the amount of $5 \times 10^{12}$ kg $y^{-1}$, of which perhaps less than half may be available for energy purposes, although many are currently used as primary energy resources in developing countries [5].

The components of biomass include cellulose, hemicelluloses, lignin, lipids, proteins, simple sugars, starches, water, hydrocarbons, ash, and other compounds. The concentration of each class of compound varies, depending on species, type of plant tissue, stage of growth, and growing conditions. Cellulose is a linear polysaccharide of $\beta$-D glucopyranose units linked with (1–4) glycosidic bonds, represented by the empirical composition $CH_{1.67}O_{0.83}$ (or more commonly as $C_6H_{10}O_5$). Hemicelluloses are polysaccharides of variable composition including both five- and six-carbon monosaccharide units. The lignin is an irregular polymer of phenylpropane units, having the approximate composition $CH_{1.23}O_{0.38}$ [2, 6–10]. Plants producing large amounts of free sugars, such as sugar cane and sweet sorghum, are attractive as feedstocks for ethanol fermentation, as are starch crops such as maize (corn) and other grains. Cellulose and hemicellulose are more difficult to ferment, and lignin is considered to be largely unfermentable. Thermal methods are less selective in application and can be used to produce heat, fuel gas, and chemicals from lignocellulosic materials. The synthesis of methanol from biomass by thermal gasification is an example. The structure of cellulosic materials and the chemistry of biomass pyrolysis have previously been reviewed in depth [11–13].

Roughly half of plant organic matter is carbon, the rest being made up of 5%–7% hydrogen, 30%–50% oxygen, along with nitrogen, sulfur, chlorine, and other elements [14]. The elements C, H, and O make up more than 99% of clean, dry, pine wood, the rest being N, S, Cl, and ash. The molar composition can be expressed on an empirical basis as $CH_{1.46}O_{0.68}$ [15]. By comparison, C, H, and O make up 81% or less of rice husk, represented as $CH_{1.26}O_{0.66}$ (the rest being primarily silica ash, which may range as high as 25% of dry weight).

The ash or inorganic fraction of biomass is highly variable and can account for more than 30% of the dry weight (principally animal manures and other waste materials, although rice straw and rice husks typically contain more than 20% ash). The ash is of two forms, inherent ash and adventitious ash, the former arising from the physiological processes of the plants, the latter representing entrained materials added during handling and processing. Temperate region woods typically have inherent ash contents below 1%, whereas some tropical woods have bulk ash contents above 5% [6]. Leaves and young tissues contain more ash than mature wood. Cereal grain straws, husks, and other agricultural biomass frequently have high concentrations of ash. The amount of ash in a biomass material and the composition of the ash are important for the selection of the conversion technology. Alkali metals (especially the essential macronutrient potassium) in the ash can cause slagging, surface fouling, and corrosion as a result of reactions with silica and other minerals. Chlorine is an important facilitator in the alkali reactions leading to deposition in boilers. It also contributes to corrosion and is involved in the formation of certain toxic emissions (e.g., dioxins and furans). Biochemical conversion methods (e.g., fermentation) avoid most of the fouling and corrosion phenomena suffered by the thermal methods because of the lower temperatures involved, although they are not yet as well developed for lignocellulosic feedstocks. Other methods, such as supercritical water oxidation, in which the oxidation is carried out at a relatively low temperature but at high pressure, may suffer less from ash reactions, but are also not yet commercial. Fouling and corrosion in thermal systems are reduced by the extraction of the offending constituents before the fuel is introduced to the combustor or reactor. Such is the case of sugar cane bagasse, commonly used as a fuel in the sugar industry. In the sugar extraction step, most of the alkali and chlorine in the cane is leached and carried off in the liquid effluent, leaving the bagasse relatively free of fouling elements. The success of this technique has also recently been demonstrated for straw [16]. The lower temperatures associated with pyrolysis compared with direct combustion lead to retention of more alkali species in the char phase. With less alkali in the volatile phase, ash fouling may be reduced for boilers fired with pyrolysis gas instead of the solid fuel directly, although the problem of even very low concentrations of alkali in pyrolysis gas remains critical to the control of blade fouling in gas turbines intended for advanced power generation cycles.

The energy content or heating value of biomass can be partially correlated with ash concentration. Woods with less than 1% ash typically have heating values near 20 MJ $kg^{-1}$. Each 1% increase in ash translates roughly into a decrease of 0.2 MJ $kg^{-1}$ [15]. Although the ash does not contribute substantially to the overall heating value, certain elements in the ash, including the alkali species, are catalytic to the thermal decomposition. Heating values can also be correlated to carbon concentration, with each 1% increase in carbon elevating the heating value by approximately 0.39 MJ $kg^{-1}$, a result identical to that found by Shafizadeh [17] for woods and wood pyrolysis products. The heating value relates to the amount of oxygen required for complete combustion, with 14,022 kJ $kg^{-1}$ of oxygen consumed [17]. Cellulose has a smaller heating value (17.3 MJ $kg^{-1}$) than lignin (26.7 MJ $kg^{-1}$) because of its higher degree of oxidation. Other constituents, such as hydrocarbons, with lower degrees of oxidation, tend to raise the heating value of biomass.

Fuel moisture is a limiting factor in both the combustion and the gasification of biomass without an external energy input. The autothermal limit for biomass combustion is typically ~65% moisture content wet basis (mass of water as moisture per mass of moist fuel). Above this point, insufficient energy is liberated by combustion to satisfy the requirements for evaporation of moisture and heating of reaction products. Practically, most combustors require a supplemental fuel, such as natural gas, when burning biomass in excess of 50%–55% moisture wet basis. Atmospheric-pressure direct gasifiers (in which partial oxidation of the fuel in the gasifier reactor provides the heat necessary to generate the pyrolysis gas) operate better with fuels below 30% moisture content. At high pressure the moisture can be used to some advantage in enhancing the hydrogen content of the gas. Updraft gasifiers (described below) can handle higher moisture fuels than can downdraft units.

*Reaction Stoichiometry*

Reaction stoichiometry determines the product distribution and serves to distinguish gasification from combustion. Viewed simply, the complete combustion of biomass in air transforms the organic fraction into carbon dioxide and water, as below:

$$C_xH_yO_z + n_1H_2O + n_2(1+e)(O_2 + 3.76N_2) = n_3CO_2 + n_4H_2O + n_5N_2 + n_6O_2, \tag{3.26}$$

where the stoichiometric coefficients depend on the concentrations of carbon, hydrogen, and oxygen in the original fuel [here expressed on a moisture-and-ash-free (maf) basis] and on the amount of excess oxygen $e$ added to the reaction. The simple reaction shown ignores the contribution from other organic constituents such as nitrogen, sulfur, and chlorine. Although these are typically present in much smaller concentrations than C, H, and O, they play an important role in biomass combustion and gasification from a pollutant emission standpoint. Fuel moisture is accounted for in the reaction by the coefficient $n_1$, and air is approximated as consisting of 21% by volume oxygen and 79% equivalent nitrogen (nitrogen, argon, and all other air species). With $n_1$ and $e$ given by the reaction conditions and $n_2$ found at $e = 0$ (the stoichiometric limit), Eq. (3.26) contains only four unknowns that are solved by the four element balances (C, H, O, N) such that

$$n_1 = (W_{\text{fo}}/W_w)M_{\text{db}}/X_o, \tag{3.27}$$

$$n_2 = x + y/4 - z/2, \tag{3.28}$$

$$n_3 = x, \tag{3.29}$$

$$n_4 = y/2 + n_1, \tag{3.30}$$

$$n_5 = 3.76n_2(1 + e), \tag{3.31}$$

$$n_6 = n_2 e, \tag{3.32}$$

where $X_o$ is the mass-basis maf organic fraction (in kilograms per kilogram dry matter) of the fuel (in this simple case, the sum of carbon, hydrogen, and oxygen mass fractions from the dry-basis elemental or ultimate analysis), $W_{\text{fo}}$ is the molecular weight of the fuel (in kilograms per kilomole, again determined for the maf organic fraction from the elemental analysis), $W_w$ is the molecular weight of water (18.0154 kg kmol$^{-1}$), and $M_{\text{db}}$ is the dry-basis moisture content (in kilograms per kilogram), viz., mass of water per mass of dry fuel. The maf air–fuel ratio $\text{AF}_o$ (again ignoring N, S, and Cl), the dry-basis air–fuel ratio $\text{AF}_d$, and the wet-basis as-fired air–fuel ratio $\text{AF}_w$ (in kilograms per kilogram) are, respectively,

$$\text{AF}_o = n_2(1 + e)(W_{O_2} + 3.76W_{N_2})/W_{\text{fo}}, \tag{3.33}$$

$$\text{AF}_d = \text{AF}_o(X_o), \tag{3.34}$$

$$\text{AF}_w = \text{AF}_d(1 - M_{\text{wb}}), \tag{3.35}$$

where $W_{O_2}$ and $W_{N_2}$ are the molecular weights of $O_2$ and $N_2$ (31.9988 and 28.0134 kg kmol$^{-1}$, respectively), and $M_{\text{wb}}$ is the wet-basis moisture content of the fuel (in kilograms per kilogram), related to the dry-basis moisture content as

$$M_{\text{wb}} = M_{\text{db}}/(1 + M_{\text{db}}). \tag{3.36}$$

The so-called stoichiometric or theoretical air–fuel ratio $\text{AF}_{\text{stoich}}$ (on maf, dry, or as-fired basis as desired) is computed for $e = 0$. Typically, for biomass fuels, $3 \le \text{AF}_{\text{o, stoich}} \le 6$.

The equivalence ratio $\phi$ is defined as

$$\phi = AF_{stoich}/AF = 1/(1+e), \tag{3.37}$$

where AF is the actual air–fuel ratio (maf, dry, or as-fired, but at the same condition as $AF_{stoich}$). In some literature, the equivalence ratio is defined as the inverse of $\phi$ so that care must be exercised in the interpretation of published results. The value $1/\phi = 1 + e$ is sometimes termed the air factor.

The equivalence ratio serves to classify thermochemical conversion regimes. Nominally, conditions for which $\phi \leq 1 (e \geq 0)$, with the actual air–fuel ratio greater than or equal to the stoichiometric air–fuel ratio, define the combustion regime. Conditions for which $\phi > 1(-1 < e < 0)$ define the pyrolysis, gasification, and fuel-rich combustion regimes, whereby the fuel is only partially or incompletely oxidized and a much larger number of reaction products are generated. These limits are not strictly defined, in that combustion may be carried out under fuel-rich conditions, up to $\sim\phi = 1.5$, but complete oxidation of the fuel is not possible and products of incomplete combustion (pollutants) are produced. For the most part, the direct gasification reactions are carried out with something less than ~30% of theoretical air with $\phi > 3$. Staged combustion is used as a means to control certain pollutant emissions, especially oxides of nitrogen ($NO_x$), and in this case the primary combustion may be carried out under fuel-rich conditions and later stages carried out under increasingly fuel-lean conditions. The advantage to this is that combustion temperature is reduced throughout, which reduces the formation of thermal $NO_x$ from atmospheric oxygen and nitrogen and may also assist in the control of ash deposition on combustor surfaces. Gasification can be thought of in some respects as a staged combustion process, with the initial solid-to-gas fuel conversion carried out under extremely fuel-rich conditions or, in the case of indirect systems, without any added oxygen beyond what is present in the fuel.

When the value of $e$ in Eq. (3.26) is substantially less than zero (to a limit of $-1$, indicating no oxygen added to the reaction) and the equivalence ratio exceeds a value of ~3, the reaction products consist not only of carbon dioxide and water, but of large amounts of carbon monoxide, hydrogen, gaseous hydrocarbons, and condensable compounds (tars and oils), along with char. Steam, in addition to air, may be supplied to the reaction to promote the production of hydrogen. Reaction conditions can be varied to maximize the production of fuel gases, fuel liquids, or char (as for charcoal). The term gasification is applied to processes that are optimized for the production of fuel gases (principally CO, $H_2$, and light hydrocarbons). Commonly $CO_2$ and $N_2$ are present in the gas mixture as diluents, depending on the type of gasifier used. If oxygen is used instead of air, the dilution from atmospheric nitrogen does not occur and the heating value of the gas is increased. Indirect gasifiers or pyrolyzers are also free of dilution from atmospheric nitrogen. Nitrogenous species in the gas arise only from nitrogen in the fuel in such cases. The term pyrolysis is more commonly applied to those processes optimized for fuel liquid production, in which a large fraction of condensable materials is desired from the reactions. Catalysts are sometimes used as promoters, especially in the cracking of high molecular weight hydrocarbons and tars produced during gasification.

A simplified global reaction for the gasification of the organic portion of biomass may be written in a manner similar to that of the reaction (3.26):

$$\begin{aligned} C_xH_yO_z + n_1H_2O + n_2(1+e)(O_2 + 3.76N_2) \\ = n_3CO_2 + n_4H_2O + n_5N_2 + n_6O_2 \\ + n_7CO + n_8H_2 + n_9CH_4 + n_{10}C(s) \\ + \text{other gases} \\ + \text{tars, oils, other condensable species.} \end{aligned} \tag{3.38}$$

Participating in the gasification are a large number of reactions along with a large number of intermediate species. A simplified reaction system based on carbon (as charcoal) gasification includes (but is by no means limited to)

$$C + O_2 = CO_2, \tag{3.39}$$

$$C + 2H_2 = CH_4, \tag{3.40}$$

$$CO + H_2O = CO_2 + H_2, \tag{3.41}$$

$$C + CO_2 = 2CO, \tag{3.42}$$

$$C + H_2O = CO + H_2, \tag{3.43}$$

$$C + 2H_2O = CO_2 + 2H_2. \tag{3.44}$$

In addition there are a number of pyrolytic reactions that directly produce hydrocarbons and other volatile and condensable species. In the forward direction, reactions (3.39)–(3.41) (the latter being the water–gas shift reaction) are exothermic at standard temperature (298 K), while reactions (3.42) (the Boudouard reaction), (3.43) (the water–gas reaction), and (3.44) are endothermic. In a direct-heat gasifier, the energy to drive the endothermic gasification reactions comes from the exothermic reactions. Ideally, oxygen (or air) gasification of rice husk, for example, with the composition noted above, would appear as

$$CH_{1.26}O_{0.66} + 0.17O_2 = CO + 0.63H_2,$$

but unfortunately the reaction is endothermic and not self-sustaining. As Reed [2] has shown for adiabatic gasification, energy equivalent to combustion of approximately a third of the CO and $H_2$ needs to be supplied to provide the heat necessary for direct gasification.

*Gas Products*

A typical producer gas composition from wood fuel in an air-blown (air as oxidizer) direct gasifier is (after particulate matter and most condensables are removed from the gas)

| Element or Compound | % by volume |
|---|---|
| $CO$ | 22 |
| $H_2$ | 14 |
| $CH_4$ | 5 |
| $H_2O$ | 2 |
| $CO_2$ | 11 |
| $N_2$ | 46 |

The higher heating value of the gas shown is ~5.9 MJ m$^{-3}$. The large concentrations of $CO_2$ and $N_2$ lead to the low heating value of this gas. The small concentration of methane (as well as other hydrocarbons that are commonly present) contributes substantially to the heating value of the gas because of its own much higher heating value (36.1 MJ m$^{-3}$) compared with CO (11.5 MJ m$^{-3}$) and $H_2$ (11.5 MJ m$^{-3}$).

For oxygen-blown gasifiers, there is no dilution from atmospheric $N_2$, and the heating value is increased into the range of 10 to 14 MJ m$^{-3}$, dry basis. Pyrolysis gas from indirect gasifiers typically has a heating value in the range of 17 to 20 MJ m$^{-3}$. The standard densities and heating values of some producer gas components are listed in Table 3.39.

The efficiency of the gasifier is reported on either a hot-gas or cold-gas basis, depending on the application. The hot-gas efficiency is computed as the ratio of the total enthalpy of the gas to the fuel heating value. The total gas enthalpy includes both the chemical energy in the combustible constituents (CO, $H_2$, and hydrocarbons) and the sensible energy associated with the gas at reactor exit temperature. The hot-gas efficiency is commonly applied to units firing furnaces. The cold-gas efficiency is computed as the ratio of chemical energy in the gas to the fuel heating value, on the assumption that the gas is cooled to standard condition (usually 298 K). The cold-gas efficiency is commonly used with gasifiers fueling engines, in which case the gas is normally cooled to near-ambient conditions as part of the gas cleaning operation and to improve the mass intake rate of fuel by the engine (i.e., increase delivery ratio). Cold-gas efficiencies generally range from 50% to 90%.

**Table 3.39. Properties of selected gases at 300 K and 101.3-kPa (1-atm) pressure [15]**

| | | Higher Heating Value | | Lower Heating Value | |
|---|---|---|---|---|---|
| Gas | Density (kg m$^{-3}$) | (MJ kg$^{-1}$) | (MJ m$^{-3}$) | (MJ kg$^{-1}$) | (MJ m$^{-3}$) |
| $CH_4$ | 0.650 | 55.56 | 36.11 | 50.04 | 32.53 |
| $CO$ | 1.137 | 10.11 | 11.49 | 10.11 | 11.49 |
| $CO_2$ | 1.787 | | | | |
| $C_2H_2$ | 1.056 | 49.95 | 52.74 | 48.22 | 50.92 |
| $C_2H_4$ | 1.137 | 50.34 | 57.24 | 47.15 | 53.61 |
| $C_2H_6$ | 1.219 | 51.93 | 63.30 | 47.51 | 57.92 |
| $C_3H_8$ | 1.787 | 50.40 | 90.06 | 46.38 | 82.88 |
| $H_2$ | 0.081 | 141.93 | 11.50 | 119.98 | 9.72 |
| $O_2$ | 1.300 | | | | |
| $N_2$ | 1.137 | | | | |

Gasifier capacity is commonly expressed on the basis of the fuel consumption per unit cross-sectional area of reactor or specific gasification rate (in kilograms per square meter per hour). The equivalent power per unit area is also sometimes used (in watts times inverse square meters). Consideration of the direct gasification reaction system suggests a general optimization. At low fuel conversion rates, gas production is low and heat loss relatively high. At extremely high fuel conversion rates, the gas heating value generally declines as a result of an increasing air–fuel ratio. Between these two extremes, the gasifier efficiency is expected to be parabolic, reaching a maximum at some intermediate level of fuel conversion rate. Tiangco, *et al.* [18] observed this with small rice husk gasifiers used to fuel spark-ignited engines. An optimum in the specific gasification rate was located at ~200 kg $m^{-2}$ $h^{-1}$, giving the maximum cold gas efficiency. Manurung and Beenackers [19] optimized the specific gasification rate at slightly lower values.

*Other Products*

***Char.*** Complete combustion of biomass leaves a solid residue of nonvolatile ash. Gasification produces solid residues, or chars, containing ash, carbon, and organic materials of variable concentration. Direct gasifiers in which the oxidant stream flows countercurrent to the fuel stream (as is the case of updraft gasifiers noted below), tend to produce low-carbon chars. In many cases such reactors produce an ash or ash slag (resolidified molten ash). Other types of gasifiers, designed to reduce the formation of liquids (e.g., downdraft reactors), tend to produce chars of higher carbon concentration. The compositions of several chars are listed in Table 3.40. As mentioned above, low temperatures and low heating rates tend to favor the production of char. These conditions are exploited in the production of charcoal. For small-scale gasifier applications, char disposal or utilization may represent a difficult and costly component of the system. Adverse environmental impacts result from improper char handling or disposal. The char can have economic value as an activated carbon, although further processing may be required for meeting technical standards for commercial-grade carbon. Some dual-reactor gasifiers combust the char to produce heat for the gasifier stage. This allows for the production of a

**Table 3.40. Heating values and compositions of biomass chars [20]**

| Type | Higher Heating Value (MJ $kg^{-1}$) | Elemental Analysis (% by weight, dry basis) | | | | | |
|---|---|---|---|---|---|---|---|
| | | C | H | O | N | S | Ash |
| Fir bark char | 19.2 | 49.9 | 4.0 | 24.5 | 0.1 | 0.1 | 21.4 |
| Rice hull char | 14.2 | 36.0 | 2.6 | 11.7 | 0.4 | 0.1 | 49.2 |
| Grass straw char | 19.3 | 51.0 | 3.7 | 19.7 | 0.5 | 0.8 | 24.3 |
| Municipal solid waste char | 18.7 | 54.9 | 0.8 | 1.8 | 1.1 | 0.2 | 41.2 |
| Redwood charcoal (694–822 K) | 28.8 | 75.6 | 3.3 | 18.4 | 0.2 | 0.2 | 2.3 |
| Redwood charcoal (733–1214 K) | 30.5 | 78.8 | 3.5 | 13.2 | 0.2 | 0.2 | 4.1 |
| Oak charcoal (711–914 K) | 24.8 | 67.7 | 2.4 | 14.4 | 0.4 | 0.2 | 14.9 |

**Table 3.41. Composition (% by weight) of tars from an air-blown downdraft gasifier**

| Fuel | C | H | O | N | S | Cl | Ash |
|---|---|---|---|---|---|---|---|
| Wood | 38.17 | 5.49 | 38.85 | 0.09 | 0.08 | 0.12 | 17.2 |
| Walnut shell | 30.85 | 4.97 | 36.37 | 0.35 | 0.09 | 0.17 | 27.2 |
| Cotton gin trash | 64.02 | 6.64 | 24.37 | 0.65 | 0.26 | 0.06 | 4.00 |

higher-energy gas with less nitrogen dilution on air-blown units. One disadvantage of this technique is the potential ash fouling and slagging associated with the char combustion stage.

***Liquids.*** Gasifiers produce condensable vapors in varying amounts. These consist of a large number of organic compounds, including acids. When the gas is intended for use in internal combustion engines, the presence of condensable materials, including water and tars, represents a significant problem because of condensation in the intake system. They are less of a problem when the gas is burned straight out of the reactor for process heat. Updraft gasifiers (described below) produce large amounts of tars and are used principally with boilers or, if a high-quality vapor is produced, for liquid fuel production. Tar yields may be as much as a third of feed mass and energy. Downdraft gasifiers produce lower amounts of tars and are more commonly applied to engines. Tar yields from downdraft units are typically a few percent or fewer of feed. High-residence-time fluidized-bed and entrained-bed gasifiers also produce lower yields of tars [2]. Pyrolysis oils and tars are generally highly oxygenated, which adds to their corrosiveness. Compositions for tars from three different biomass fuels are listed in Table 3.41.

*Equilibrium*

The gas composition can be predicted by assuming thermodynamic equilibrium of reaction products at the product temperature. A number of computer codes are available that can handle the full complement of reaction products.

A simplification of reaction (3.38) leads to a rough approximation of the equilibrium gas composition. Excluding the other gases, tars, and oils, the product side of reaction (3.38) contains eight unknowns. In a properly functioning gasifier, the amount of oxygen in the products should be near zero, and can here be disregarded, leaving seven unknowns. By further disregarding the concentrations of $CH_4$ and C(s) (solid residual char) in the products as small (although the importance of methane to the energy content of the gas has already been noted, and complete conversion of the carbon is often difficult to achieve), a straightforward equilibrium solution is obtained for the five remaining unknowns: the stoichiometric coefficients for $CO_2$, $H_2O$, $N_2$, CO, and $H_2$. Four element balances can be written as

$$\begin{aligned}
\text{C}: &\quad x = n_3 + n_7, \\
\text{H}: &\quad y + 2n_1 = 2n_4 + 2n_8, \\
\text{O}: &\quad z + n_1 + 2n_2(1+e) = 2n_3 + n_4 + n_7, \\
\text{N}: &\quad 3.76n_2(1+e) = n_5.
\end{aligned}$$

The law of mass action provides a fifth equation:

$$\prod_t (p_i)^{n_i} = K_p(T), \tag{3.45}$$

where $p_i$ is the equilibrium partial pressure of species $i$ in the system, $n_i$ is the stoichiometric coefficient for species $i$, $K_p(T)$ is the equilibrium constant, a known function of temperature, and the pi symbol denotes the product operation.

The four products, CO, $H_2$, $CO_2$, and $H_2O$, can be related through the water–gas shift reaction, Eq. (3.41) above. Note that

$$p_i/p = n_i/n = x_i, \tag{3.46}$$

where $p$ is the total pressure, the total number of moles $n = n_3 + n_4 + n_5 + n_7 + n_8$, and $x_i$ is the mole fraction of gas species $i$, then

$$K_p(T) = (x_{CO_2} x_{H_2})/(x_{CO} x_{H_2O}) = (n_3 n_8)/(n_7 n_4). \tag{3.47}$$

The value of the equilibrium constant $K_p$ is found from the reaction temperature $T$. An expression for $K_p$ in Eq. (3.47) for temperatures up to 2100 K is given by Gumz [21]:

$$\log_{10} K_p = 36.72508 - (3994.704/T) \\ +4.462408 \times 10^{-3} T - 0.671814 \times 10^{-6} T^2 - 12.220277 \log_{10} T$$

for $T$ in degrees Kelvin.

The reaction temperature can be determined by means of energy balance. For adiabatic gasification, the first law yields

$$H_P = H_R, \tag{3.48}$$

where $H_P$ and $H_R$ are the enthalpies of the products and the reactants, respectively, and are nonlinear functions of temperature.

Solving the four element balance equations with the law of mass action yields the value of $n_3$ in quadratic form, from which the other coefficients are determined:

$$n_3 = [-b \pm (b^2 - 4ac)^{1/2}]/(2a), \tag{3.49}$$

where

$$a = K_p^{-1} - 1,$$
$$b = (z - x - y/2) + n_2(1 + e) - [z + n_1 + 2n_2(1 + e)],$$
$$c = K_p^{-1}[x(z - x) + xn_1 + 2xn_2(1 + e)].$$

Iteration on $T$ yields the proper value of $K_p$. Alternatively, a linearization of the enthalpy functions [14] allows the value of $T$ and hence $K_p$ to be determined directly. This simple model can be used to investigate approximate compositions for various feedstocks, although agreement with experimental data has been shown to decline with increasing fuel moisture. Many gasifiers do not operate well at higher fuel moistures in any case. Nor do all gasifiers produce product compositions close to equilibrium. For

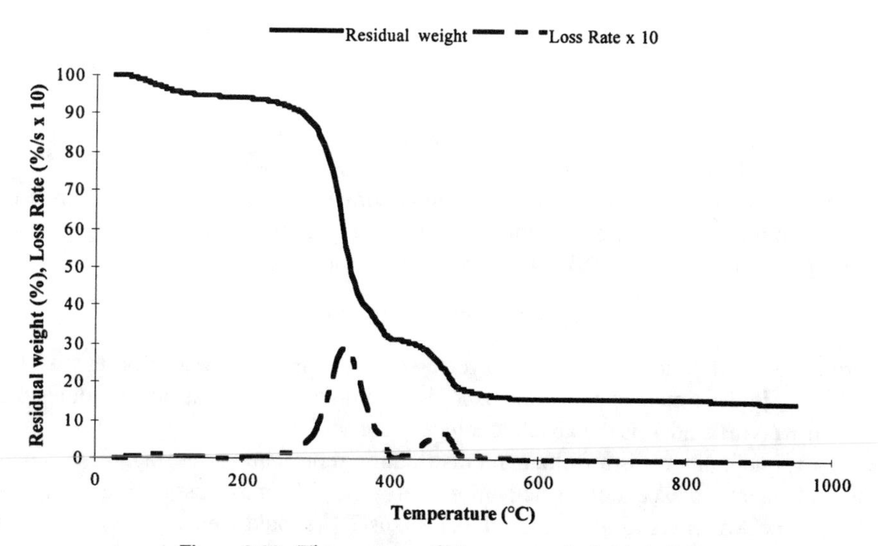


**Figure 3.32. Thermogram of biomass pyrolysis in air [22].**

more accurate estimates of the equilibrium composition, more robust computer codes are routinely available.

*Kinetics*

The kinetics of biomass pyrolysis are important to the design and scaling of reactors and for optimizing the desired product stream. The pyrolysis rate is typically measured by the weight loss occurring as a function of heating rate and atmosphere. Heating rate is important to the product distribution, as noted above. Fast pyrolysis produces high yields of tar vapor, whereas charcoal production is normally carried out slowly. A typical thermogram for a biomass fuel (in this case, rice straw) heated in air is shown in Figure 3.32. Also shown is the rate of weight loss. In the figure, the initial weight loss between 50 and 150°C is due to drying. Beginning at 200°C, there is a sharp drop in weight due to pyrolysis, with slower char oxidation reactions peaking between 450 and 500°C. The residual weight beyond 600°C is that of ash.

The kinetics are sometimes modeled by an Arrhenius form. The rate of reaction $\mathrm{d}x/\mathrm{d}t$ during heating can be written as

$$\mathrm{d}x/\mathrm{d}t = kf(x), \tag{3.50}$$

where $k$ is the reaction rate constant (in inverse seconds), $f(x)$ is a function that depends on the extent of reaction $x$, and

$$x = (m_o - m)/(m_o - m_f), \tag{3.51}$$

where $m$ is the mass (in grams) at any time $t$ (in seconds), $m_o$ is the initial mass (in grams), and $m_f$ is the final mass (in grams) obtained on heating to the final temperature.

Typically, the function in $x$ is taken to be

$$f(x) = (1 - x)^n \tag{3.52}$$

where $n$ is a dimensionless reaction order. Many times the value of $n$ is set to unity, representing a first-order reaction.

The value of $k$ is obtained from a thermogravimetric analysis giving the mass of a heated sample over time, either under constant temperature or under a programmed temperature ramp, as in Fig. 3.32. The Arrhenius formulation for $k$ gives

$$k = A_0 \exp[-E/(R_0 T)], \tag{3.53}$$

where $A_0$ is the frequency factor (in inverse seconds), $E$ is the activation energy for reaction (in joules times inverse moles), $R_0$ is the universal gas constant (in joules per mole per Kelvin), and $T$ is the absolute temperature (in Kelvin).

The exponential term is a Boltzmann distribution representing the fraction of molecules with energy in excess of the activation energy $E$. If the mass loss during heating is strictly of the Arrhenius form, a plot of $\ln(k)$ versus $T^{-1}$ should yield a straight line. If a constant heating rate $q = \mathrm{d}T/\mathrm{d}t(\mathrm{K\ s^{-1}})$, is used, Eq. (3.50) can be modified as follows:

$$\mathrm{d}x/\mathrm{d}T = (A_0/q)\exp[-E/(R_0 T)](1 - x)^n. \tag{3.54}$$

The integration of the above equation is done either numerically or by use of approximate solutions [22]. For biomass, a single-component model of the type of Eq. (3.54) is not normally adequate to model the full volatilization range. The reaction can be described in stages, or multicomponent models can be constructed of the type

$$\mathrm{d}x/\mathrm{d}t = \sum_i (\mathrm{d}x_i/\mathrm{d}t), \tag{3.55}$$

with each component $i$ having its own frequency factor and activation energy. Equation (3.55) is sometimes applied as a linear superposition of the separate decompositions of the main structural components of biomass (cellulose, hemicellulose, lignin). However, interactions among components do occur, so such assumptions are only approximate [13]. The values of $A_0$ and $E$ found from experiments vary widely for different biomass fuels exposed to different heating rates and atmospheres, although $E$ generally varies less than $A_0$. For cellulose, $E$ is typically $\sim$240 kJ mol$^{-1}$ [23]. Much work has been done recently in trying to characterize the rate coefficients for biomass pyrolysis [2, 12, 13, 24].

*Catalysis*

Selectivity in the partitioning among biomass pyrolysis and gasification products can be enhanced through the use of catalysts. Biomass naturally contains some catalytic material in the form of alkali compounds that promote the yield of gases and increase reaction rates. Leaching of cereal straws to remove alkali metals has recently been shown to retard the pyrolysis rate [25]. The catalyst $K_2CO_3$ has been shown to be effective in enhancing the yield of gases from moist wood, and certain nickel catalysts have also been shown to enhance gas yields and are effective in methanol synthesis [13]. Cost of materials and the need to recover catalysts in the process may preclude their use with smaller systems.

### *Gasification Reactors*

Gasification reactors can be classified into three principal types: 1, fixed- or moving-bed, 2, fluidized-bed, and 3, entrained-bed gasifiers. Gasifier types for coal have been reviewed in [26] and for biomass in [27].

#### *Fixed or Moving Beds*

These reactors are distinguished by a stationary or mechanically moved fuel bed. Design types range from simple pipes (such as the throatless open-core reactors) to sophisticated automatic units. Open-core gasifiers are generally small (10–1000 kW thermal), usually batch-operated devices in which the fuel bed remains stationary (except for fuel and char settling) while the reaction front passes through it. These are, for the most part, designed as suction-type gasifiers attached to engines. The engine intake vacuum induces air flow through the reactor and gas cleaning train. Open-core gasifiers such as those developed by Kaupp [28] and Reed [2] are of this type and have proved quite successful on rice husk, although problems remain in gas cleanup and waste handling for small engine units.

More sophisticated gasifiers in this classification maintain a fuel bed above a grate and include active ash discharge through the grate and in some cases automated fuel loading. Reactors may be indirectly or directly heated. Particle size is important to the operating performance. The particles must be large enough ($>\sim$3 mm) that they are not entrained on the gas flow and so that the pressure drop across the bed does not become excessive as a result of high packing density. The particle size generally does not exceed $\sim$50 mm; otherwise gas bypassing becomes a problem (depending on gasifier design) and gas quality deteriorates.

These gasifiers are further classified into updraft, downdraft, and cross-draft types, although more practically they are classified as either cocurrent or countercurrent reactors, depending on the flow directions of the gas and the solid in the reactor. In an updraft or countercurrent reactor (Fig. 3.33), oxidant is admitted through the grate at the bottom into the fuel bed, where char oxidation releases heat for gasification, pyrolysis, and drying. The char production from updraft gasifiers is reduced relative to downdraft units, but the gas is more heavily laden with tars, and this makes use of the gas for engines more difficult. Downdraft or cocurrent reactors (Fig. 3.33) admit oxidant at the top of the fuel bed or somewhere within it, such as through nozzles or tuyerres, so that the pyrolysis tars, in principle, pass through the oxidation zone and are thermally cracked to simpler compounds. Outlet tar yields are thereby reduced. Char, rather than ash, is discharged from the reactor. Downdraft gasifiers are more difficult to scale up than updraft reactors [2] because of the need to supply air to the interior of the fuel bed above the grate. One difficulty lies in generating a uniformly distributed high-temperature cracking zone in the reactor; hence a constriction, or throat, is often added inside the reactor below the level of the tuyerres. While the throat may help to reduce bypassing of the cracking zone by tars, it also impedes the fuel and char flow and can lead to troublesome bridging and irregular operation. Open-core reactors often do

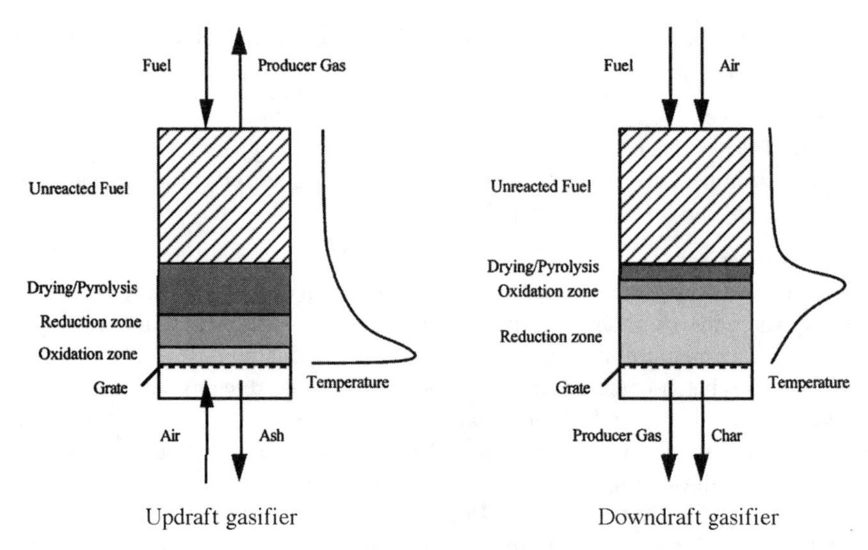


**Figure 3.33. Schematic illustration of updraft (counter-current) and downdraft (co-current) gasifier types.**

not incorporate a constricted throat in the design. Cross-draft reactors employ a mixed flow regime.

Gas cleaning is the principal problem when gasifiers of any type are used with engines. Removal of particulate matter and tars from the gas is critical to maintaining engine life and acceptable operation. Unfortunately, gas cleaning represents one of the biggest obstacles to the application of the technology. Systems are available to produce acceptable gas quality, but generally rely on some combination of wet and dry scrubbing and filtering. The compounds that are removed from the gas into the waste stream include hazardous materials (such as polycyclic aromatic hydrocarbons) that require proper disposal or treatment. Environmental and health impacts have often been neglected in proposals to implement gasifier technology in remote or developing areas. Gas cleaning and waste handling remain the critical engineering challenges to the adoption of the technology at small scale and low cost.

*Fluidized and Entrained Beds*

The fluidized bed takes its name from the suspension of solid bed particles created by the oxidant or gas flowing through the bed at a velocity high enough to overcome the particle weight by means of the drag force. The reactor typically consists of a distributor (a porous plate or set of nozzles) and a bed of sand or refractory particles (media) above the distributor that is fluidized when an oxidant (usually air) is blown into the bed through the distributor. This is shown schematically in Fig. 3.34.

The bed is preheated above the fuel ignition temperature on startup. Two types of fluidized-bed combustors are currently used: bubbling beds and circulating beds. In the bubbling bed, air delivered through the distributor into the bed at a sufficiently high velocity (0.1 to 3 m/s) causes a bubble phase to form, and the bed resembles a boiling

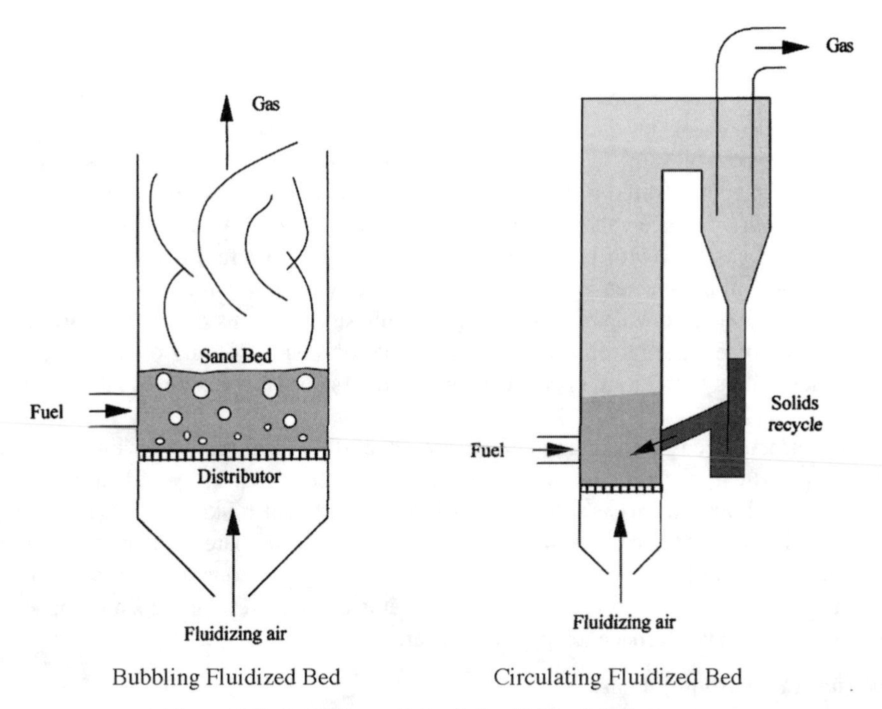


**Figure 3.34. Bubbling and circulating fluidized-bed concepts.**

liquid with turbulent mixing of particles in the bed. Small particles become entrained on the flow and exit the reactor. This is a principal means of removing fuel ash from the reactor, although bed draining is also used. The circulating bed operates at a higher gas velocity than the bubbling bed and uses a particle separation device (such as a cyclone) either within the primary reactor or external to it. At higher velocities, a greater fraction of the bed is entrained on the gas flow and exits the reactor. The larger particles, including the incompletely reacted fuel particles, are returned to the bed to increase fuel conversion efficiency. The fluidized bed is rather homogeneously mixed and has a more uniform bed temperature distribution. Localized high-temperature zones are largely eliminated, thus reducing undesirable bed media agglomeration and slagging; the latter is more common with moving-bed reactors (sometimes by design). Multiple beds may be used, such as the twin reactor designs developed both for coal and for biomass [2, 13, 26]. Char combustion in a secondary fluidized bed provides the necessary heat for fuel pyrolysis and gasification in the primary fluidized bed. A gas free of nitrogen dilution is thereby produced. Char from the first-stage gasifier is circulated to the second-stage char combustor.

Entrained-bed gasifiers operate at higher velocities than fluidized beds and are similar in nature to suspension burners, with fuel particles carried on the oxidant stream. Entrained-bed gasifiers require small particle sizes and have been developed primarily for coal.

### *Gas Utilization*

*Heating and Steam Raising*

The most straightforward use of producer gas is in furnaces for heating and steam raising. In such cases, tar and particulate matter are of lesser concern than for engine applications, as long as the gas can be kept hot up to the flame. A gasifier may be used to convert a gas- or oil-fired furnace to solid fuel without a major retrofit of the furnace proper (although furnace volume may no longer be optimized). Advantages also accrue to the use of gasifiers with high fouling fuels. Gasifying the fuel may allow greater retention of fouling species in the solid char phase, with lower fouling rates on heat exchange equipment downstream of the gas combustor. Systems of this type are now under development for retrofitting existing biomass boilers. Cofiring of coal or other fossil fuel boilers with biomass can also be accommodated if a gasifier is added to the plant.

Gas burners need to accommodate the lower air–fuel ratio required with producer gas compared with natural gas and other hydrocarbon fuels [29]. The wood gas with the composition shown above has a stoichiometric air–fuel ratio of 1.5 kg $kg^{-1}$. Additionally, the adiabatic flame temperature of the gas is only 1960 K, compared with natural gas at 2330 K. The lower flame temperature can be an advantage in reducing the formation of thermal $NO_x$, but may also require a greater heat exchange area for the same capacity because of the lower furnace temperature differential.

*Internal-Combustion Engines*

A common desire is to use gasifiers to fuel internal-combustion engines, especially reciprocating spark-ignited or compression-ignited engines. Much effort has gone into the development of engines for remote or developing country applications for which other forms of power are unavailable or expensive and biomass exists as an indigenous resource. Small-scale units are often sought to supply much needed power for milling, water pumping, lighting, and other uses. Two main problems arise in the application of gasifiers in such circumstances: 1, gas cleaning and disposal of waste and 2, derating of engine power with simple air-blown gasifiers. The former problem is by far the most critical; the latter is solved by increasing design engine capacity at a somewhat higher cost.

***Gas Cleaning.*** For internal-combustion engines, gas cleaning is mandatory for long engine life. Reduction in tar yields has been of key design interest for gasifiers that fuel engines. The downdraft reactor, with its lower tar production compared with that of updraft units, has been the design of choice for reciprocating engine applications. However, most downdraft units still produce sufficient amounts of tar and particulate matter to require rather substantial gas cleaning ahead of the engine. In many cases, the gas cleaning train is composed of various wet and dry scrubbers and filters that generate contaminated waste streams requiring proper handling and disposal. Some progress has been made with the development of reversed flow reactors for postgasifier tar cracking [27]. Attempts have also been made to crack the tars internally in the gasifier by adding secondary air downstream of the main air supply to produce a hot zone ($>$900°C) for further tar cracking. When secondary oxygen is added, part of the primary gas and tar are burned, releasing heat with the intent of promoting tar-cracking reactions, yielding

**Table 3.42. Recommended limits for tar, particulate matter, and alkali metals in producer gas engine fuels (adapted from [27])**

| | Internal-Combustion Engines | Gas Turbines |
|---|---|---|
| Particulate matter (ppm)[a] | <30 (preferrably <5) | <30 (preferrably <2) |
| Particle size ($\mu$m) | <10 (preferrably <5) | <5 |
| Tar (mg $Nm^{-3}$) | <100 (preferrably <5) | |
| Alkali metals (ppm) | | <0.2 (preferrably <0.05) |

[a] ppm, parts in $10^6$.

noncondensable fuel gases. Such designs are so far not well developed. Other designs provide secondary oxidation external to the main reactor. Particle concentrations are typically reduced through hot-gas filtration and inertial separation. Cold filtration leads to the blinding of filter surfaces by condensing tars and other vapors, including water. Recommended limits for tar and particulate matter in producer gas engine fuels are listed in Table 3.42.

***Engine Derating.*** Power derating occurs primarily as a result of changes to the air capacity of the engine. Whereas a gasoline- or diesel-fueled engine operates with a stoichiometric mass air–fuel ratio of ~15, the stoichiometric air–fuel ratio for producer gas typically is ~1. With this much fuel in the engine cylinder and in the form of gas, there is less volume available for combustion air needed to burn the fuel; hence less fuel can be burned per cycle and the engine power is derated. This is a particular problem with naturally aspirated spark-ignited engines; supercharged engines can offset the deficiency somewhat by increasing boost pressure. Derating may also occur as a result of reductions in mechanical efficiency of the engine when operating on producer gas. To some extent, derating can be offset in spark-ignited engines by an increase in the spark advance and the compression ratio of the engine to take advantage of the higher octane rating (~120) of producer gas fuel compared with that of gasoline [30].

If an assumption of equal indicated efficiency and equal volumetric efficiency is made for the naturally aspirated engine, the derating of the engine is easily found in terms of the charge energy derating and the friction power. The brake work per cycle $W_b$ (in joules) is the product of the brake thermal efficiency ($\eta_b$) and the charge energy added on each cycle $Q_a$ (in joules):

$$W_b = \eta_b Q_a, \tag{3.56}$$

or, with the mechanical efficiency $\eta_m$ and the indicated efficiency $\eta_i$,

$$W_b = \eta_m \eta_i Q_a. \tag{3.57}$$

The theoretical intake mixture mass flow rate to the cylinder $m_t$ (in kg $s^{-1}$), based on piston displacement rate, is

$$m_t = \rho_i D / t_c, \tag{3.58}$$

where $\rho_i$ is the intake mixture density (in kilograms per cubic meter), $D$ is the displacement volume (in cubic meters), and $t_c$ is the time for one cycle (=120/rpm for four-stroke cycle engines). The mixture density is written from the air density $\rho_a$ and the fuel density $\rho_f$, at the intake conditions, and the mass air–fuel ratio AF as

$$\rho_i = \rho_a(1 + \mathrm{AF})/(\rho_a/\rho_f + \mathrm{AF}). \tag{3.59}$$

The brake mean effective pressure, $p_m = W_b/D$ (in pascals), can be written from the indicated mean effective pressure $p_i$, and the friction mean effective pressure $p_f$:

$$p_m = p_i - p_f = \eta_i \eta_v \rho_a Q_f/(\rho_a/\rho_f + \mathrm{AF}) - p_f. \tag{3.60}$$

The brake mean effective pressure is the average pressure in the cylinder yielding the same brake work output as the actual, time-varying cylinder pressure. Similarly, the indicated mean effective pressure is the equivalent average pressure giving the same indicated work (that developed on top of the piston by the combustion gases), and the friction mean effective pressure expresses the equivalent friction work. In tests of an engine operating on gasoline and producer gas, the friction work has been found to be similar for both fuels at the same engine speed [31]. The correlation of Bishop [32] has been found to model adequately the friction mean effective pressure. Bishop's correlation yields

$$p_f = 1.3 \times 10^5 u_p \mu_o/b + 1.7 \times 10^8 \mu_o^2(r + 15)/(\rho_o b^2), \tag{3.61}$$

where the mean piston speed (in meters per second) $u_p = 120\, s/n$, $\mu_o$ is the lubricating oil viscosity (in pascal seconds), $\rho_o$ is the oil density (in kilograms per cubic meter), $b$ is the cylinder bore (in meters), $s$ is the piston stroke (in meters), $n$ is the engine speed (in revolutions per minute), and $r$ is the compression ratio.

The ratio of $p_m$ of the engine fueled with producer gas to $p_m$ of the engine fueled with gasoline gives the derating of the engine. Note that Eq. (3.60) is generally valid and explains the drop in mechanical efficiency, equal to the ratio of brake work to indicated work, when a producer gas fuel is used. The friction mean effective pressure remains constant for both fuels at the same speed, but the charge energy with producer gas is substantially lower. Of the indicated work developed, a lower fraction of it appears as brake work; the remainder is needed to overcome the friction.

The actual derating of a small, spark-ignited gasoline engine rated 3.7 kW at 3600 rpm with a 6.2 compression ratio was observed for the engine operated on producer gas from rice husk [31]. Details of the gasifier and engine performance are included in Tables 3.43 and 3.44. The engine was operated on gasoline at the same speed for comparison.

The spark advance and air–fuel ratio were optimized for each fuel to give maximum brake power. Indicator diagrams giving engine cylinder pressure as a function of cylinder volume during the compression and expansion strokes are shown in Fig. 3.35. The producer gas achieves approximately two-thirds of the peak pressure on gasoline. The difference in charge energy per cycle is apparent from Table 3.44, with the producer gas supplying approximately half that of gasoline. Although the indicated efficiency of the engine is slightly higher on producer gas than gasoline, the friction power is roughly the same, such that the mechanical efficiency of the engine is substantially reduced

**Table 3.43. Composition and properties of producer gas from rice husk and gasifier performance at 3600 engine rpm [31]**

| Species | Volume Concentration (%) |
|---|---|
| CO | 15.81 |
| $H_2$ | 10.36 |
| $CH_4$ | 2.37 |
| $H_2O$ | 3.05 |
| $CO_2$ | 12.14 |
| $N_2$ | 55.06 |
| $O_2$ | 0.82 |
| $C_2H_2$ | 0.02 |
| $C_2H_4$ | 0.34 |
| $C_2H_6$ | 0.03 |
| Molecular weight (kg $kmol^{-1}$) | 26.69 |
| Lower heating value (MJ $kg^{-1}$) | 3.5 |
| Stoichiometric air–fuel ratio (kg $kg^{-1}$) | 0.94 |
| Lower heating value of stoichiometric air–gas mixture (MJ $kg^{-1}$) | 1.82 |
| Reactor cold-gas efficiency (%) | 56 |
| Specific gasification rate (kg $m^{-2}$ $h^{-1}$) | 172 |
| Brake specific fuel consumption (kg $kWh^{-1}$) | 2.5 |

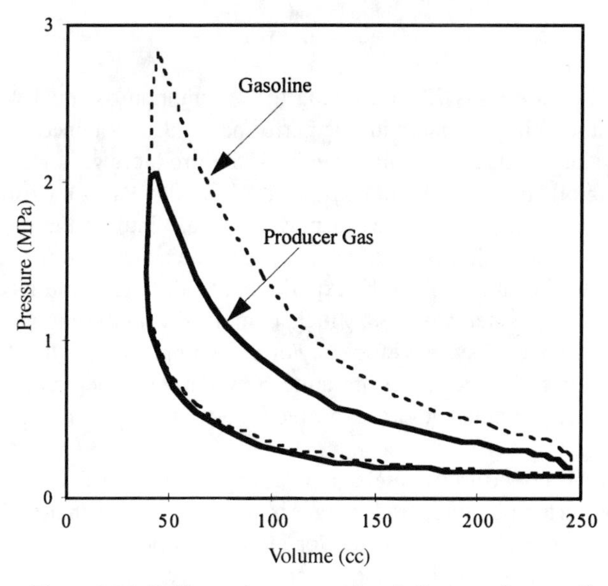


**Figure 3.35. Indicator (pressure–volume) diagrams for a small (3.7-kW) spark-ignited engine operated on producer gas and gasoline at 3600 rpm [31].**

**Table 3.44. Brake performance at 3600 rpm for a spark-ignited internal-combustion engine fueled with rice husk and gasoline[a] [31]**

| Parameter | Producer Gas | Gasoline |
|---|---|---|
| Air–fuel capacity | | |
| Mass flow rate of fuel (g $s^{-1}$) | 2.51 | 0.37 |
| Mass flow rate of combustion air (g $s^{-1}$) | 2.27 | 5.16 |
| Total flow rate of fresh mixture (g $s^{-1}$) | 4.79 | 5.53 |
| Air–fuel ratio | 0.91 | 14.1 |
| Equivalence ratio | 1.03 | 1.05 |
| Volumetric efficiency | 0.74 | 0.70 |
| Charge energy added per cycle (J) | 295 | 542 |
| Power, mep, and efficiency | | |
| Brake power (W) | 1335 | 3216 |
| Brake mean effective pressure (kPa) | 216 | 520 |
| Brake thermal efficiency (%) | 15 | 20 |
| Indicated work per cycle (J) | 85.2 | 149.4 |
| Compression work per cycle (J) | 39.7 | 46.7 |
| Expansion work per cycle (J) | 124.9 | 196.1 |
| Indicated power (W) | 2557 | 4481 |
| Indicated mean effective pressure (kPa) | 414 | 725 |
| Indicated thermal efficiency (%) | 29 | 28 |
| Friction power (W) | 1222 | 1265 |
| Friction mean effective pressure (kPa) | 198 | 205 |
| Mechanical efficiency (%) | 52 | 72 |
| Overall gasifier/engine efficiency (%) | 8 | |

[a] Totals may not agree because of rounding.

on producer gas (52% versus 72%). Making the assumption of equal volumetric and indicated thermal efficiencies and with the performance data obtained on gasoline, the lower heating value and the molecular weight of the producer gas, and the properties of the lubricating oil ($\rho_o = 886$ kg $m^{-3}$, $\mu_o = 8.86 \times 10^{-3}$ Pa s), the estimated power ratio (ratio of power on producer gas to power on gasoline) of the engine is 0.4 at 3600 rpm for 6.2 compression ratio (60% derating). The actual power ratio was 0.415. The predicted value is within 4% of the experimental value. The cold-gas efficiency of the simple open-core gasifier was 56%, which, combined with the engine brake thermal efficiency, yields an overall system efficiency of 8%, compared with the 20% efficiency on gasoline. Low overall efficiencies are quite common with simple, small systems of this type. Much higher efficiencies are expected for advanced power systems now under development.

Producer gas can also be used to fuel compression-ignited (diesel) engines. Because of its high self-ignition temperature, producer gas is normally used in dual-fuel applications. A small amount of diesel fuel is injected for ignition purposes and control of ignition timing. For dual-fuel diesel engines, the gas can generally supply up to 70% of the total fuel energy, sometimes more. At higher levels, the engine becomes more susceptible to severe knock resulting from the long ignition delay associated with producer gas [33, 34]. This long ignition delay is the same property that gives the gas its high octane rating for spark-ignited engines.

*Chemical Synthesis*

If the reactor uses enriched or pure oxygen or if the reactor is indirectly heated, the gas produced can be used for chemical synthesis. Such gas is referred to as synthesis gas, or syngas, and is not diluted with large amounts of atmospheric nitrogen. The cost of producing oxygen is high, however, and such systems are generally proposed for larger scales or for producing higher-value commodities, such as chemicals and premium liquid fuels. Methanol, an alcohol fuel, can be produced from syngas, for example, although it is more commonly produced from natural gas. Methanol, $CH_3OH$, is produced by the catalytic reaction

$$CO + 2H_2 = CH_3OH. \tag{3.62}$$

This reaction is favored at a low temperature (400°C) but high pressure (30–38 MPa). Zinc oxide and chromic oxide are common catalysts. With copper as the catalyst, the reaction temperature and pressure can be reduced (260°C, 5 MPa), but the copper is sensitive to sulfur poisoning and requires good gas scrubbing [26]. Fischer–Tropsch reactions can be utilized to produce a spectrum of chemicals including alcohols and aliphatic hydrocarbons. Temperature and pressure requirements are reduced and greater selectivity can be obtained by choice of catalyst.

Liquids, such as gasolines, can be produced by means of indirect routes involving gasification or pyrolysis to produce reactive intermediates that can be catalytically upgraded [35–37]. Liquids produced directly by pyrolysis are usually corrosive and cannot be directly used as engine fuels. Many products are also carcinogenic. Refining of some sort is generally necessary to produce marketable compounds. Liquid fuels can also be produced by direct thermochemical routes, such as by hydrogenation in a solvent with a catalyst present [38–40]. As mentioned above, however, the use of catalysts in biomass gasification systems may be limited for economic reasons by the relatively small scales of biomass systems [13].

*Advanced Power Generation Systems*

Gasification is currently thought to be an important technology for increasing the efficiency of biomass power generation. Conventional Rankine (steam) cycle power plants utilizing direct combustion of biomass currently operate on average at only ~20% station efficiency. Advanced systems having the potential to increase generation efficiency are under development.

One concept for advanced power generation includes the use of a pressurized biomass gasifier to produce fuel gas for a gas turbine engine (topping cycle). Pressurizing the gasifier avoids the need for a compressor to handle hot, dirty gas. Waste heat from the turbine can be recovered for use in a steam cycle (bottoming cycle) to produce additional power. An integrated gasifier combined cycle power generation concept of this type is schematically illustrated in Fig. 3.36. The efficiencies of these systems are calculated to be higher (>30%) than those of conventional Rankine cycles. Higher efficiencies offer key economic advantages by reducing the fuel requirements and hence fuel costs. Major engineering challenges include hot-gas cleaning to provide a gas of adequate quality to the turbine, and the development of reliable high-pressure reactors and fuel feeding systems. To avoid fouling, only very low alkali concentrations (generally well below 1 ppm) are tolerated in the turbine inlet gas, as shown in Table 3.42. Although

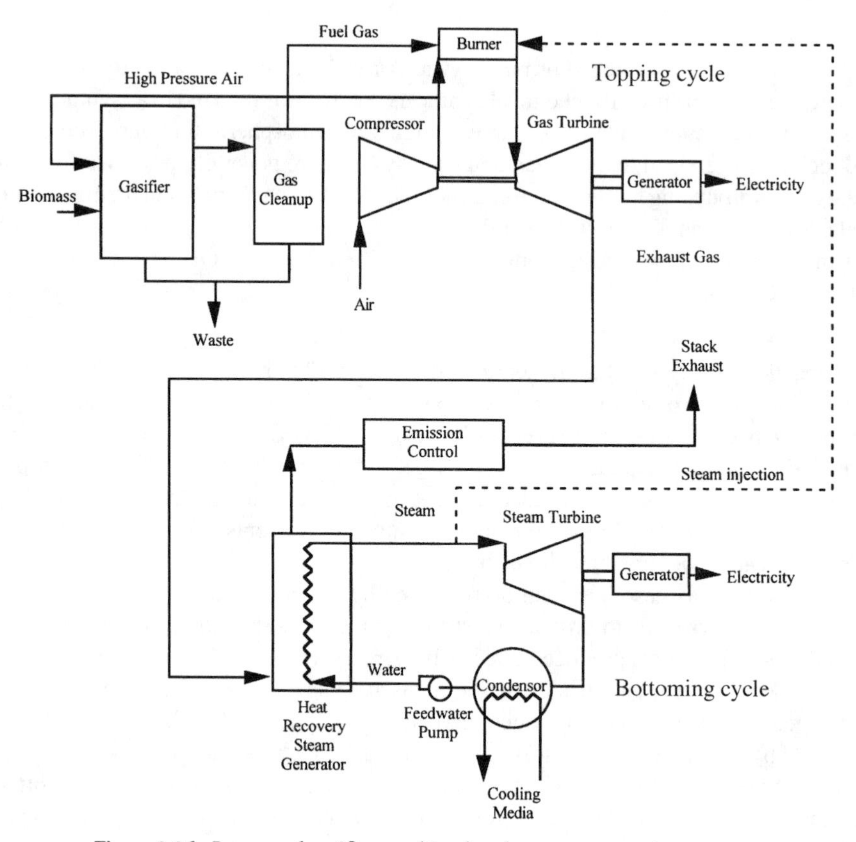


**Figure 3.36. Integrated gasifier combined cycle power generation concept.**

operation at gasifier temperature may retain more of the biomass fuel alkali in the solid char phase, sufficient amounts volatilize to require extensive separation ahead of the turbine. A number of strategies are proposed to do this, including condensing alkali compounds on particulate matter and separating the particulate matter on ceramic or other high-temperature filters. The use of a gasifier is thought to be an advantage over a direct combustor because heat loss in the gas cleaning system is of lesser concern, most of the fuel enthalpy being in the form of chemical energy in the fuel gas. Systems of this type are currently under development and several large-scale demonstration projects are proceeding [41], but commercial operating data are not yet available and many perceived advantages are still rather speculative.

Figure 3.36 also illustrates the possible use of steam injection for reducing thermal $NO_x$ emissions and enhancing the power output of the gas turbine. The high heat capacity of the steam compared with that of the combustion products leads to a power increase because of the larger enthalpy change for a given temperature drop [42]. The addition of steam reduces the flame temperature, which is beneficial for reducing thermal $NO_x$ formation.

Contaminant fouling of the turbine is eliminated if hot air, rather than hot combustion gas, is put through the turbine. In the indirect or downheated cycle, the combustion gas is passed through a heat exchanger to transfer heat to the compressor outlet air. In this case, there is no need to pressurize the gasifier, but the system efficiency is reduced because of the heat exchange, and contaminant fouling of the heat exchanger becomes important. A fully closed Brayton (gas turbine) cycle need not use air as the working fluid for the turbine. Helium and other fluids can be used to increase efficiency, although they are more expensive than air and leaks must be avoided. The turbine can expand to pressures below atmospheric to enhance efficiency further. Cogeneration systems combine electrical generation with thermal energy production and utilization (such as steam generation for food processing or district heating) and also enhance the overall fuel utilization efficiency.

Stirling engines are also under development for use with biomass [43]. In theory, the Stirling engine can operate at high efficiency. The Stirling engine utilizes external combustion with heat transfer at the engine head rather than internal combustion to supply the heat needed to operate the cycle. Most engines are coupled with direct-combustion furnaces, but gasifiers can be used as well. As an external-combustion engine, the Stirling avoids many of the problems with tars and particulate matter associated with the use of producer gas in internal-combustion engines. Aside from the conventional problems of sealing associated with Stirling engines, there is the need to avoid excessive head fouling and to reduce the dead volume in the heat exchanger to maintain high engine efficiency. The use of a gasifier in place of the direct-combustion furnace may prove of benefit in such circumstances, but limited development has occurred to date. Of the direct-combustion units operating on biomass, only small, low-power, low-efficiency engines have so far been developed commercially.

Another potential high-efficiency conversion device for utilizing producer gas is the fuel cell. The fuel cell is an electrochemical energy converter, theoretically unconstrained in efficiency by the Carnot cycle heat engine limitation. Two main classes of fuel cell are considered for producer gas applications: the molten carbonate fuel cell, operating at temperatures between 550 and 650°C, and the solid oxide fuel cell, operating at 1000–1100°C [44]. These cells are internally reforming with good tolerance of the main constituents in pyrolysis gas. Alkaline fuel cells are intolerant of $CO_2$, and phosphoric acid fuel cells tolerate CO only in concentrations below $\sim$1%. The solid polymer electrolyte fuel cell (also called the proton-exchange membrane cell) is also sensitive to CO. Molten carbonate and solid oxide cells are still under active development for more conventional fuels (e.g., natural gas and alcohols), and limited development has been pursued so far on the use of biomass-derived pyrolysis gases. The advantages of fuel cells in power system design make them attractive technologies for future development.

## References

1. Kaupp. A. and J. R. Goss. 1984. *Small Scale Gas Producer-Engine Systems*. Braunschweig/Wiesbaden, W. Germany: Priedr. Vieweg & Sohn.

2. Reed, T. B. 1985. Principles and technology of gasification. *Advances in Solar Energy*, eds. Boer, K. W. and J. A. Duffie, vol. 2, Chap. 3. Boulder, CO: American Solar Energy Society.
3. Hubbert, M. K. 1971. The energy resources of the earth. *Sci. Am.* 224(3):31–40.
4. Gibbons, J. H., P. D. Blair, and H. L. Gwin. 1989. Strategies for energy use. *Sci. Am.* 261(3):136–143.
5. Jenkins, B. M. 1992. Agricultural residues as energy conversion feedstocks. *Renewable Energy in Agriculture and Forestry* ed., Vanstone, B. J., pp. 145–188. Location: International Energy Agency.
6. Chum, H. L. and Baizer, M. M. 1985. The electrochemistry of biomass and derived materials. ACS Monograph 183. Washington, DC: American Chemical Society.
7. Schultz, T. P. and F. W. Taylor. 1989. Wood. *Biomass Handbook*, eds. Kitani, O. and C. W. Hall, New York: Gordon & Breach, pp. 133–141.
8. Sudo, S., F. Takahashi, and M. Takeuchi. 1989. Chemical properties of biomass. *Biomass Handbook*, eds. Kitani, O. and C. W. Hall, pp. 892–933. New York: Gordon & Breach.
9. Lynd, L. R. 1990. Large-scale fuel ethanol from lignocellulose. *Appl. Biochem. Biotechnol.* 24/25:695–719.
10. Wyman, C. E. and N. D. Hinman. 1990. Ethanol, fundamentals of production from renewable feedstocks and use as a transportation fuel. *Appl. Biochem. Biotechnol.* 24/25:735–753.
11. Shafizadeh, F. 1968. Pyrolysis and combustion of cellulosic materials. *Adv. Carbon Chem.* 23:419–474.
12. Antal, M. J. 1983. Biomass pyrolysis: a review of the literature. Part 1—carbohydrate pyrolysis. *Advances in Solar Energy*, eds. Boer, K. W. and J. A. Duffie, vol. 1, pp. 61–111. Boulder, CO: American Solar Energy Society.
13. Antal, M. J. 1985. Biomass pyrolysis: a review of the literature. Part 2—lignocellulose pyrolysis. *Advances in Solar Energy*, eds. Boer, K. W. and J. A. Duffie, vol. 2, Chap. 4, pp. 175–255. Society, Boulder, CO: American Solar Energy Society.
14. Jenkins, B. M. and J. M. Ebeling. 1985. Correlation of physical and chemical properties of terrestrial biomass with conversion. *Proceedings Energy from Biomass and Wastes IX*, ed. Klass, D. L., pp. 371–400. Chicago: Institute of Gas Technology.
15. Jenkins, B. M. 1989. Physical properties of biomass. *Biomass Handbook*, eds. Kitani, O. and C. W. Hall, pp. 860–891. New York: Gordon and Breach.
16. Jenkins, B. M., R. R. Bakker, and J.B. Wei. 1996. On the properties of washed straw. Biomass Bioenergy 10(4):177–200.
17. Shafizadeh, F. 1981. Basic principles of direct combustion. *Biomass conversion processes for energy and fuels*, eds. Sofer, S. S. and O. R. Zaborsky, pp. 103–124. New York: Plenum.
18. Tiangco, V. M., B. M. Jenkins, and J. R. Goss. 1996. Optimum specific gasification rate for static bed rice hull gasifiers. Biomass Bioenergy 11(1):51–62.
19. Manurung, R. K. and A. A. C. M. Beenackers. 1994. Modeling and simulation of an open-core downdraft moving bed rice husk gasifier. *Advances in Thermochemical*

*Biomass Conversion*, ed. Bridgwater, A.V., pp. 288–309. London: Blackie Academic and Professional.

20. Graboski, M. and R. Bain. 1981. Properties of biomass relevant to gasification. *Biomass Gasification, Principles and Technology*, ed. Reed, T. B., pp. 41–71. Park Ridge, NJ: Noyes Data Corp.
21. Gumz, W. 1950. *Gas producers and blast furnaces, Theory and Methods of calculation*. New York: Wiley.
22. Bining, A. S. and B. M. Jenkins. Thermochemical reaction kinetics for rice straw from an approximate integral technique. ASAE Paper 926029. St. Joseph, MI: ASAE.
23. Ebeling, J. M. and B. M. Jenkins. 1987. Thermogravimetric analysis and kinetic reaction rates for rice hulls and rice straw. ASAE Paper 876028. St. Joseph, MI: ASAE.
24. Gaur, S. and T. B. Reed. 1994. Prediction of cellulose decomposition rates from thermogravimetric data. Biomass Bioenergy 7(1-6):61–67.
25. Jenkins, B. M., R. R. Bakker, L. L. Baxter, J. H. Gilmer, J. B. Wei. 1996. Combustion characteristics of leached biomass. *Developments in Thermochemical Biomass Conversion*, eds. Bridgwater, A. V. and D. G. B. Boocock, pp. 1316–1330. London: Blackie Academic and Professional.
26. Probstein, R. F. and R. E. Hicks. 1982. *Synthetic Fuels*. New York: McGraw-Hill.
27. Beenackers, A. A. C. M. and K. Maniatis. 1996. Gasification technologies for heat and power from biomass. *Biomass for Energy and the Environment*, eds. Chartier, P. *et al.*, vol. 1, pp. 228–259. New York: Pergamon/Elsevier Science.
28. Kaupp, A. 1983. Gasification of rice hulls, theory and praxis. Unpublished Ph.D. dissertation, Davis, CA: University of California.
29. Schaus, O. O., J. C. K. Overall, and G. K. Lee. 1985. Development of an industrial gas burner for the utilization of low calorific value gases produced by biomass and waste utilization processes. *Energy from Biomass and Wastes IX*, pp. 405–436. Chicago: Institute of Gas Technology.
30. National Research Council. 1983. *Producer Gas: another Fuel for Motor Transport*. Washington, DC: National Academy.
31. Jenkins, B. M. and Goss, J. R. 1988. Performance of a small spark-ignited internal combustion engine on producer gas from rice hulls. *Research in Thermochemical Biomass Conversion*, eds. Bridgwater, A.V. and J. L. Kuester, pp. 1057–1070. London: Elsevier.
32. Bishop, L. N. 1964. Effects of design variables on friction and economy. SAE paper 812A. Warrendale, PA: Society of Automotive Engineers.
33. Chancellor, W. J. 1980. Alternate fuels for engines. *Proceedings 49th Rural Energy Conference*. Davis CA: University of California.
34. Ogunlowo, A. S., W. J. Chancellor, and J. R. Goss. 1981. Dual-fuelling a small diesel engine with producer gas. *Trans. ASAE* 24(1):48–51.
35. Kuester, J. L., C. M. Fernandez, T. Wang, and G. Heath. 1985. Liquid hydrocarbon fuel potential of agricultural materials. *Fundamentals of Thermochemical Biomass Conversion*, pp. 875–895. London: Elsevier.

36. Prasad, B. V. R. K. and J. L. Kuester. 1988. Process analysis of a dual fluidized bed biomass gasification system. *Ind. Eng. Chem. Res.* 27:304–310.
37. Kuester, J. L. 1991. Conversion of guayule residues into fuel energy products. *Bioresource Technol.* 35:217–222.
38. Chornet, E. and R. P. Overend. 1987. Liquid fuels from lignocellulosics. *Biomass, Regenerable Energy*, eds. Hall, D. O. and R. P. Overend. Chichester: Wiley.
39. Elliott, D. C., D. Beckman, A. V. Bridgwater, J. P. Diebold, S. B. Gevert, and Y. Solantausta. 1991. Developments in direct thermochemical liquefaction of biomass: 1983–1990. *Energy Fuels* 5:399–410.
40. Bridgwater, A. V. and S. A. Bridge. 1991. A review of biomass pyrolysis and pyrolysis technologies. *Biomass Pyrolysis Liquids Upgrading and Utilization*, eds. Bridgwater, A. V. and G. Grassi, London: Elsevier.
41. Bain, R. L., R. P. Overend, and K.R. Craig. 1996. Biomass-fired power generation. *Proceedings Biomass Usage for Utility and Industrial Power*. New York: Engineering Foundation.
42. Weston, K. C. 1992. *Energy Conversion*. St. Paul, MN: West Publishing Company.
43. Carlsen, H. 1996. Stirling engines for biomass: state-of-the-art with focus on results from Danish projects. *Biomass for Energy and the Environment*, eds. Chartier, P. *et al.*, pp. 278–283. Oxford: Pergamon/Elsevier Science.
44. Blomen, L. J. M. J. and M. N. Mugerwa. 1993. *Fuel Cell Systems*. New York: Plenum.

## 3.3 Solid Fuels

*P. C. Badger*

Because of their ease of use and availability, biomass fuels have traditionally been, and in the near future will continue to be, used in solid fuel form. Solid fuels are also easy to store, transport, and handle manually—features that are especially attractive to developing countries where bioenergy use is highest. The solid fuels reviewed in this section include—in their natural order of use because of cost, convenience, and availability—firewood and charcoal, organic wastes, and energy crops.

### 3.3.1 Fuel Wood and Charcoal

Fuel wood, because of its widespread availability and ease of use, was the first fuel used regularly by humans. For these same reasons, fuel wood, along with charcoal, continues to be a fuel of choice in lesser developed countries. In addition to the primary use of these fuels for cooking, they are essential in developing countries for many small- to medium-scale trades and industries such as brick making, textile manufacture, baking, and brewing. In rural communities, fuel wood and charcoal can have important other uses such as crop drying or curing. Where other fuels are scarce and charcoal is available, it is also used in industrial metallurgy applications [1].

These fuels are also important to developing countries because they do not have to be imported with foreign currency, can theoretically be available throughout most parts of a country, and are renewable. These fuels also provide a means of employment and income for many people in many situations for which there are few or no job opportunities [1].

***Fuel Wood***

Worldwide in 1990, roughly 1.4 billion (i.e., $\times 10^9$) tons of biomass (mainly fuel wood and charcoal) were used for energy, accounting for over half of the wood harvested. This amount accounted for 70% of energy used in developing countries but, as shown in Table 3.45, was equivalent to only 14% of the world's total energy use [1].

In 1978 the Food and Agriculture Organization (FAO) of the United Nations estimated the per capita worldwide production of fuel wood, including charcoal, at 0.37 $m^3$. It also estimated that the per capita use was 0.46 $m^3$ in developing countries and 0.13 $m^3$ in developed countries. Developed countries have always used more energy per capita but have used relatively little wood and charcoal. In contrast, developing countries have used less energy per capita but have obtained a relatively large share of their energy from wood and charcoal [2].

In 1990, ~40% of the world's people, or 2 billion people, used fuel wood and charcoal as their main source of energy for residential heating and cooking—mostly the latter. This number is expected to reach 3 billion by the year 2000. Among the present users, half currently face wood supply shortages, and 5% are experiencing fuel wood famine [1]. Part of the supply problem is one of local availability since resources may be unavailable because of physical (e.g., distance, topography, transportation available) or social reasons (e.g., state reserves, customary practices, social differentiation) [3].

Fuel wood shortages create hardships for families in several ways. Wood collectors, usually women and children, have to spend increasing amounts of time foraging for and transporting wood and, in the process, taking them from more productive activities. Health can suffer when fuel-wood shortages lead to less cooking of nutritional foods, less fuel available to boil water, and less heat available in winter [3].

Fuel wood harvesting has been blamed for rapid worldwide deforestation; however, studies show that 70% of the permanent deforestation in Africa has been a result of land clearing for agricultural purposes [3]. There is a growing awareness that a high percentage of fuel wood (and wood used for a myriad of other uses) comes not from forests but from rural landscapes [4] including farmland and common property or communal land such as river banks, road sides, village commons, and fallow lands [5]. Wood from these sources is usually viewed as a common property resource, leaving it unaccounted for, unregulated, and vulnerable to overharvesting. Subsistence users are perhaps the worst offenders, using wood fuels for expediency and not out of preference or concern for environmental matters [1].

To meet the growing shortage, one FAO study estimated that ~28 $Mm^3$ of wood fuel per year would be needed in sub-Saharan Africa by the year 2000. In order to meet this demand, annual plantings would have to increase by a factor of 20 and involve the establishment of roughly 19 Mha of fuel wood plantations—over 100 times more than exists today. Throughout the world, some 50 Mha would have to be planted at a cost of $50 billion to avert a fuel wood shortage by the year 2000. The present rate of planting will meet only 5%–10% of these requirements [6].

**Table 3.45. Fuel wood and charcoal: the world picture [1][a]**

| Regions/Countries | I | II | III | IV | V | VI | VII | VIII | IX |
|---|---|---|---|---|---|---|---|---|---|
| | | Fuel wood [including biomass (all forms) charcoal] FAO | | Fuel wood Conventional | Biomass Conventional | Total Average Per Capita Energy Consumption | | Biomass: % Total Energy Consumption | |
| | Conventional (PJ) | (PJ) | (PJ) | (I + II) (PJ) | (I + III) (PJ) | (IV/pop.) (GJ/cap) | (V/pop.) (GJ/cap) | (II/IV) (%) | (III/V) (%) |
| World | 305,885 | 18,625 | 49,943 | 324,511 | 355,829 | 65 | 71 | 6 | 14 |
| Industrialized | 235,456 | 2895 | 7312 | 238,351 | 242,768 | 200 | 203 | 1 | 3 |
| N. America | 91,636 | 1319 | 4027 | 92,955 | 95,662 | 345 | 355 | 1 | 4 |
| Europe | 64,832 | 609 | 1527 | 65,441 | 66,359 | 132 | 134 | 1 | 2 |
| Russia | 57,690 | 930 | 1720 | 58,620 | 59,411 | 207 | 210 | 2 | 3 |
| Asia | 17,106 | 6 | 6 | 17,112 | 17,112 | 135 | 135 | 0 | |
| Oceania | 4192 | 31 | 31 | 4223 | 4223 | 216 | 216 | 1 | |
| Developing | 70,430 | 15,730 | 42,631 | 86,160 | 113,061 | 23 | 30 | 18 | 38 |
| Africa | 7363 | 4481 | 9160 | 11,845 | 16,524 | 20 | 28 | 38 | 55 |
| C. America | 5488 | 504 | 868 | 5992 | 6356 | 42 | 45 | 8 | 14 |
| S. America | 8328 | 2419 | 2858 | 10,748 | 11,186 | 38 | 40 | 23 | 26 |
| Asia | 49,146 | 8263 | 29,689 | 57,409 | 78,835 | 20 | 28 | 14 | 38 |
| Oceania | 105 | 62 | 56 | 167 | 161 | 29 | 28 | 37 | 35 |

[a] FAO fuel wood plus charcoal consumption is compared with a total biomass consumption for each country.

As shown in Table 3.46, the amount of land suitable for reforestation is quite large [7]. However, reforestation costs can be high and, since crops take years to mature and may be less profitable than other crops, local cooperation can be difficult to obtain [4]. Lands available may also not be readily accessible to fuel wood users.

Because of the interaction and the complexity of the issues, conflicts, and stresses involved, coupled with difficulty in obtaining accurate information, there does not appear to be an easy solution to the growing fuel-wood shortage [4]. For the present, it appears that fuel wood shortages will persist well into the future.

***Charcoal***

Charcoal was discovered in the ashes of partially burned fires and became widely adopted as a fuel because of, relative to wood fuels, its smokeless nature, more efficient burn, and ease of transportation and storage. As a result, the art of making charcoal has been practiced in virtually every part of the world for centuries. The basic principle underlying the charcoal manufacturing process is that of incomplete combustion caused by limiting air to the process.

Charcoal is carbonized carbonaceous material, often derived from wood, wood wastes, and agricultural and food processing by-products. It plays an important role in the economies and daily lives of people throughout the world, as a fuel for heat and cooking, as a major or minor metallurgical material in some countries, and in developed countries as a fuel for outdoor cooking. An estimated $400 \times 10^6$ $m^3$ of fuel wood is converted to charcoal worldwide every year according to the FAO of the United Nations [8].

Much of the charcoal produced in developing countries is used for household heating and cooking but it also finds industrial uses. In Brazil, which ranks ninth in world steel production, for example, 34% of its pig iron production and 18% of its steel production was based on charcoal in 1994 [9], and it plays a smaller but significant role in metallurgy in many other countries. In 1995, 824,000 tons of charcoal were produced in the United States [10], almost all of it for outdoor cooking.

Charcoal consists mostly of 50%–95% carbon, 5%–40% volatile matter (tar and water), and 0.5%–5% ash, depending on the feedstock and the processing conditions [8]. It has a higher heating value of ~28 kJ/kg, or if briquetted, 23 to 25 kJ/kg because of the binder and sometimes fillers added to briquettes. When emerging from the kiln, charcoal contains ~1% water but absorption of water is rapid and the product sold can contain 5%–10% water. Yields, based on air-dried weight of the feedstock, range from 50% for a 300°C carbonization to 25% for a 1000°C carbonization, with an average of ~30% for the more usual 500–600°C carbonization end point. Of the various process conditions (heating rate, final temperature, residence time, and kiln atmosphere) the carbonization or final temperature is the most important in determining quality [9].

So-called lump charcoal is produced from solid wood such as logs, end cuts, and slabs. Charcoal briquettes can be produced from charcoal fines that are combined with anything from petroleum coke to wood mill wastes [8].

Charcoal reacts with oxygen in the air at a glowing red heat to form carbon monoxide gas, which then combines with additional oxygen to burn with a barely visible blue flame. Since the fire is relatively smokeless and odorless, flues are usually not used with

**Table 3.46. Prospects for plantations in developing regions, *million hectares* [7][a]**

| Country | Cropland Measures | | | | Alternative Measures of Land Areas Potentially Available for Plantations | |
|---|---|---|---|---|---|---|
| | Present Cropland | Potential Cropland | Cropland Required in 2025[b] | Excess Potential Cropland in 2025[c] | 10% of Cropland plus Permanent Pasture plus Forests and Woodlands | Degraded Lands Suitable for Reforestation[d] |
| Latin America | 179.2 | 889.6 | 269 | 621 | 171 | 156 (+ 32) |
| Africa | 178.8 | 752.7 | 268 | 484 | 176 | 101 (+ 148) |
| Asia (except China) | 348.3 | 412.5 | 522 | −110 | 111 | 169 (+ 150) |
| Total | 706.3 | 2054.8 | 1059 | 995 | 458 | 426 (+ 330) |

[a] Except for the last two columns on the right, the data here are for the 91 developing countries for which the FAO estimated potential cropland areas.

[b] The Response Strategies Working Group of the Intergovernmental Panel on Climate Change has estimated that the area required for cropland in developing countries will increase 50% by 2025.

[c] This is the difference between the potential cropland and the cropland requirements in 2025.

[d] The data refer to all countries on these continents, including China. The first entry is the sum of the land areas in logged forests, forest fallow, and deforested watersheds, all of which are estimated to be suitable for reforestation. The number in parenthesis is 1/5 of the desertified drylands, the fraction of desertified drylands in developing countries estimated to be suitable for reforestation. Reprinted with permission from *Renewable Energy-Sources for Fuels and Electricity,* eds. Johansson, T. B., *et al.* 1993, Washington, DC and Covelo, CA: Island Press.

charcoal stoves. The fact that wood can be burned in compact portable stoves without flues partially explains charcoal's wide degree of use [4].

*Charcoal Feedstocks*

Charcoal production in developing nations is a highly refined art that makes use of local and regional resources. Although roundwood is the preferred feedstock, where it is unavailable, nonwoody materials such as nut shells and husks, straw and reeds, garbage wastes, sugar cane bagasse, bamboo, food processing wastes, bark, agricultural residues, and other organic materials are sometimes also used as feedstocks [8]. However, nonwoody materials (and wood in the form of small particles, e.g., sawdust) produce charcoal in the form of a fine powder that must be briquetted in order to be used for most applications. Briquetting also requires a binder and usually starch is used. Starch may not readily be available if food shortages occur or if agricultural sources are not nearby. The excessive use of agricultural residues can also harm the soil ecosystem and is usually not encouraged. However, in some places, crop residues such as sugar cane stalks can be excessive and must be partially removed anyway [8].

Both hardwoods and softwoods (conifers) make satisfactory charcoal; however, hardwood charcoal is usually stronger. Softwoods also tend to produce more tars and less acetic acid and similar products in the pyrolysis oil fraction. When pine wood is used, it is also usually more cost effective to chip the wood, extract the rosin, and use the residue for charcoal feedstock [8].

Usually roundwood is the preferred feedstock and requires ~5 tons per ton of charcoal and as a rule involves cutover and degraded forests [11]. An unmanaged but controlled cutover forest can maintain a mean annual increment of 2–4 $m^3$/ha per year on a 40-year rotation. Savannah forests, typically providing 35 $m^3$/ha, are the preferred type of forest, rather than a rain forest, because the wood is usually suitable only for fuel wood and thus cheap and high in lignin (which produces better charcoal). The long, dry seasons promote drying, thus increasing the charcoal yield and minimizing insect and fungal damage. The classic charcoal production sources of Africa, South America, and Asia are nearly all Savannah forests [11].

Population pressures and diminishing resources, however, are forcing advances into rain-forest use, which provides a high yield: up to 100 $m^3$/ha even after removal of marketable wood. The wood of rain forests, however, is high in moisture and rapidly deteriorates, decreasing yields and increasing production costs. The United Nations, among many organizations, has addressed the issue of diminishing forest resources through promotion of managed plantations of fast-growing species such as eucalyptus to compensate for these problems [11]. These efforts, beset by political and ethnic upheavals and population pressures combined with diminishing natural resources, may reflect a change in the way traditional charcoal is made, but the ancient craft of charcoal production, so vital a part of the culture of developing countries, is unlikely to be lost [11].

*Charcoal Production Processes*

Charcoal is produced by the heating of biomass materials under conditions of limited oxygen availability, a process called carbonization in which complex carbon-containing chemicals are broken down into simpler ones such as elemental carbon and carbon-containing tars. The chemical term for the process is called pyrolysis.

When carbonaceous material is heated in an atmosphere containing insufficient oxygen to ignite the material, it passes through several stages. Initially, it remains slightly above 100°C until the wood is bone dry. When the free moisture is evaporated, the temperature rises and at ~270°C, the wood begins to decompose, emitting a gas composed mainly of carbon monoxide, carbon dioxide, acetic acid, and methanol. At this point the reactions are still endothermic. Between 270 and 290°C the reaction becomes exothermic and proceeds spontaneously. Between 290 and 400°C the breakdown continues, producing a mixture of noncondensable, combustible gases called wood gas and condensable gases that form a complex liquid called pyroligneous acid (more commonly called pyrolysis oil) [12]. At 400°C the transformation to charcoal is essentially complete, but the charcoal still contains considerable tar and the temperature is usually taken to 500°C to drive off some of these tars. At 300°C, for example, charcoal contains 31% volatiles; at 700°C it contains 7% volatiles [11].

Originally the process of manufacturing charcoal was to stack wood as densely as possible into pits, cover the wood with dirt except for holes for air access, and then ignite the pile. The earth sealed the wood from oxygen and insulated the process against excessive heat loss. The operator controlled the admission of air until charring was complete. Air flow was then cut off and the contents allowed to cool. Later, instead of constructing a pit, the operator carefully piled the wood on the ground, covered it with dirt, and ignited it. In spite of technological advancements, and because of their simplicity, low cost, and capability of producing good-quality charcoal with proper management, pit-and-pile processes are still used in some parts of the developing world where labor is relatively inexpensive. Another advantage to pit-and-pile technologies is that they are amenable to small-scale production and to the lifestyle of a small farmer in developing countries [11].

Eventually masonry and steel retorts were developed to increase the yield efficiency further, speed up the process, and reduce labor requirements [11]. As used here, a kiln or furnace is defined as a device that burns part of the charge in order to carbonize the remainder and from which the by-product heat is capable of being recovered. A retort is defined as a container in which heat is applied from an external source. Both systems typically burn the volatiles produced in the charring process to provide the bulk of the heat necessary for the process. Retorts remove them before they are burned. This provides the retort operator with the options of condensing some or all of the gases to remove chemicals or of burning some or all of them [13]. The charcoal yield from retorts is 50% higher than that from kilns [12].

The production of charcoal in the 19th century industrialized world was closely associated with the production of chemicals such as acetic acid, acetone, and methanol that were collected in the off gas from sealed cast iron retorts. As more efficient means of producing these products from petrochemicals evolved in the 1920's, wood distillation declined, but the demand for activated charcoal in the food and chemical industries increased and the demand for charcoal remained high [14]. As a result, new, faster, more efficient, and less labor-intensive methods of charcoal production evolved.

Currently in the industrialized world, charcoal is most commonly manufactured by a batch process either in metal or concrete kilns operating on 7- to 14-day cycles or in continuous furnaces producing a minimum of 1 ton/h. According to one industrial source, 50%–75% of the charcoal produced in the United States is produced in these batch kilns [15], mostly in Arkansas, Missouri, Tennessee, Kentucky, and Texas, where mesquite charcoal is becoming a kind of cult product [16] for outdoor chefs.

Kilns are made from brick, concrete, and metal. Shapes used for brick and concrete included beehive and rectangular boxes with curved roofs. Metal kilns consisted of a wider variety of shapes because of the greater flexibility of metal.

The Reichert process, which uses inert or reducing gas to prevent ignition, was the first to heat successfully the charge inside the retort directly instead of through metal walls [12]. A further improvement was the continuous Lambiotte process [12] developed in the 1940's (US patent 2,289,917). It is a vertical retort that produces lump charcoal from slab and roundwood. The wood is continuously added at the top and progresses downward to emerge at the bottom as charcoal. Carbonization is achieved by collecting the off gases near the top of the retort, burning part of them in an external furnace, and injecting the resulting inert gas into the center of the retort. The rest of the gas is scrubbed to remove tars and cooled and injected at the retort bottom to cool the charcoal before it exits the retort. Alternatively, the retort off gases can be used to produce heat and power, and fossil fuel can be used to provide the carbonization heat.

All the above methods require chunk wood, logs, or processed wood in pieces such as slab wood to provide the proper gas and charcoal flow characteristics. Part of the reason for this is that for carbonization of small particles, the particles must be kept moving [12]. The increasing production of small-sized wood such as sawdust, shavings, bark, and sanding wastes that could not be processed in these kilns led to the development of a kiln adapted from metallurgy called the Herreshoff furnace [12]. The Herreshoff furnace consists of a vertical cylinder 6–8 m in diameter containing 4–12 vertically spaced, circular hearths with material introduced at the top. The hearths contain drop holes alternately on the outside rim and at the shaft. As the material moves down through the kiln, rakes on a centrally mounted slowly turning shaft move the material toward the drop holes in the hearths until it discharges at the bottom as charcoal, where it is cooled in a screw conveyor sprayed with water. The kiln is started by gas or oil burners on each hearth to raise the temperature to $\sim$600°C, causing the feedstock to ignite. Once the hearths reach 900–1000°C, the burners are shut off and the kiln is operated continuously at 24 h/day.

Other continuous processes that use fluidized beds, rotary kilns, and screw conveyors have also been suggested for making charcoal from small-sized wood. There are also pyrolysis processes carried out in rotary kilns; examples in the US include Angelo Industries in Arkansas and Industrial Boiler in Georgia [12].

Whereas charcoal is produced by use of internal heating of the charged wood by burning part of it, the gases usually pass into the atmosphere as smoke, creating atmospheric pollution. The recovery of these gases can prevent pollution and potentially improve the process economics if the gases are used as an energy source for the charcoal process or other applications or by the recovery of chemicals [8].

A steel retort facilitates the capture of the gas that, once flowing adequately, can be used to fuel the process or can be used for other applications [17]. Typically wood gas contains 23% carbon monoxide, 2% hydrogen, 17% methane, 38% carbon dioxide, 18% nitrogen, and 2% oxygen. It has a heating value of ~10.8 $MJ/m^3$ and can be used as a substitute for natural gas in many applications [11]. The presence of toxic gases, especially carbon monoxide, requires safety precautions such as locating kilns and retorts outdoors.

The pryolysis oil, if allowed to stand, separates into a water layer, a layer of insoluble tars, and a solution tar layer containing a host of organic chemicals. The composition varies depending on the feedstock and the carbonization conditions. For Northern Hemisphere hardwoods, for example, 1000 kg of air-dried wood produces 50 kg of acetic acid, 16 kg of methanol, 8 kg of acetone and methyl acetone, 190 kg of soluble tars, and 50 kg of insoluble tars. The pyrolysis oil will ignite if heated to 100°C and it can be used as a substitute for fuel oil. However, the calorific value of pyrolysis oil produced as a by-product of charcoal production is relatively low, and the tars can make it difficult to handle [12].

Wood distillation was once an important source of a variety of chemicals but it has been all but displaced by other sources. The number of by-product recovery plants in the US dwindled from over 100 in the 1920's until the last one closed in 1969. Some insoluble tars are used as a veterinarian disinfectant. A few specialty chemicals can be extracted from the soluble tars and some can be used in brick making to increase porosity, but its most promising use is as a fuel [14].

*Briquettes*

Beginning in the 1950's, production of briquettes, which can be made from bark and wood waste products such as sawdust and mill shavings, became an important product [10]. Briquettes also appeal to many because they are relatively free of dust and uniform in size. Briquettes can be many shapes and sizes but the usual barbecue briquette is a pillow-shaped object ~50 mm × 50 mm and 25 mm thick.

Briquettes are composed of fine charcoal plus a binder and sometimes a filler. Wood ash, coal or petroleum solids, sawdust, or calcium carbonate (in the form of ground limestone, chalk, or ground sea shells) are sometimes used as fillers, ostensibly to provide a slower burn, but they are also significantly cheaper than charcoal. Sodium nitrate is also a common additive to facilitate burning. A binder, usually a starch of some type, is used in a quantity equal to ~8%–10% of the weight of charcoal [8]. Waxes, clay, tar, and pitch have also been used as binders; cornstarch is commonly used in the US [11].

In the manufacture of briquettes, the charcoal is first pulverized, usually in a hammer mill, and screened to remove gross impurities. Charcoal produced from bark, twigs, and leaves usually contains soil, mineral sand, and clay particles that increase the ash content. Most such contaminants can be separated by screening the pulverized charcoal and rejecting the material passing a 2–4-mm screen. Starch in the form of a paste and a filler are then added and the mixture is blended in a ribbon blender and metered to a briquetting machine [8]. This device is essentially a roll press consisting of two steel rollers with matching indentations. The most difficult control process is controlling the feed rate to fill each indentation of the rapidly rotating rolls which are pressed together by several bars of pressure. The briquettes fall from the rolls onto a moving belt and enter a tunnel dryer where they are dried at ~80°C from ~30%–5% water. Drying sets the starch, forming a briquette that can be used in place of lump charcoal for residential applications. Generally, only briquettes that use pitch or tar are suitable for industrial uses [12]. In developed countries where barbecuing is the principal use, sales are seasonal and storage can become a considerable expense.

*Uses of Charcoal*

The principal worldwide use for charcoal is for preparing food, and the preferred form is in lumps because of its ease of handling and ignition [8]. The general public in the United States directly encounters charcoal only in the ubiquitous bags of briquettes used for outdoor cooking, for which ~95% of US charcoal is produced [11]. The remaining US charcoal is used for minor uses in metallurgy, feed additives, a raw material for production of activated carbon, and in horticulture [12]. In the developing world, heating and cooking is usually the primary use, although in countries with steel industries, few coal reserves, and abundant wood resources charcoal often substitutes for coke [9].

The barbecue fuel market has little competition; however, industrial users can usually substitute coal, coke, petroleum coke, or lignite fuels for charcoal. The following properties are of significance to industrial markets [8]:

- low sulfur content,
- high ratio of carbon to ash,
- relatively few and unreactive inorganic impurities,
- stable pore structure with high surface area,
- good reduction ability,
- almost smokeless.

Most specifications for charcoal have come from the steel or the chemical industries, but these specifications tend to be used even if the charcoal is intended for the barbecue market [11]. Common specifications are 5%–15% moisture, a volatile content of ~30%, and an ash content of 3%. Low-volatilty charcoal burns cleanly but is difficult to light; high-volatility charcoal lights easily but may burn with a smoky flame. High-volatilty charcoal is also less friable, another important property in the barbecue market.

### 3.3.2 Organic Residues

As discussed here, organic residues are materials derived from plants that have low or negative values (disposal costs), a feature that can make them an attractive source of low-cost energy. In addition to their low cost, in many cases, they have the added advantages of being already collected and, within each type, possessing relatively uniform characteristics. Depending on the intended use, they may be already preprocessed in terms of particle size and moisture content. Some potential disadvantages can include high moisture contents, availability in only limited quantities, and seasonal availability.

In industrialized countries, organic residues can play an important role for biomass energy by (1) serving as a source of low-cost fuels during the developmental phase of new biomass energy conversion technologies and (2) serving as a temporary fuel source until energy crop plantings can mature and provide a sustainable resource. In developing countries, organic residues can be a significant fuel source, especially as forest reserves dwindle.

Information on the following residues is provided in this section:

- forestry crop and processing residues,
- agricultural crop and processing residues,
- food processing wastes,
- animal manures,
- municipal solid waste (MSW) including landfills and urban wood waste,
- sewage sludge.

***Forestry Crop Residues***

Forestry crop residues, as defined here, include low-value materials resulting from harvesting, thinning, and land-clearing operations for replanting (harvesting and logging residues are considered the same here). They may include limbs, tops, stumps, roots, and undersized trees. Residues from harvesting can be recovered either before, during, or after harvesting of the desired or economical tree portions [18]. Normally these residues are not recovered because their economic value is less than their value for fuel or other markets. However, there are certain silviculture operations, such as thinning, land clearing for replanting, and residue removal for fire prevention, for which the residues must be handled anyway, and their recovery and use can help offset the cost of these operations.

The amount of residue generated depends on several factors, including tree species and size, the type and method of silviculture operation conducted that results in residue, and the machinery used. For example, a study in the Southern US found that whole-tree harvesting resulted in approximately 60% more residue than would be recovered after a conventional (selective) harvest [18]. The amount available must take into account soil nutrient requirements, the efficiency of recovery, and the accessibility to the site due to physical or social limitations.

Estimates of the amount of harvesting residues generated in the US range from 20% to 60% of the total original biomass. A study of seven mixed pine hardwood sites typical of the Southern US found residue levels ranging from 25 to 134 $m^3$/ha with ~60%, on average, left standing after conventional harvest. Harvesting residues in Canada averaged 65 $m^3$/ha with the western coastal regions averaging 126–168 $m^3$/ha with amounts up to 483 $m^3$/ha [19]. Table 3.57 in the urban wood residues subsection lists fuel-related properties for woody materials.

***Forestry Processing Residues***

Forestry processing residues are usually classified as either from primary (e.g., pulp, paper, lumber) or secondary (e.g., furniture, composite boards, wooden handles) manufacturing operations. The various types of residues can include wood flour, sawdust, shavings, sander dust, pole and post peelings, chip rejects, end cuts, slabs, flawed dimension lumber, and other residues.

**Table 3.47. Mill residue production factors for wood processing operations [20]**

| Type of Operation | Residue Type | Typical Residue Production Factors (% of Input) | (kg per Unit of Product) |
|---|---|---|---|
| Sawmill | Bark | 10[a] | 175.6/m$^3$ |
| | Chips | 30[a] | 623.9/m$^3$ |
| | Sawdust | 15[a] | 311.9/m$^3$ |
| Chip mill | Bark | 10[a] | 76.9/t |
| Veneer mill and plywood mill | Bark | 10[a] | 172.5/m$^3$ |
| | Chips | 30[a] | 671.8/m$^3$ |
| Oriented Strand Board mill | Bark | 10[a] | 185.3/m$^3$ |
| Pole mill | Bark | 10[a] | 80.61/m$^3$ |
| | Shavings | 5[a] | 48.41/m$^3$ |
| Pulp mill | Bark | 10[a] | 337.1/t |
| Pallet mill | Chips | 20[b] | n/a |
| | Sawdust | 10[b] | n/a |
| Planing mill | Shavings | 15[b] | n/a |
| Furniture and cabinet plants | Chips | 25[b] | n/a |
| | Sawdust | 10[b] | n/a |
| Wood components | Chips | 25[b] | n/a |
| | Sawdust | 10[b] | n/a |

[a] Percentage of log volume.
[b] Percentage of lumber volume.

*Availability*

It is difficult to determine the amounts of forestry processing residues accurately; partially because mill owners are reluctant to report true quantities for fear of potential regulatory problems. Thus estimates and published values are frequently conservative.

Although not providing information as to quantities available (because of competing uses), factors to estimate the amounts generated by some types of primary forest processing plants have been developed based on the type of equipment used to process the wood or, like the factors in Table 3.47, based on the product produced [20]. Although pulp and paper mills generate large quantities of wood residues (i.e., bark, fines, and unusable chips), this waste is typically used in house and is not available.

*Characteristics and Fuel Value*

Primary forestry processing residues, which include sawdust, chip rejects, end cuts, slabs, and peelings, depending on the product, can be processed when either dry or green and thus have moisture contents ranging from 5% to 60%. Primary residues frequently consist of bark or include portions of bark. Bark inherently has a higher ash content; it also may contain ash in the form of soil picked up from being dragged along the ground during harvesting operations.

Secondary forestry processing residues are typically bark free and relatively dry as many of the latter residues come from materials that must be glued, which usually

**Table 3.48. Identifying the unit weight of various fuels [21]**

| | Weight per Gravity-packed unit (kg/m$^3$) | | | |
|---|---|---|---|---|
| Moisture(%) | Hog Fuel | Shavings | Bark | Sawdust |
| 0 | 160 | 96 | 208 | 152 |
| 10 | 177.6 | 106.4 | 231.2 | 168 |
| 20 | 200 | 120 | 260 | 190.4 |
| 30 | 227.2 | 136 | 296 | 216 |
| 40 | 265.6 | 159.2 | 345.6 | 252 |
| 50 | 320 | 192 | 416 | 304 |

**Table 3.49. How fuel moisture content affects efficiency [21]**

| Moisture Content (As-Received)(%) | Gross Heating Value[a] (As-Fired) (MJ/kg) | Net Heating Value[b] (Heat Available for Steam Production) (MJ/kg) | Boiler Efficiency[c] (%) |
|---|---|---|---|
| 15 | 16.92 | 13.03 | 77 |
| 30 | 13.40 | 9.78 | 73 |
| 40 | 10.70 | 7.49 | 70 |
| 50 | 8.91 | 5.79 | 65 |
| 60 | 6.67 | 3.87 | 58 |

[a] Assumes a higher heating value of 20.3 MJ/kg for wood and bark.
[b] Assumes a stack temperature of 260°C and excess air (%) equal to the percentage of moisture contained in the fuel.
[c] Overall efficiency is found by subtraction of other losses, such as radiation, blowdown, etc.

requires a moisture content in the range of 5%–6%. Residues from secondary sources may contain glues, sander dust, surface coverings (e.g., paint, plastic laminates), and other nonwood materials that can cause undesirable emissions when thermochemically converted or can harm processing and conversion equipment.

The particle size and shape of residues can vary from fine powder (wood dust) to small particles (sawdust) to shavings, chips, and slivers (pole peelings) to large particles (blocks and slabs). Bark can range from friable to tough and stringy, the latter posing special handling problems. Particle density can also vary widely, as shown in Table 3.48, and have an impact on transportation and handling [21]. For solid fuel applications, fuel quality depends on particle size and calorific value, ash, and debris content. As shown in Table 3.49, the most significant factor affecting calorific value is moisture content [21]. Moisture content can have a large economic impact by requiring more material to be handled, thus requiring larger storage areas and increasing operating and maintenance costs. During the first stages of combustion and gasification processes, fuel-contained moisture must be driven off. The amount of energy in the fuel that is consumed to provide fuel drying increases proportionally as moisture content increases, thus decreasing combustion efficiency and increasing costs. Debris in the form of tramp metal, dirt, and rocks can damage processing and conversion equipment. To minimize the risk of picking up these materials, residues should be kept off the ground on concrete pads or, ideally, be placed immediately into enclosed containers (e.g., overhead bins, tractor trailer vans). (See the subsection on urban wood wastes for further discussion on wood-fuel quality.)

***Agricultural Crop Residues***

Agricultural crop residues—materials left on the land after harvest of the economically valuable part of the crop—may be collected during harvest of the crop or left for later harvest. It includes orchard and vineyard prunings and field residues from horticultural and agronomic crops.

The annual global crop residue production has been estimated at 2.3 billion tons, with the US production estimated at 400 million tons. US recovery potential, taking into account environmental factors, ranges from 17% to 45% [22].

The measure of crop residue is the harvest index (HI), which is defined as the mass of harvested or usable plant product as a fraction of total plant biomass produced per unit of land area (or per plant). The HI depends on crop variety, moisture and nutrient availability, sunlight intensity and duration, and harvesting machinery settings [23].

In almost all cases, some percentage of the field residues must be left in the field for erosion control and to maintain soil quality. The former depends on soil characteristics, slope, wind, and rainfall. Soil quality factors include soil tilth, nutrients, water holding and cation exchange capacities, and reducing rainfall impact, thereby reducing soil crusting [24].

The amount that can be economically removed will depend on collection costs, the residue's value, and if planting must occur immediately after harvest (double cropping), which may require residue removal. Collection costs partially depend on residue density on the field, the passes needed for residue harvest, and equipment efficiency and requirements. Residue value partially depends on its physical and chemical characteristics relative to its conversion and use.

Because of their bulk, residues may interfere with planting following crops, especially if residues are in large particles or in large quantities or if double cropping is desired. In some cases, crop residues must be removed from the field to mitigate accidental fires [24] and the carry-over of insects and disease to nearby or following crops [23]. Such removal can reduce or eliminate the need for chemicals and their application costs and, by reducing insects and disease, potentially improve productivity of future crops. Where it is necessary to remove field residues, their value-added use for energy can at least partially offset the cost of removal and disposal.

As shown in Table 3.50, estimates on the availability of crop residues are usually based on residue-to-crop ratios [6]. These factors should be applied to crops harvested, not to crops planted. The availability of crop residues is partially controlled by the time of crop harvest, although in some cases, depending on climate and other factors, residues can be left in the field until the field is readied for planting again. The availability of residues not collected at harvest will partially depend on weather and field conditions for later harvest and the rate of decomposition of residues. Seasonal availability limits the use of residues unless storage systems are used. Otherwise, cases must be developed to supply fuel when residues are not available. Since collection and transportation costs are factors that limit use, crop residues will be used in the US in the future only as a result of government mandates or increased energy prices [23].

**Table 3.50. Relation of main product to by-product for different agricultural crops [6]**

| Product | Main Product | By-Product | Ratio of Main Product to By-Product |
|---|---|---|---|
| Cereals | | | |
| Wheat | Grain | Straw | 1:1.3 |
| Barley | Grain | Straw | 1:1.2 |
| Maize | Grain | Straw | 1:1 |
| Oats | Grain | Straw | 1:1.3 |
| Rye | Grain | Straw | 1:1.6 |
| Rice | Grain | Straw | 1:1.4 |
| Millet | Grain | Straw | 1:1.4 |
| Sorghum | Grain | Straw | 1:1.4 |
| Pulse | | | |
| Pea | Grain | Straw | 1:1.5 |
| Bean | Grain | Straw | 1:2.1 |
| Soya | Grain | Straw | 1:2.1 |
| Tuber and root crops | | | |
| Potatoes | Tuber | Stalk | 1:0.4 |
| Feedbeet | Root | Stalk | 1:0.3 |
| Sugarbeet | Root | Stalk | |
| Jerusalem artichoke | Tuber | Stalk | 1:0.8 |
| Peahut | Nut | Stalk | 1:1 |
| Cocoa | Nut | Shell and outer fiber | 1:0.2 |
| Sugar Cane | Sugar | Bagasse | 1:1.16 |

*Biomass Regenerable Energy*, edited by D. O. Hall and R. P. Overend. Copyright John Wiley & Sons Limited, Reproduced with permission.

*Straw*

Among major worldwide crops, the largest sources of crop residues are wheat and rice straw, each at approximately 21%, followed by maize (corn) at 14% [6]. Primarily because of government policy, the use of straw for energy is rapidly growing in Europe. In 1991, Denmark used 730,000 tons of straw for energy including 450,000 tons in farm-scale boilers, 265,000 in central heating plants, and 67,000 tons in power plants. The 3 million tons of unused surplus straw annually produced in Denmark could provide 45 PJ corresponding to 5.8% of Denmark's gross energy consumption in 1988 [25]. In the US, a significant portion of wheat and rice straw is burned in the field for disease control. This practice generates smoke, causing air pollution and traffic hazards and has resulted in legislation preventing such burning [23].

*Cotton Stalks*

Legislation in some US states has mandated the removal of cotton stalks from the field for disease control, resulting in research on cotton stalk harvesters. Tests over 4 years in Arizona found that the amount of recoverable cotton stalks varied from year to year because of crop variety, harvest method, and length of time elapsed from cotton harvest to stalk harvesting. Recovered amounts from mechanical collection ranged from

2354 to 4014 kg/ha of dry matter compared with those from the hand harvest, which ranged from 4197 to 4604 kg/ha. Moisture content ranged from 23.3% to 41.4% with a 34.9% average [26].

*Sugar Cane*

Most cane fields in developing countries are burned before harvest to remove leaves and tops. This practice removes an estimated equivalent of 6–10 barrels of oil per hectare. Research in Puerto Rico indicates that whole-plant harvesting (tops and leaves harvested with stalks) could provide 25% more total biomass than that from the traditional harvesting of stalks only [27].

*Characteristics of Crop Residues*

Table 3.51 includes heating values and compositional data for selected crop residues [28]. Crop residues contain chemical elements that can cause slagging and fouling of combustion systems. See the subsection on slagging and fouling concerns for a discussion of these factors and how to mitigate them.

***Agricultural Crop and Food Processing Residues***

Agricultural crop and food processing residues as defined here are materials left after crops removed from the field are processed into feed, food, or other products. These residues may include sugar cane bagasse, rice husks, cotton gin trash, grain dust and chaff, corn cobs, weed seeds, hulls from nuts (e.g., peanuts, almonds, walnuts), nut shells (e.g., pecans, almonds), fruit pits and stones (e.g., olives, cherries, peaches, plums, apricots), fruit pomace (e.g., apple, citrus, and grape), and other materials. For energy applications, these residues have an advantage of already being collected at the processing site.

Quantities of these residues can be significant and can represent major disposal costs for their associated industries. For example, in 1993, over 45,000 tons of fruit stones from cherries, peaches, and plums were produced in the US [29].

Because of the variety of ways to process foods and unavailability of good data, limited information on commercial food processing residues is included here. Also, a high percentage of residues from commercial food processing is in the form of slurries and, although suitable for anaerobic digestion, is unsuitable for thermochemical conversion without drying, which is usually not practical.

*Sugar Cane Bagasse*

Sugar cane bagasse is one of the major agricultural processing residues in the world, with most of it used to fuel sugar cane processing plants [30]. India is the leading sugar cane producer in the world, with over 450 mills producing over 11 million tons of refined sugar and accounting for 60% of sugar cane cultivated. At 30% of cane by weight and 50% moisture content, these mills in aggregate produce 40 million tons of bagasse per year [31].

Bagasse quality and quantity depend on variety and age of the cane and methods of harvesting and processing. It usually contains an average of 46%–52% moisture (depending on grinding speed and mill efficiency), 43%–52% fiber, and 2%–6% solubles (mostly sucrose) [32] with a higher heating value of 17.3 MJ/kg [28].

Table 3.51. **Heating value and compositional data for crop residues and energy crops [28]**

| Type of Biomass | Heating Values (MJ/kg, Dry Basis) | | Approximate Analysis (% by Weight, Dry Basis) | | Ultimate Elemental Analysis (% by Weight, Dry Basis) | | | | | | | |
|---|---|---|---|---|---|---|---|---|---|---|---|---|
| | Higher | Lower | Volatile | Ash | F. Carbon | Carbon | Hydrogen | Oxygen | Nitrogen | Sulfur | Chlorine | Residue |
| Crop Residues | | | | | | | | | | | | |
| Alfalfa seed straw | 18.45 | 17.33 | 72.6 | 7.25 | 20.15 | 46.76 | 5.4 | 40.72 | 1 | 0.02 | 0.03 | 6.07 |
| Almond prunings | 20.01 | 18.91 | 76.83 | 1.63 | 21.54 | 51.3 | 5.29 | 40.9 | 0.66 | 0.01 | 0.04 | 1.8 |
| Barley straw | 17.31 | 16.22 | 68.8 | 10.3 | 20.9 | 39.92 | 5.27 | 43.81 | 1.25 | | | 9.75 |
| Bean straw | 17.46 | 16.3 | 75.3 | 5.93 | 18.77 | 42.97 | 5.59 | 44.93 | 0.83 | 0.01 | 0.13 | 5.54 |
| Corn cobs | 18.77 | 17.55 | 80.1 | 1.36 | 18.54 | 46.58 | 5.87 | 45.46 | 0.47 | 0.01 | 0.21 | 1.4 |
| Corn stover | 17.65 | 16.5 | 75.17 | 5.58 | 19.25 | 43.65 | 5.56 | 43.31 | 0.61 | 0.01 | 0.6 | 6.26 |
| Cotton stalks | 18.26 | 17.15 | 73.29 | 5.51 | 21.2 | 47.05 | 5.35 | 40.77 | 0.65 | 0.21 | 0.08 | 5.89 |
| Grape Prunings | | | | | | | | | | | | |
| Cardinal | 19.21 | | 78.17 | 2.22 | 19.61 | | | | | | | |
| Chenin blanc | 19.13 | 17.91 | 77.28 | 2.51 | 20.21 | 48.02 | 5.89 | 41.93 | 0.86 | 0.07 | 0.1 | 3.13 |
| Gewürztraminer | 19.16 | | 77.27 | 2.47 | 20.26 | | | | | | | |
| Merlot | 18.84 | | 77.47 | 3.04 | 19.49 | | | | | | | |
| Pinot noir | 19.05 | 17.84 | 76.83 | 2.71 | 20.46 | 47.14 | 5.82 | 43.03 | 0.86 | 0.01 | 0.13 | 3.01 |
| Riber | 19.12 | | 76.97 | 3.03 | 20 | | | | | | | |
| Thompson seedless | 19.35 | 18.15 | 77.39 | 2.25 | 20.36 | 47.35 | 5.77 | 43.32 | 0.77 | 0.01 | 0.01 | 2.71 |
| Tokay | 19.31 | 18.1 | 76.53 | 2.45 | 21.02 | 47.77 | 5.82 | 42.63 | 0.75 | 0.03 | 0.07 | 2.93 |
| Zinfandel | 19.06 | | 76.99 | 3.04 | 19.49 | | | | | | | |
| Rice straw (fresh) | 16.28 | 15.32 | 69.33 | 13.42 | 17.25 | 41.78 | 4.63 | 36.57 | 0.7 | 0.08 | 0.34 | 15.9 |
| Rice straw (weathered) | 14.56 | 13.74 | 62.31 | 24.36 | 13.33 | 34.6 | 3.93 | 35.38 | 0.93 | 0.16 | | 25 |
| Russian thistle | 18.46 | | | 15.4 | | | | | | | | |
| Safflower straw | 19.23 | 18.08 | 77.05 | 4.65 | 18.3 | 41.71 | 5.54 | 46.58 | 0.62 | | | 5.55 |
| Sorghum stalks | 15.4 | 14.32 | | | | 40 | 5.2 | 40.7 | 1.4 | 0.2 | | 12.5 |
| Wheat straw | 17.51 | 16.47 | 71.3 | 8.9 | 19.8 | 43.2 | 5 | 39.4 | 0.61 | 0.11 | 0.28 | 11.4 |

| | | | | | | | | | | | | |
|---|---|---|---|---|---|---|---|---|---|---|---|---|
| Energy Crops | | | | | | | | | | | | |
| Cattails | 17.81 | 16.72 | 71.57 | 7.9 | 20.53 | 42.99 | 5.25 | 42.47 | 0.74 | 0.04 | 0.38 | 8.13 |
| Eucalyptus camaldulensis | 19.42 | 18.2 | 81.42 | 0.76 | 17.82 | 49 | 5.87 | 43.97 | 0.3 | 0.01 | 0.13 | 0.72 |
| Eucalyptus, leaves only | 20.31 | | 77.78 | 4.83 | 17.39 | | | | | | | |
| Eucalyptus Globulus | 19.23 | 18 | 81.6 | 1.1 | 17.3 | 48.18 | 5.92 | 44.18 | 0.39 | 0.01 | 0.2 | 1.12 |
| Eucalyptus Grandis | 19.35 | 18.13 | 82.55 | 0.52 | 16.93 | 48.33 | 5.89 | 45.13 | 0.15 | 0.01 | 0.08 | 0.41 |
| Guayule | 25.16 | | | 5.2 | | | | | | | | |
| Jute stick | 19.34 | | 92.55 | 0.9 | 6.55 | 49.04 | 6.69 | 40.19 | 2.92 | 0.25 | | 0.91 |
| Leucaena | 19.07 | 17.81 | 80.94 | 1.53 | 17.53 | 49.2 | 6.05 | 42.74 | 0.47 | 0.03 | | 1.51 |
| Sudan grass | 17.39 | 16.28 | 72.75 | 8.65 | 18.6 | 44.58 | 5.35 | 39.18 | 1.21 | 0.08 | 0.13 | 9.47 |
| Water hyacinth | 16.02 | | | 22.4 | | 41.1 | 5.29 | | 1.96 | 0.41 | | |

*Rice Husks*

In most rice-growing developing countries—approximately 90% of the world's rice is grown in Asia [6]—rice husks are the most abundantly available crop residue. The estimated annual production of rice husks in millions of tons is India, 18; Indonesia, 8; Bangladesh, 5; Thailand, 4; Burma, 3; Philippines, 2; and ~1 in Pakistan. Although large quantities of rice are grown, approximately half is consumed within a few kilometers of the fields where it was grown, indicating that most husks are produced in very small quantities [33].

At least 0.2 ton of rice husks are generated from each ton of paddy rice. One ton of husks, at 16.16 MJ/kg, is capable of providing the energy for drying 15.6 tons of paddy rice. Rice husk ash is characterized by its amount (20% of the initial weight) and its silica content (roughly 65%) [34].

*Cotton Gin Trash*

Cotton gin trash is also produced worldwide. Eighty percent of US or roughly 1.4 million tons per year of cotton gin trash are produced in Arizona, California, New Mexico, Oklahoma, and Texas over a 3- to 4-month ginning season. Cotton is picked either by pickers that use rotating spindles to remove only the cotton lint or strippers that remove the entire boll from the plant. For every 227-kg bale of *ginned* cotton, spindle-picked cotton generates approximately 68 kg of gin trash while stripper-picked cotton produces 318 kg [35].

*Grain Cleanings*

Grain cleanings (dust, chaff, weed seeds) have been researched for energy. A Canadian study found that grain cleanings account for 1% of grain receipts at a major grain elevator [36]. In the US, elevator losses of 3% are estimated for the handling and milling of oil seeds [23].

*Characteristics*

Table 3.52 shows heating value and compositional data for selected food and agricultural crop processing residues [28]. Because of potential slagging and fouling problems, virtually all pose special problems for combustion applications.

*Slagging and Fouling Concerns*

One of the biggest concerns for agricultural residues or other high-ash feedstocks to be used in thermochemical conversion processes is their propensity to cause chemical deposits on combustion chamber walls (slagging) or on heat transfer surfaces (fouling). These deposits can significantly reduce heat transfer rates with a resulting reduction in boiler efficiency, can grow to the extent that air flow through the boiler is severely reduced (again reducing boiler efficiency), and can contribute to corrosion problems. Additionally, large deposits have been known to break off and cause severe damage from falling onto combustion chamber surfaces. The seriousness of these problems can require frequent and costly shutdowns for removing deposits [37].

In his research, Miles [37] found that most slagging and fouling problems can be attributed to the presence of high levels of alkalis and alkali earth metals in the fuel, with potassium and sodium the worst offenders. Potassium and sodium are necessary for the

**Table 3.52. Heating value and compositional data for food/crop processing residues and MSW [28]**

| | Heating Values (MJ/kg, Dry Basis) | | Approximate Analysis (% by Weight, Dry Basis) | | | Ultimate Elemental Analysis (% by Weight, Dry Basis) | | | | | | |
|---|---|---|---|---|---|---|---|---|---|---|---|---|
| Type of Biomass | Higher | Lower | Volatile | Ash | F. Carbon | Carbon | Hydrogen | Oxygen | Nitrogen | Sulfur | Chlorine | Residue |
| Food/crop processing residues | | | | | | | | | | | | |
| Almond hulls | 18.22 | 17.11 | 71.33 | 5.78 | 22.89 | 45.79 | 5.36 | 40.6 | 0.96 | 0.01 | 0.08 | 7.2 |
| Almond shells | 19.38 | 18.14 | 73.45 | 4.81 | 21.74 | 44.98 | 5.97 | 42.27 | 1.16 | 0.02 | | 5.6 |
| Babassu husks | 19.92 | 18.81 | 79.71 | 1.59 | 18.7 | 50.31 | 5.37 | 42.29 | 0.26 | 0.04 | | 1.73 |
| Cherry pits | 21.75 | | 84.2 | 1 | 14.8 | | | | | | | |
| Cocoa hulls | 19.04 | 17.95 | 67.95 | 8.25 | 23.8 | 48.23 | 5.23 | 33.19 | 2.98 | 0.12 | | 10.25 |
| Coconut fiber | 19.24 | | 69.6 | 5.1 | 25.3 | | | | | | | |
| Coconut shell | 20.56 | | 81.8 | 0.5 | 17.6 | | | | | | | |
| Cotton gin trash | 16.42 | 15.33 | 67.3 | 17.6 | 15.1 | 39.59 | 5.26 | 36.38 | 2.09 | | | 16.68 |
| Grape pomace | 20.34 | 19.11 | 68.54 | 9.48 | 21.98 | 52.91 | 5.93 | 30.41 | 1.86 | 0.03 | 0.05 | 8.81 |
| Macadamia shells | 21.02 | 19.97 | 75.92 | 0.4 | 23.68 | 54.41 | 4.99 | 39.69 | 0.36 | 0 | | 0.56 |
| Olive pits | 21.39 | 10.1 | 78.64 | 3.16 | 18.19 | 48.81 | 6.23 | 43.48 | 0.36 | 0.02 | | 1.1 |
| Peach pits | 20.82 | 19.6 | 79.12 | 1.03 | 19.85 | 53 | 5.9 | 39.14 | 0.32 | 0.05 | | 1.59 |
| Peanut hulls | 18.64 | 17.51 | 73.02 | 5.89 | 21.09 | 45.77 | 5.46 | 39.56 | 1.63 | 0.12 | | 7.46 |
| Pineapple waste | 18.47 | 17.22 | 78.82 | 4.39 | 16.79 | 47.3 | 6.03 | 40.93 | 1.13 | 0.21 | | 4.4 |
| Pistachio shells | 19.26 | 18.03 | 82.03 | 1.13 | 16.84 | 48.79 | 5.91 | 43.41 | 0.56 | 0.01 | 0.04 | 1.28 |
| Plum pits | 21.14 | | 85.3 | 0.1 | 14.6 | | | | | | | |
| Prune pits | 23.28 | 22.06 | 76.99 | 0.5 | 22.51 | 49.73 | 5.9 | 43.57 | 0.32 | | | 0.48 |
| Rice hulls | 16.14 | 15.25 | 65.47 | 17.86 | 16.67 | 40.96 | 4.3 | 35.86 | 0.4 | 0.02 | 0.12 | 18.34 |
| Rice (wild) hulls | 19.14 | | 80.57 | 4.91 | 14.52 | 47.12 | 6.06 | 40.05 | 1.21 | 0.42 | 0.04 | 5.1 |
| Sugar cane bagasse | 17.33 | 16.22 | 73.78 | 11.27 | 14.95 | 44.8 | 5.35 | 39.55 | 0.38 | 0.01 | 0.12 | 9.79 |
| Tomato pomace | 23.77 | | 85.1 | 4.1 | 10.8 | | | | | | | |
| Walnut shells | 20.18 | 18.99 | 78.28 | 0.56 | 21.16 | 49.98 | 5.71 | 43.35 | 0.21 | 0.01 | 0.03 | 0.71 |
| Wheat dust | 16.2 | 15.14 | 69.85 | 13.68 | 16.47 | 41.38 | 5.1 | 35.19 | 3.04 | 0.19 | | 15.1 |

(Continues)

**Table 3.52. (Continued)**

| | Heating Values (MJ/kg, Dry Basis) | | Approximate Analysis (% by Weight, Dry Basis) | | | Ultimate Elemental Analysis (% by Weight, Dry Basis) | | | | | | |
|---|---|---|---|---|---|---|---|---|---|---|---|---|
| Type of Biomass | Higher | Lower | Volatile | Ash | F. Carbon | Carbon | Hydrogen | Oxygen | Nitrogen | Sulfur | Chlorine | Residue |
| MSW | 16.57 | | 72.7 | 16.57 | 10.73 | | | | | | | |
| Paper | | | | | | | | | | | | |
| Corrugated boxes | 17.28 | 16.1 | | | | 43.73 | 5.7 | 44.93 | 0.09 | 0.21 | | 5.34 |
| Brown | 17.92 | 16.66 | | | | 44.9 | 6.08 | 47.84 | 0 | 0.11 | | 1.07 |
| Food cartons | 17.98 | 16.71 | | | | 44.74 | 6.1 | 41.92 | 0.15 | 0.16 | | 6.93 |
| Waxed milk cartons | 27.28 | 25.36 | | | | 59.18 | 9.25 | 30.13 | 0.12 | 0.1 | | 1.22 |
| Plastic coated | 17.91 | 16.63 | | | | 45.3 | 6.17 | 45.5 | 0.18 | 0.08 | | 2.77 |
| Newspaper | 19.72 | 18.45 | | | | 49.14 | 6.1 | 43.03 | 0.05 | 0.16 | | 1.52 |
| Refuse-derived fuel | 17.4 | | | | | 42.5 | 5.84 | 27.57 | 0.77 | 0.48 | 0.57 | 22.17 |

growth of green plants and are thus commonly found in agricultural crops and residues and dirty wood residues. However, it is a combination of the volatility and the quantity of these elements in some plant materials, especially potassium, that causes the problem. All faster-growing plants (such as grasses or fast-growing trees) or faster-growing plant parts (such as twigs and leaves) contain relatively high levels of alkali compounds. To illustrate, Miles found that alkali levels in willow tops were almost three times higher than in butt-end samples from the same plants [37].

Fuels high in alkali are also frequently relatively rich in chlorine, silicon, and sulfur, which in combination with potassium represent the primary fouling agents. Chlorine is an important facilitator, leading to the condensation of potassium chloride salt onto surfaces. These salts are readily attacked by sulfur oxides to form potassium sulfate, which leads to the creation of sticky coatings that increase buildup from particle impingement [37].

The high levels of alkali create problems in two ways. Depending on their chemical content, different compounds have different ash fusion temperatures (the temperature when ash turns sticky and starts to melt). Because of their relatively low melting points, alkali materials will combine with chemicals present in the fuel to form compounds with ash fusion temperatures below those typically found in industrial boiler systems. Thus, as combustion occurs, the resulting ash melts and coats (slags onto) the combustion chamber surfaces [37].

Deposits also occur as a result of the boiler design and operation. Usually conventionally designed combustion systems are not suitable for burning high-alkali fuels. Designs should include (1) adequate water-wall surface area or parallel heat exchange surfaces; (2) combustion air control to maintain exit gas temperatures below 850°C; (3) grates suitable for removing large quantities of ash, and (4) soot blowers to remove fly-ash deposits [37].

Commercially available chemical additives that form compounds with higher ash fusion temperatures may be effective in reducing slagging but not deposit formation. To date, no systematic study has been performed on the effectiveness of chemical additives [37].

A method of predicting slagging tendency has been developed with techniques from the coal industry. The method [37] involves calculating the weight in alkali oxides (potassium, $K_2O$, and sodium, $Na_2O$) per heat unit (kilograms per gigajoule) in the fuel with the higher heating value (HHV):

$$\text{kg alkali/GJ} = 1 \times 10^6/\text{HHV kJ/kg(dry)} \times \%\ \text{ash} \times \%\ \text{alkali in ash.}$$

This method combines all the necessary fuel-related information into one index number. It has been determined that deposits rarely occur at values below 0.17 kg/GJ, with the likelihood of deposit formation increasing from 0.17 to 0.34 kg/GJ. Deposit formation is almost certain to occur at values above 0.34 kg/GJ. However, "...while an indicator of potential problems this information must be combined with field experience and boiler operating conditions to evaluate the impact of a fuel on a particular boiler" [37].

It is important to note that classical analyses of feedstocks and hence handbook information can provide misleading information to a designer. For example, ash content and characteristics can be changed significantly by the harvesting, handling, and storage processes used. Also, ash chemical composition can vary depending on chemicals present in the soil where the crop was grown. Ash characteristics that are determined in the laboratory by standard techniques and temperatures may differ significantly from the ash produced in the conversion process because of incomplete combustion and the presence of higher temperatures that volatilize more of the chemicals in the ash. Combustion field tests can also provide misleading results if not performed for sufficient time since deposits usually form slowly. To mitigate these potential variances partially, samples should be selected for testing that have been handled in the same manner as feedstocks delivered to the processing plant. Also, analyses can be performed with temperatures and conditions similar to those of the conversion process.

In spite of these limitations, handbook information can provide an indication of anticipated characteristics and performance. Unfortunately, with the exception of wood, limited handbook data are available on physical and chemical characteristics for many biomass materials and even less are available on ash characteristics. Table 3.53 lists some biomass residues for which data are available [37].

***Animal Manures***

Animal excreta (manures) are another potential biomass resource available for energy. Raw manure (without bedding) availability depends on the amount generated over

**Table 3.53. Agricultural crop processing residue [37]**

| Fuel Type | Sugar cane bagasse HI (Dry) | Straws, Rice (Dry) | Straws, Wheat, Imperial (Dry) | Straws/Husks Rice Hulls (Dry) | Almond Hulls (Dry) | Olive Pits (Dry) |
|---|---|---|---|---|---|---|
| Ash (%) | 2.44 | 18.67 | 9.55 | 20.26 | 6.13 | 1.72 |
| HHV (MJ/kg) | 18.85 | 18.85 | 16.78 | 15.81 | 18.84 | 21.54 |
| Chlorine (%) | 0.03 | 0.58 | 2.06 | 0.12 | 0.02 | 0.04 |
| Water-soluble Alkalis (%) | | | | | | |
| $Na_2O$ | | | | 0.022 | | |
| $K_2O$ | | | | 0.665 | | |
| CaO | | | | 0.008 | | |
| Elemental Composition | | | | | | |
| $SiO_2$ | 46.61 | 74.67 | 37.06 | 91.42 | 9.28 | 30.82 |
| $Al_2O_3$ | 17.69 | 1.04 | 2.23 | 0.78 | 2.09 | 8.84 |
| $TiO_2$ | 2.63 | 0.09 | 0.17 | 0.02 | 0.05 | 0.34 |
| $Fe_2O_3$ | 14.14 | 0.85 | 0.84 | 0.14 | 0.76 | 6.58 |
| CaO | 4.47 | 3.01 | 4.91 | 3.21 | 8.07 | 14.66 |
| MgO | 3.33 | 1.75 | 2.55 | <0.01 | 3.31 | 4.24 |
| $Na_2O$ | 0.79 | 0.96 | 9.74 | 0.21 | 0.87 | 27.8 |
| $K_2O$ | 4.15 | 12.3 | 21.7 | 3.71 | 52.9 | 4.4 |
| $SO_3$ | 2.08 | 1.24 | 4.44 | 0.72 | 0.34 | 0.56 |
| $P_2O_5$ | 2.72 | 1.41 | 2.04 | 0.43 | 5.1 | 2.46 |
| $CO_2$/other | | | | | 20.12 | |
| Undetermined | 1.39 | 2.68 | 14.32 | | −2.89 | −0.7 |
| Total | 100 | 100 | 100 | 100.64 | 100 | 100 |
| Alkali (kg/GJ) | 0.35 | 7.08 | 9.63 | 2.72 | 9.42 | 1.39 |

time per animal, the percentage of dry matter, the percentage of animals confined, and the percentage collectable. Bedding will increase the amount of waste and change its characteristics.

Data on animal manure characteristics from the American Society of Agricultural Engineers Standards are presented in Table 3.54 [38]. The values in the table represent fresh (as-voided) feces and urine. Actual values may vary because of differences in animal diet, age, usage, productivity, and management. Site-specific data should be used whenever possible. Factors for estimating manure availability were developed for the Southeast US and are shown in Table 3.55 [39].

Ability to combust animal manures may be limited because of potential slagging and fouling problems (see the subsection on slagging and fouling concerns) and in some cases, high moisture contents. Anaerobic digestion processes may thus be a better option for converting resources.

Animal manures also have significant nutrients and thus can have higher value as fertilizers and soil conditioners for crops and, in some cases, feed for animals [23]. Excessive manure quantities, limited markets, salt content, and presence of weed seeds are reasons why combustion is considered [40]. At least two power plants that directly combust cattle manure have been built in the US [41].

***Municipal Solid Waste (MSW) and Refuse-Derived Fuels***

MSW is defined here as raw, unprocessed residues typically in waste streams of large municipalities. Typical percentages for MSW in the US are paper, 40; yard wastes, 17.6; metals, 8.5; glass, 7.0; plastics, 8.0; other 11.6; and food wastes, 7.4 [42]. Refuse-derived fuels (RDF's) are defined as fuels derived from MSW by removal of the noncombustibles. Table 3.52 provides information on MSW and RDF characteristics [28].

*Municipal Solid Waste*

Table 3.56, published in 1984, shows some typical amounts and volumes of MSW generated in various countries of the world [43]. The amount of MSW generated in the US ranges from 1.1 to 1.8 kg per person per day [39] with a HHV ranging from 13.13 to 19.87 MJ/kg [28].

Between RDF and MSW, unprocessed MSW is currently the most available fuel form. Its advantage is that, with the exception of removing tires (because of their high sulfur content), with large, noncombustible items (e.g., appliances) and infectious medical wastes little front-end sorting is required before combustion. A disadvantage is that the significant amount of noncombustible materials (metals, glass, rocks, etc.) limits the type of combustion system that can be used [44].

For political and environmental reasons, the trend in the US has gone from disposing of MSW in landfills, to waste-to-energy systems, to recovering and recycling materials. It is anticipated that MSW available for fuel will decline as communities strive to meet government mandated recycling goals (fuel use is not considered recycling by many political entities). A small segment of MSW will always be available for fuel since certain items (e.g., cardboard residues used in shipping raw meat) cannot by law be recycled or can be reused only a limited number of times [45].

**Table 3.54. Fresh manure production and characteristics per 1000-kg live animal mass per day [38]**

| Parameter | Units* | | Typical Live Animal Masses | | | | | | | | | | |
|---|---|---|---|---|---|---|---|---|---|---|---|---|---|
| | | | Dairy 640 kg† | Beef 360 kg | Veal 91 kg | Swine 61 kg | Sheep 27 kg | Goat 64 kg | Horse 450 kg | Layer 1.8 kg | Broiler 0.9 kg | Turkey 6.8 kg | Duck 1.4 kg |
| Total Manure‡ | kg | mean§ | 86 | 58 | 62 | 84 | 40 | 41 | 51 | 64 | 85 | 47 | 110 |
| | | std. deviation | 17 | 17 | 24 | 24 | 11 | 8.6 | 7.2 | 19 | 13 | 13 | ** |
| Urine | kg | mean | 26 | 18 | ** | 39 | 15 | ** | 10 | ** | ** | ** | ** |
| | | std. deviation | 4.3 | 4.2 | ** | 4.8 | 3.6 | ** | 0.74 | ** | ** | ** | ** |
| Density | $kg/m^3$ | mean | 990 | 1000 | 1000 | 990 | 1000 | 1000 | 1000 | 970 | 1000 | 1000 | ** |
| | | std. deviation | 63 | 75 | ** | 24 | 64 | ** | 93 | 39 | ** | ** | ** |
| Total solids | kg | mean | 12 | 8.5 | 5.2 | 11 | 11 | 13 | 15 | 16 | 22 | 12 | 31 |
| | | std. deviation | 2.7 | 2.6 | 2.1 | 6.3 | 3.5 | 1 | 4.4 | 4.3 | 1.4 | 3.4 | 15 |
| Volatile solids | kg | mean | 10 | 7.2 | 2.3 | 8.5 | 9.2 | ** | 10 | 12 | 17 | 9.1 | 19 |
| | | std. deviation | 0.79 | 0.57 | ** | 0.66 | 0.31 | ** | 3.7 | 0.84 | 1.2 | 1.3 | ** |
| Ammonia nitrogen | kg | mean | 0.079 | 0.086 | 0.12 | 0.29 | ** | ** | ** | 0.21 | ** | 0.08 | ** |
| | | std. deviation | 0.083 | 0.052 | 0.016 | 0.1 | ** | ** | ** | 0.18 | ** | 0.018 | ** |
| Total phosphorus | kg | mean | 0.094 | 0.092 | 0.066 | 0.18 | 0.087 | 0.11 | 0.071 | 0.3 | 0.3 | 0.23 | 0.54 |
| | | std. deviation | 0.024 | 0.027 | 0.011 | 0.1 | 0.03 | 0.016 | 0.026 | 0.081 | 0.053 | 0.093 | 0.21 |
| Potassium | kg | mean | 0.29 | 0.21 | 0.28 | 0.29 | 0.32 | 0.31 | 0.25 | 0.3 | 0.4 | 0.24 | 0.71 |
| | | std. deviation | 0.094 | 0.061 | 0.1 | 0.16 | 0.11 | 0.14 | 0.091 | 0.072 | 0.064 | 0.08 | 0.34 |
| Calcium | kg | mean | 0.16 | 0.14 | 0.059 | 0.33 | 0.28 | ** | 0.29 | 1.3 | 0.41 | 0.63 | ** |
| | | std. deviation | 0.059 | 0.11 | 0.049 | 0.18 | 0.15 | ** | 0.11 | 0.57 | ** | 0.34 | ** |
| Magnesium | kg | mean | 0.071 | 0.049 | 0.033 | 0.07 | 0.072 | ** | 0.057 | 0.14 | 0.15 | 0.073 | ** |
| | | std. deviation | 0.016 | 0.015 | 0.023 | 0.035 | 0.047 | ** | 0.016 | 0.042 | ** | 0.0071 | ** |
| Sulfur | kg | mean | 0.051 | 0.045 | ** | 0.076 | 0.055 | ** | 0.044 | 0.14 | 0.085 | ** | ** |
| | | std. deviation | 0.01 | 0.0052 | ** | 0.04 | 0.043 | ** | 0.022 | 0.066 | ** | ** | ** |
| Sodium | kg | mean | 0.052 | 0.03 | 0.086 | 0.067 | 0.078 | ** | 0.036 | 0.1 | 0.15 | 0.066 | ** |
| | | std. deviation | 0.026 | 0.023 | 0.063 | 0.052 | 0.027 | ** | ** | 0.051 | ** | 0.012 | ** |
| Chloride | kg | mean | 0.13 | ** | ** | 0.26 | 0.089 | ** | ** | 0.56 | ** | ** | ** |
| | | std. deviation | 0.039 | ** | ** | 0.052 | ** | ** | ** | 0.44 | ** | ** | ** |
| Iron | mg | mean | 12 | 7.8 | 0.33 | 16 | 8.1 | ** | 16 | 60 | ** | 75 | ** |
| | | std. deviation | 6.6 | 5.9 | ** | 9.7 | 3.2 | ** | 8.1 | 49 | ** | 28 | ** |
| Cadmium | mg | mean | 0.003 | ** | ** | 0.027 | 0.0072 | ** | 0.0051 | 0.038 | ** | ** | ** |
| | | std. deviation | ** | ** | ** | 0.028 | ** | ** | ** | 0.032 | ** | ** | ** |
| Lead | mg | mean | ** | ** | ** | 0.084 | 0.084 | ** | ** | 0.74 | ** | ** | ** |
| | | std. deviation | ** | ** | ** | 0.012 | ** | ** | ** | ** | ** | ** | ** |

Table 3.55. Factors for Estimating Manure Availability for the Southeast US [39]

| Animal | kg/Manure/ Animal/Year | Percentage Confined | Percentage Collectible | Percentage Dry Matter | Factor |
|---|---|---|---|---|---|
| Dairy cows | 1772 | 0.5 | 0.85 | 0.14 | 105.4 |
| Cattle on feed | 564 | 0.1 | 0.65 | 0.15 | 5.5 |
| Pigs/hogs | 64 | 0.85 | 1 | 0.13 | 7.07 |
| Chickens | 1.14 | 1 | 0.8 | 0.25 | 22.8 |

Table 3.56. Some typical refuse generation rates [43]

| Place | kg/Person/Day | Volume/Day (liters) |
|---|---|---|
| India | 0.25 | 1 |
| Ghana | 0.25 | 1 |
| Aden | 0.25 | 1 |
| Egypt | 0.3 | 1.25 |
| Syria | 0.3 | 1.25 |
| Sri Lanka | 0.4 | 1.6 |
| Phillipines | 0.5 | 2 |
| Turkey | 0.6 | 2.4 |
| Malaysia | 0.7 | 3.5 |
| Singapore | 0.85 | 4.25 |
| Arabian Gulf State | 1 | 5 |
| Europe | 1 | 8 |
| United States | 1.25 | 12 |

*Managing Solid Wastes in Developing Countries*, edited by John R. Holmes. Copyright John Wiley & Sons Limited, Reproduced with permission.

*Residue-Derived Fuel*

MSW can be processed by manual or mechanical means such as screens and air classifiers to remove most of the noncombustibles to make RDF, which can then be pelletized, cubed, or used as fluff for fuel. RDF has the advantage in that, once proper emission control equipment is in place, it can be used in existing solid fuel combustion systems with little or no retrofit requirements.

***Urban Wood Waste***

As used here, urban wood waste is defined as wood waste that has traditionally been in the landfill waste stream. Examples include wooden pallets, packaging materials, furniture, and toys; land-clearing, construction, and demolition debris; and dimension lumber, tree trimmings, forestry processing wastes, utility poles, and other wooden materials.

Fortunately, a significant proportion of wood is brought to landfills in monoloads or loads that are either all wood waste or all wood waste of one kind. Examples of monoloads can include tree trimmings, land-clearing debris, and pallets. Since this wood is already separated from the landfill waste stream, recovery is greatly simplified and the wood can be kept cleaner, making it easier to reach high-value markets. If sorting is performed, experts believe that sorting out the higher-quality material from the waste stream rather than the opposite approach provides much cleaner material [46].

Since value closely follows the degree of waste wood cleanliness and the moisture content, separating waste wood according to its degree of cleanliness and moisture content provides the best means to reach the highest-value markets. Pallets and wooden packaging materials that have been kept inside will have an approximate moisture content of 10%–12%. Wood freshly cut from living plants will have a moisture content of approximately 50% in the summer and 45% in the winter [46].

Cleanliness is the absence of fines and nonwood materials, with nonwood-derived materials such as metals or chemicals being the most undesirable. Dirt or other nonwood materials can decrease fuel value by increasing ash content, reducing combustion efficiency, causing equipment problems, and potentially contributing to slagging problems [46] (see the subsection on slagging and fouling concerns for a discussion of slagging aspects).

The primary environmental concerns involve residues containing or mixed with nonwood materials. These include materials that are physically separable (e.g., nails, rocks) and those that are not physically separable (e.g., paints, preservatives, fire retardents, surface laminates). In addition to magnets and nonferrous metal detectors for metals, float tanks are sometimes used to remove dirt, rocks, glass, and nonwood materials [46].

Relative to fuel use, primary concerns are chemicals in the form of paint pigments; preservatives such as creosote, pentachlorophenol, and chromated copper arsenate (CCA); and fire retardents that can create toxic air or ash emissions. Paint pigments can contain lead, titanium, mercury, and aluminum; preservatives can contain arsenic, copper, chromium, lead, and mercury [47].

Usually, creosote and pentachlorophenol treated materials can be burned in the US if combustion-zone temperatures are high enough and residence times long enough in the combustion device. CCA materials can be burned only in specially permitted solid waste disposal burners and are not desirable fuels. Methods of removing chemicals to allow use for fuel or other applications are being developed [48]. Table 3.57 shows fuel-related properties of woody materials [28].

***Landfills***

Landfills, as described here, are the typical MSW landfills in the US. The energy available from a landfill depends on the size, depth, composition, water retention capacity, and other physical characteristics of the landfill. Weather conditions, especially temperature and rainfall, are also factors, as is the age of the landfill. In general, landfill gas is generated at the rate of 0.012 to 0.031 $m^3$/kg of waste in place per year, and the gas calorific value ranges from 15 to 22 J/$cm^3$ [45].

***Sewage Sludge***

Sewage sludge, also called biosolids, as defined here, is the solid residue left after municipal wastewater treatment. Many industrial processes, such as paper manufacturing, also generate organic sludges that, depending on their characteristics, may be used for energy. Estimates of sewage sludge generation in the US are ~0.11 kg per person per day [39].

Although organic in nature, biosolids may contain chemical elements from the waste stream and some, like cadmium, are potentially hazardous [23]. These elements may cause toxic air or ash emissions if combusted.

**Table 3.57. Heating values and compositional data for forestry crops and residues [28]**

| Type of Biomass | Heating Values (MJ/kg, Dry Basis) | | Approximate Analysis (% by Weight, Dry Basis) | | Ultimate Elemental Analysis (% by Weight, Dry Basis) | | | | | | | |
|---|---|---|---|---|---|---|---|---|---|---|---|---|
| | Higher | Lower | Volatile | Ash | F. Carbon | Carbon | Hydrogen | Oxygen | Nitrogen | Sulfur | Chlorine | Residue |
| Alder (red) | 19.3 | 18.04 | 87.1 | 0.4 | 12.5 | 49.55 | 6.06 | 43.78 | 0.13 | 0.07 | | 0.41 |
| Black locust | 19.71 | 18.52 | 80.94 | 0.8 | 18.26 | 50.73 | 5.71 | 41.93 | 0.57 | 0.01 | 0.08 | 0.97 |
| Black oak | 18.65 | 17.4 | 85.6 | 1.4 | 13 | 48.97 | 6.04 | 43.48 | 0.15 | 0.02 | | 1.34 |
| Black walnut | 19.83 | 18.62 | 80.69 | 0.78 | 18.53 | 49.8 | 5.82 | 43.25 | 0.22 | 0.01 | 0.05 | 0.85 |
| Canyon live oak | 18.98 | 17.78 | 88.2 | 0.5 | 11.3 | 47.84 | 5.8 | 45.76 | 0.07 | 0.01 | | 0.52 |
| Cedar (western red) | 20.56 | | 86.5 | 0.3 | 13.2 | | | | | | | |
| Cedar (western red) bark | 20.84 | | 77.6 | 1.2 | 21.2 | | | | | | | |
| Douglas fir | 20.37 | 10.09 | 87.3 | 0.1 | 12.6 | 50.64 | 6.18 | 43 | 0.06 | 0.02 | | 0.01 |
| English walnut | 19.63 | 18.46 | 80.82 | 1.08 | 18.1 | 49.72 | 5.63 | 43.14 | 0.37 | 0.01 | 0.06 | 1.07 |
| Hemlock (western) | 19.89 | | 87 | 0.3 | 12.7 | | | | | | | |
| Maple (big leaf) | 18.86 | 17.6 | 87.9 | 0.6 | 11.5 | 49.89 | 6.09 | 43.27 | 0.14 | 0.03 | | 0.58 |
| Ponderosa pine | 20.02 | 18.78 | 82.54 | 0.29 | 17.17 | 49.25 | 5.99 | 44.36 | 0.06 | 0.03 | 0.01 | 0.3 |
| Poplar | 19.38 | 18.17 | 82.32 | 1.33 | 16.35 | 48.45 | 5.85 | 43.69 | 0.47 | 0.01 | 0.1 | 1.43 |
| Redwood (comb. sample) | 20.72 | 19.48 | 79.72 | 0.36 | 19.92 | 50.64 | 5.98 | 42.88 | 0.05 | 0.03 | 0.02 | 0.4 |
| Redwood (heartwood) | 21.14 | | 80.28 | 0.17 | 19.55 | | | | | | | |
| Redwood (sapwood) | 20.31 | | 80.12 | 0.67 | 19.21 | | | | | | | |
| Tan oak (comb. sample) | 18.93 | 17.7 | 80.93 | 1.67 | 17.4 | 47.81 | 5.93 | 44.12 | 0.12 | 0.01 | 0.01 | 2 |
| Tan oak (sapwood) | 19.07 | | 83.61 | 1.03 | 15.36 | | | | | | | |
| White fir | 19.95 | 18.71 | 83.17 | 0.25 | 16.58 | 49 | 5.98 | 44.75 | 0.05 | 0.01 | 0.01 | 0.2 |
| White oak | 19.42 | 18.3 | 81.28 | 1.52 | 17.2 | 49.48 | 5.38 | 43.13 | 0.35 | 0.01 | 0.04 | 1.61 |

Because of their high moisture content and potentially hazardous chemicals, biosolids may be better suited to anaerobic digestion processes. Approximately 75% of biosolids are volatile and available for digestion. For typical digester efficiencies of 30% 60% (40%–50% is more typical), most digesters will yield between 0.6 to 1.5 $m^3$ of biogas per kilogram of volatile solids digested, with a calorific value of 18 to 22 $J/cm^3$ [45].

### 3.3.3 Energy Crops

If biomass energy is to have a large, scale, sustainable future, then resources in the form of energy crops must be available. The term *energy crops* as used here refers to crops grown specifically for energy purposes. Usually these crops are bred or genetically engineered to maximize biomass productivity and to optimize plant characteristics for energy applications. Energy crops as defined in this section do not include commodity crops that are used for a variety of markets that may include energy applications [such as maize (corn) for ethanol].

Energy crops are typically perennial crops and thus distinguish themselves from conventional agricultural crops that are usually annual. Energy crops reviewed here include short-rotation wood crops (SRWC) and herbaceous energy crops (HEC). Aquatic crops, especially microalgae for oil production, have also been researched for energy crops, but are not included in this review, nor are oil seed crops. The actual choice of crop will depend on several factors including the end use, ease of processing, suitability for the climate and soil conditions, delivered costs, harvest timing, and equipment and production requirements.

Energy crops have been researched in the United States for over 20 years and longer in other parts of the world. The development of energy crops in the US has been hampered by low energy prices but continued research is slowly improving their competitiveness. Relatively small reductions in the cost of producing and harvesting energy crops together with slightly higher fossil energy prices could make biomass energy cost competitive in the US [49]. Under present conditions, the drivers for continued research include the long lead time to develop crops, the development of niche opportunities such as for wastewater treatment, interest in the use of some energy crops for other markets (e.g., fiber), and regulatory drivers such as global climate change mitigation.

Outside of the US, there is more interest in energy crops for a variety of reasons, including lack of indigenous fossil fuels, higher energy prices, scarcity of foreign currency, biomass availability, and government policies that support the development of biomass fuels. For example, Sweden, with higher energy prices and more favorable policies, has commercialized systems to produce and harvest fast-growing willows for energy applications and had over 70,000 ha planted in 1995 [50].

Like other biomass resources, energy crops can be converted into virtually any energy form. However, energy crops can be developed to optimize key characteristics for energy applications and their sustained production can better ensure long-term, large-scale supplies with uniform characteristics. Energy crops also have significantly higher yields per unit of land area than natural stands. These higher yields improve their cost effectiveness over conventional crops and minimize land requirements, associated chemical use, hauling requirements, and other negative environmental impacts [51]. Energy crops, especially trees, can also improve the environment by acting as biological filters when planted along waterways where they intercept and remove herbicides, pesticides, and excess fertilizers from water flows [52]. Frequently energy crops can better tolerate limited flooding or drought conditions and thus be more productive than conventional agricultural crops when grown on sites that are subject to such conditions. As the world's population, environmental concerns, and land use pressures grow, these and other factors become increasingly important for the development of energy resources and the use of energy crops.

Much of the following information is general in nature since the technology is rapidly evolving and recommendations are rapidly changing. Additionally, the worldwide diversity of crops, soil and climate conditions, production capabilities, and other factors would require volumes to cover all aspects. The sections below cover production and harvesting issues, relevant physical and chemical characteristics, and environmental factors.

***Production of SRWC***

A wide variety of short rotation woody crops (SRWC) have been researched around the world. (Sometimes SRWC are called short-rotation intensive crops or SRIC.) These include willows, eucalyptus, hybrid popular, black locust, sycamore, sweetgum, silver maple, and other species. SRWC's have the distinguishing features of being fast-growing hardwoods, grown under intensive production practices, and harvested in cycles that are 3 to 10 years in duration versus 20 to 30 years for conventional silviculture practices [53]. SRWC's can be harvested repeatedly and resprout (coppice) from the stump, thus eliminating the need for replanting after each harvest. However, repeated harvest cycles significantly reduce plant vigor and productivity and it is usually recommended that a new crop be planted after three harvest cycles. Because of the current rapid development of new varieties with improved attributes, it is currently recommended that replanting occur after every harvest and that new varieties be planted [51].

Because they are clones, SRWC's are planted from cuttings and not seedlings (although cuttings may be rooted beforehand). The optimum SRWC site is a deep, medium-texture soil, where root penetration to at least 1 m is not interrupted by bedrock, water table, hardpans, or gravel layers. SRWC hardwoods will grow well on both upland and bottomland sites, provided the soil is well drained and either has a good moisture-holding capacity or has a water table that is within reach of the roots during the growing season [51].

Site preparation can improve soil structure and aeration, reduce existing plant competition and some pests, and facilitate accessibility. Depending on prior vegetative cover and general site conditions, preparation may range from simple to complex and expensive. For crops established on open, cleared land, site preparation is relatively simple and similar to agronomic practices for conventional crops. Tillage, no-till, or a combination of both systems may be used [51].

To obtain optimum performance, topography, soils, climate, and location must match the genetic requirements of the trees. Although there is a growing body of SRWC knowledge in the world, site-specific conditions can make the matching of feedstocks with sites difficult. Carefully managed, scientifically planned trials at one or two representative sites will help define specific trees that will give optimum performance for the conditions present. These trials can also help define management practices needed for optimum crop performance. Although the cost and the land required for these trials are relatively minor, the information provided can be the most important factor in establishing a successful SWRC operation [54].

Proper establishment and the timing of planting are critical factors in the eventual success of the SRWC crop. As a rule of thumb for areas where frost occurs, planting can occur anytime after frost leaves the ground, usually March through June in the Northern Wemisphere [50]. Planting may be performed manually with any number of devices, including shovels, dibbles, and hoedads. However, manual labor is expensive and time consuming; therefore most SRWC plantings are done by machine. In industrialized countries, the most frequently used device is a partially mechanized, tractor-towed machine that opens a continuous furrow. Unrooted cuttings, small rooted cuttings, or seedlings are hand fed into the planting mechanism, which inserts the plant into the furrow and tamps the surrounding earth with press wheels [51]. The Swedes have commercialized an automated, tractor-drawn and -powered, four-row planter for willow SRWC [50].

The selection of plant spacings can have an impact on several factors. Spacing affects the cost of establishment, required duration of weed control, prevalence of disease, final size of trees, biomass distribution among tree components, nutrient requirements, type of harvesting machine, harvesting frequency and costs, yield, and wildlife diversity. Thus spacing selection must take into account the species grown, management objectives, and market requirements. Normally, spacings range from 1 m × 1 m to 2 m × 3 m [51].

Even with superior stock, it is necessary to manage SRWC's at the same level as that of an agricultural crop in order to obtain high yields. Weed control is important during the first 2 years to prevent shading, limit competition for water and nutrients, and to eliminate plant phytotoxins. Applications of fertilizer may be required for obtaining the best results. Fertilization will not affect all species and clones in the same way, so nutrient additions should be tailored to specific tree requirements [51].

Worldwide, SRWC average annual growth rates start at 10 $m^3$/ha/year and have an upper limit considered to be in the 50–70 $m^3$/ha/year range. Average regional yields are in the range of 12 to 30 $m^3$/ha/year for *eucalyptus* SRWC species, and 10 to 20 $m^3$/ha/year for *pinus* species. Although more research is needed, preliminary estimates for the *populus* species commercially grown in the Northern Hemisphere range from 10 to 30 $m^3$/ha/year [54]. Table 3.58 summarizes typical growth rates for commercial SRWC plantings in various countries.

Roughly 4%–8% of the land area within a 40-km radius would be required for supplying to typical electricity or ethanol production facilities with processing capacities of up to 1000 to 2000 dry t per day [53].

Harvesting provides the critical link between crop production and end use and, for SRWC, can represent up to half the delivered cost of feedstocks [51]. In designing a harvesting system, biological requirements of the feedstock must be taken into account. An important aspect of the management strategy for SRWC plantations is the reliance on coppice regeneration for stand reestablishment (where desired) for one or more generations beyond the initial planted rotation. This imposes a number of constraints and

**Table 3.58. Typical growth rates at commercial plantations in various countries [54]**

| Region | Main Species in Plantations | Rotation period (yrs) | Number of thinnings | Normal Growth Rates, (MAI $m^3$/ha/yr), |
|---|---|---|---|---|
| Argentina | *Pinus taeda, P. eliotti* | 20–25 | 2–4 | 20–25 |
| | *Populus spp., Eucalyptus spp.* | 10–14 | — | 15–20 |
| Brazil | *Eucalyptus spp.* | 3 × 7 | — | 16–60 |
| | *Pinus eliotti, P. taeda* | 20–25 | 2–4 | 20–30 |
| | *Pinus carigaea, P. oocarpa* | 20–25 | 2–4 | 12–20 |
| Chile | *Pinus radiata* | 20–25 | 2 | 15–30 |
| US South | *Pinus taeda, P. eliotti, P. palustris, P. echinata* | 20–45 | 0–2 | 12–15 |
| | *Populus spp., Salix, spp., Liquidambar spp., etc.* | 10–100 | | 4–35 |
| Portugal | *Pinus pinaster* | 40 | 2–3 | 10 |
| | *Eucalyptus globulus* | 3 × 8 to 10 | | 10–18 |
| Spain | *P. pinaster, P. radiata, Eucalyptus globulus* | 25–40 | 2–3 | 4–15 |
| | *E. camaldulensis* | 3 × 8 to 20 | | 4–20 |
| Australia | *P. radiata, P. pinaster, p. elliotti, P. caribea* | 20–35 | 1–3 | 15–17 |
| | *Eucalyptus spp.* | 3 × 10 | — | 15–20 |
| New Zeland | *Pinus radiata* | 25–35 | 2–3 | 15–30 |
| South Africa | *Pinus radiata and other pine species* | 15 | 3 | 18–20 |
| | *Eucalyptus spp.* | 11 | — | 20–22 |

requirements on harvesting methods and equipment as they relate to tree size and form and to stump–root system management [51].

Most SRWC plantations are likely to be established at planting densities ranging from 400 to 800 trees per hectare. At time of harvest, individual stems will likely have diameters of 10–20 cm near ground level and stand 7.5–15 m tall. In the later generations, coppice stools will produce a number of shoots, several of which are likely to survive to time of harvest. Although total stool biomass may be greater, each of the shoots will be smaller in diameter and height than the stems of the initial generation, and the overall form will be of several outwardly and upwardly spreading shoots rather than a single straight stem. Harvesting methods must be capable of accommodating these differences [51].

The time of harvest affects coppice success and vigor with the degree of sensitivity to this factor highly variable between species, varieties, and even individual clones. In general, however, it can be stated that dormant season harvesting (during the period October–April in the Northern Hemisphere) is far preferable to harvesting during the growing season [51]. Harvesting productivity and therefore the cost of harvesting, are influenced by a number of factors, including site conditions, the distance between trees, and tree size or volume. Harvesting methods for SRWC range from manual methods to sophisticated machines that sever the stem from the stump and then chip or bundle the material. Manual methods include brush saws and chainsaws. In addition to being labor intensive, the central difficulty with the manual approach to SRWC harvesting is one of efficiency and economics. SRWC trees are small diameter and low volume;

hence many trees must be processed to obtain a ton of material. In the relatively homogeneous physical environment of a SRWC plantation, the principal factor becomes tree size [51].

A study by the Tennessee Valley Authority [55] that reviewed the state-of-the-art of forest harvesting machines concluded with the following specific recommendations for SRWC harvester design:

- Use of saws as the severance mechanism: Circular saws, chainsaws, and bandsaws have several advantages over shears: They require less power, can cut continuously, are capable of storing energy in the form of inertia, and leave a cleanly cut and undamaged stump.
- Minimize power requirements: Avoiding large power-consuming functions such as in-field chipping will reduce the physical bulk and the mass of the machine, thereby reducing soil compaction, and significantly reducing machine cost.
- Design harvesting equipment as attachments to existing carriers: Again, there is an analogy in agriculture in which many farm implements are designed as tractor attachments. This approach reduces costs by allowing the farmer to buy only the harvesting function and furnish his own motor function (existing farm tractor).
- Produce piles, bundles or bales: Aggregation to units of sufficient size and proper orientation facilitates handling, storage, and processing, and thereby reduces costs.

Over the past few years, several prototypical machines have been developed around the world. Commercially available machines designed specifically to harvest SRWC are made in Europe, where two types of machines capable of harvesting willows are available. The first type is the whole-stem harvester–chipper that cuts, chips, and blows the chips into a wagon following or pulled by a harvester. The second type is the harvester stacker machine. This cuts whole stems and then dump bundles of them in the field, which are moved by grappling equipment for on-site storage, direct transport, or chipping and transport [50].

***Herbaceous Energy Crops***

Herbaceous Energy Crops (HEC's) include a large variety of nonwoody plants that are harvested by removal of their aboveground biomass. Most attention has been given to perennial grasses and a few annual grasses. Annual crops can pose problems with soil and wind erosion, require greater weed control efforts, and have higher production costs because of the need to reestablish the crop annually. Perennial crops usually have lower input requirements, lower crop risk, and are better suited to environmentally sensitive lands. Once a site is planted to a perennial crop, it usually does not require any additional tillage or herbicide [49].

Grass energy crops are typically managed the same as similar types of agricultural crops. In general researchers have focused on thin-stemmed perennials for temperate climates and thick-stemmed grasses for tropical climates [49].

*Thin-Stemmed Grasses*

Although thin-stemmed grasses are or could be used for forage, dual use is not recommended. It is desirable for energy crops to have low nutrient values at harvest to minimize crop nitrogen requirements and to minimize plant nitrogen content, which can cause emissions when thermochemically converted into energy [49].

After a number of cool-season and warm-season grasses were screened, the HEC currently recommended for North America temperate zones is switchgrass (*Panicum virgatum*). Most HEC research has focused on the development of monocultures; however, mixtures can provide better wildlife habitat, mitigate crop risk, and, if mixed with leguminous crops, can reduce nitrogen requirements [49].

Establishing and maximizing productivity of grass crops include the following steps: soil testing, use of live seed and optimum seeding rates, and following recommended practices for seedbed preparation, planting techniques, and crop management [49]. Where little or no information is available for a desired energy crop, growth trials will have to be performed on representative sites in small plots to determine sound crop production practices.

Crop establishment may include conventional tillage, conservation tillage, or no-till practices. The first method requires that the site be free of residues. Weed control can be by means of cultivation, herbicide application, or a combination of both. Conservation tillage or no-till methods are recommended for erosion sensitive sites. "*Conservation tillage* is defined as any tillage and planting system that maintains a residue cover of at least 30% of the soil surface covered after planting. *No-till* is defined as tillage in which the soil is undisturbed before planting, planting is done in a narrow seedbed or slot created by a planter or drill, and weeds are controlled primarily with herbicides" [49].

Although hardy once established, perennials take 1–3 years to establish and can be difficult to establish. Factors that make perennials difficult to establish include weather and soil conditions, competing weeds, hard rains after seeding, and soil crusting. To ensure good yields, these factors must be considered carefully before site and crop selection and planting [49].

Harvesting systems for thin-stemmed grasses have a significant advantage in that they can usually use traditional farm machinery and the traditional haying methods of cutting, field drying, raking, and baling. Although balers that produce large round bales are readily available in industrialized countries, large square bales are usually preferred for high-volume material handling. Bales increase hay density, reduce transportation costs, and facilitate handling. However, bales are bulky and usually require special equipment at the processing plant for handling bales.

Simultaneous harvesting and in-field pelletizing or cubing of HEC's by use of machines similar to combines can retain the advantages of densification similar to bales but eliminate the bulky handling problem of bales. A prototypical mobile, in-field pelletizing machine developed in Germany [56] produces 60 mm × 15 mm × 40 mm pellets with a bulk density similar to that of grain. The pellets can be easily handled with grain-handling equipment and readily burned in most combustion systems. Mobile, in-field hay cubers that produce 25-mm cubes have been commercially available in the US for over 20 years.

*Thick-Stemmed Grasses*

Thick-stemmed grasses for energy crops include sugar cane and energy cane—both hybrids of cane or *Saccharum* species and napiergrass (elephantgrass or *Pennisetum purpureum*). The cropping practices for these grasses are similar to those for thin-stemmed perennials. However, the thick-stemmed grasses maintain stand viability ~5 years (versus 10 years for the thin-stemmed) and need to be vegetatively propagated and planted by hand [49].

Other thick-stemmed grasses that have been studied for energy crops are corn (*Zea mays*), sorghum (*Sorghum bicolor*), sorghum × sudangrass (*S. bicolor*), and sudangrass (*S. sudanensis*) [49]. Bamboo demonstrates good yields—especially under tropical conditions—but is difficult to propagate [57]. Equipment for large-scale bamboo harvesting is also not available.

Current processes used for harvesting cane for energy crop applications are not cost effective, and additional research to develop harvesting equipment for thick-stemmed grasses is needed. Forage-type harvesting systems and silage storage systems seem the most promising [49].

*Leguminous Crops*

Leguminous crops such as alfalfa have been studied but have not demonstrated high yields, lack stand viability, and require insecticide applications and high inputs of potassium. Leguminous crops are also rich in nitrogen and are relatively low in hemicellulose and cellulose content [49]. However, crops such as alfalfa that have high protein contents or other potentially valuable coproducts may be a key to cost- effective energy crops.

***Physical and Chemical Characteristics***

One of the biggest concerns for feedstocks from fast-growing plants to be used in thermochemical conversion processes is their propensity to cause slagging or fouling [37]. See the subsection on slagging and fouling causes and methods of mitigation for a detailed discussion. Obviously, in addition to breeding for yield, hardiness, and other traditional factors, developers of new energy crops must take into account the crop's physical and chemical characteristics that will facilitate conversion and end use.

Unfortunately, only limited handbook data are available on physical and chemical characteristics for energy crops and even less are available on ash characteristics. Tables 3.51 and 3.59 [37] list some energy crops for which data are available.

***Environmental Considerations***

Although energy crops offer several energy and economic benefits, there are environmental trade-offs that must be carefully considered before launching into any full-scale project. Environmental considerations include water, soil, and air quality, greenhouse gas emissions, habitat change and biodiversity, and long-term site productivity. Fortunately energy crops can provide significant environment benefits if done properly but unfortunately can provide potentially serious damage if not done properly.

Large-scale energy crop production can require the use of tens of millions of hectares of land. The net environmental effect will depend on many factors, including the previous land use, the crop selected, production methods used, the integration of the local environment into the cropping systems, and other factors. In general, the positive environmental benefits of energy crops range from modest to significant compared with conventional agricultural row crops and the negative impacts are generally less than those of conventional row crops following typical management practices [53].

**Table 3.59. Physical and chemical characteristics for selected energy crops [37]**

| Fuel | Miscanthus Silberfeder | | Sorghastrum Avenaceum | | Wood Hybrid Poplar | | Willow SA22-3 Year | | Bana Grass HI, Immature | | Switchgrass Summer-MN, MN | |
|---|---|---|---|---|---|---|---|---|---|---|---|---|
| Type | As Rec'd | Dry | As Rec'd | Dry | As Rec'd | Dry | As Rec'd | Dry | As Rec'd | Dry | As Rec'd | Dry |
| Approximate analysis | | | | | | | | | | | | |
| Fixed Carbon | 10.73 | 12.55 | 12.57 | 14.18 | 11.63 | 12.49 | 14.47 | 16.07 | 15.93 | 16.68 | 12.47 | 14.37 |
| Volatile Matter | 72.12 | 84.4 | 72.42 | 81.67 | 78.97 | 84.81 | 74.01 | 82.22 | 70.13 | 73.44 | 71.93 | 83.94 |
| Ash | 2.61 | 3.05 | 3.68 | 4.15 | 2.51 | 2.7 | 1.54 | 1.71 | 9.43 | 9.88 | 2.33 | 2.69 |
| Moisture | 14.54 | — | 11.33 | — | 6.89 | — | 9.98 | — | 4.51 | — | 13.27 | — |
| Total | 100 | 100 | 100 | 100 | 100 | 100 | 100 | 100 | 100 | 100 | 100 | 100 |
| Ultimate analysis | | | | | | | | | | | | |
| Carbon | 40.41 | 47.29 | 41.94 | 47.3 | 46.72 | 50.18 | 44.92 | 49.9 | 43.02 | 45.06 | 41.21 | 47.51 |
| Hydrogen | 4.92 | 5.75 | 5.33 | 6.01 | 5.64 | 6.06 | 5.31 | 5.9 | 5.18 | 5.42 | 5.03 | 5.8 |
| Oxygen | 37.19 | 43.52 | 37.4 | 42.18 | 37.66 | 40.44 | 37.63 | 41.81 | 36.95 | 38.69 | 37.81 | 43.6 |
| Nitrogen | 0.28 | 0.33 | 0.28 | 0.32 | 0.56 | 0.6 | 0.55 | 0.61 | 0.8 | 0.84 | 0.31 | 0.36 |
| Sulfur | 0.05 | 0.06 | 0.04 | 0.04 | 0.02 | 0.02 | 0.07 | 0.07 | 0.11 | 0.11 | 0.04 | 0.04 |
| Ash | 2.61 | 3.05 | 3.68 | 4.15 | 2.51 | 2.7 | 1.54 | 1.71 | 9.43 | 9.88 | 2.33 | 2.69 |
| Moisture | 14.54 | — | 11.33 | — | 6.89 | — | 9.98 | — | 4.51 | — | 13.27 | — |
| Total | 100 | 100 | 100 | 100 | 100 | 100 | 100 | 100 | 100 | 100 | 100 | 100 |
| HHV (MJ/kg) | 16 | 19 | 17 | 19 | 18 | 19 | 18 | 20 | 17 | 18 | 16 | 19 |
| Chlorine (%) | 0.05 | 0.06 | 0.04 | 0.04 | 0.01 | 0.01 | <0.01 | <0.01 | 0.79 | 0.83 | <0.01 | 0.01 |

(Continues)

Table 3.59. (Continued)

| Fuel | Miscanthus Silberfeder | | Sorghastrum Avenaceum | | Wood Hybrid Poplar | | Willow SA22-3 Year | | Bana Grass HI, Immature | | Switchgrass Summer-MN, MN | |
|---|---|---|---|---|---|---|---|---|---|---|---|---|
| Type | As Rec'd | Dry | As Rec'd | Dry | As Rec'd | Dry | As Rec'd | Dry | As Rec'd | Dry | As Rec'd | Dry |
| Water-soluble alkalis (%) | | | | | | | | | | | | |
| $Na_2O$ | 0.003 | 0.003 | 0.005 | 0.005 | 0.003 | 0.003 | 0.007 | 0.007 | 0.016 | 0.017 | 0.004 | 0.004 |
| $K_2O$ | 0.288 | 0.337 | 0.295 | 0.332 | 0.545 | 0.623 | 0.242 | 0.269 | 3.52 | 3.686 | 0.134 | 0.154 |
| CaO | | | | | | | | | | | | |
| Elemental Composition | | | | | | | | | | | | |
| $SiO_2$ | | 61.84 | | 71.05 | | 5.9 | | 2.35 | | 33.65 | | 61.64 |
| $Al_2O_3$ | | 0.98 | | 1.78 | | 0.84 | | 1.41 | | 0.8 | | 1.32 |
| $TiO_2$ | | 0.05 | | 0.02 | | 0.3 | | 0.05 | | 0.07 | | 0.19 |
| $Fe_2O_3$ | | 1.35 | | 0.92 | | 1.4 | | 0.73 | | 0.63 | | 1.08 |
| CaO | | 9.61 | | 6.81 | | 49.92 | | 41.2 | | 3.57 | | 11.11 |
| MgO | | 2.46 | | 2.14 | | 18.4 | | 2.47 | | 1.71 | | 4.86 |
| $Na_2O$ | | 0.33 | | 0.24 | | 0.13 | | 0.94 | | 0.38 | | 0.64 |
| $K_2O$ | | 11.6 | | 8.7 | | 9.64 | | 15 | | 42.8 | | 8.24 |
| $SO_3$ | | 2.63 | | 1.08 | | 2.04 | | 1.83 | | 0.85 | | 0.8 |
| $P_2O_5$ | | 4.2 | | 4.3 | | 1.34 | | 7.4 | | 2.74 | | 3.09 |
| $CO_2$/other | | | | | | 8.18 | | 18.24 | | 8.97 | | |
| Undetermined | | 4.95 | | 2.96 | | 1.91 | | 8.38 | | 3.83 | | 7.03 |
| Total | | 100 | | 100 | | 100 | | 100 | | 100 | | 100 |
| Alkali, (kg/GJ) | | 0.19 | | 0.2 | | 0.14 | | 0.14 | | 2.43 | | 0.12 |

For lands that are converted to energy crops from agricultural production, many of the same factors that affect agriculture will also affect HEC and SRWC. The energy crops considered here are perennial and thus require less cultivation than row crops. Usually energy crops also are more efficient in their use of fertilizers since there is some nutrient recycling between growing seasons [53].

Each crop will have different management strategies and differing impacts on the environment. The level of knowledge on energy crop environmental issues is still in the developmental stages and the reader is encouraged to seek the latest research reports before project implementation.

## References


1. Lamb, R. 1995. *Forests, Fuels and the Future*, Forestry Topics Report No. 5, FAO Forestry Department, Food and Agricultural Organization of the United Nations, pp. 4–18.
2. FAO. 1987. *Simple Technologies for Charcoal Making*, FAO Forestry Paper 41. Rome: Food and Agricultural Organization of the United Nations.
3. Munslow, B., Y. Katerere, A. Ferf, and P. O'Keefe. 1988. *The Fuelwood Trap*, pp. 1–29, London: Earthscan Publications.
4. Soussan, J. 1991. Building sustainability in fuelwood planning, *Bioresource Technol.*: 49–56.
5. Leach, G. and R. Mearns. 1988. *Beyond the Fuelwood Crisis—People, Land, and Trees in Africa*, pp. 1–24, London: Earthscan Publications.
6. Hall, D. O. and R. P. Overend, eds. 1987. *Biomass Regenerable Energy*, pp. 3–49. New York: Wiley.
7. Johansson, T. B., H. Kelly, A. K. N. Reddy, and R. H. Williams, eds. 1993. *Renewable Energy—Sources for Fuels and Electricity*, p. 641. New York: Island Press.
8. FAO. 1985. *Industrial Charcoal Making*, FAO Forestry Paper 63, Rome: Food and Agricultural Organization of the United Nations.
9. Lemmerich, G., and C. A. Luengo 1996. Babassu Charcoal: A Sulfurless Renewable Thermo-reducing Feedstock for Steelmaking. *Biomass Bioenergy* 10(1):41–44.
10. Meyers, D. 1996. DHM, Inc., Holmdell, New Jersey, personal communication.
11. FAO. 1983. *Simple Technologies for Charcoal Making*, FAO Forestry Paper 41. Rome: Food and Agricultural Organization of the United Nations.
12. Brocksiepe, H.-G. 1986. Charcoal, *Ullmann's Encyclopedia of Industrial Chemistry*, 5th ed., pp. 157–162. VCH Publishers, New York, NY.
13. Smith, T. G. 1985. *Alaska Charcoal Production Feasibility Study*, prepared for the Pacific Northwest and Alaska Regional Biomass Energy Program, administered and published by the Bonneville Power Administration, November, p. 63.
14. Wegner T. H., A. J. Baker, B. A. Bendtsen, J. J. Brenden, W. E. Eslyn, J. F. Harris, J. L. Howard, R. B. Miller, R. C. Pettersen, J. W. Rowe, R. M. Rowell, W. T. Simpson, D. F. Zinkel, and the U.S. Department of Agriculture, eds. 1984. Wood, in: *Kirk-Othmer Encyclopedia of Chemical Technology*, 3rd ed., pp. 603–604. New York: Wiley.

15. Russart, T. 1996. Kingsford Products, Louisville, Kentucky, Personal communication.
16. *Center for Semi-Arid Forest Resources, 1996. Annual Report 1995-1996*, Kingsville, TX: Texas A&M University.
17. Product literature. The Hiller Group, Inc., Tampa, FL.
18. Stokes, B. J. 1992. Harvesting small trees and forest residues. *Biomass Bioenergy*, 2:131–147.
19. Stokes, B. J. and T. McDonald. 1996. *Logging/Harvesting Slash (Residues) Summary*. Auburn, AL: USDA Forest Service, Engineering Research Unit.
20. Muehlenfeld, K. J. 1997. *Biomass Energy Sourcebook: A Guide for Economic Developers in the Southeast*. AL: Southeastern Regional Biomass Energy Program, Tennessee Valley Authority, Muscle Shoals.
21. Schwieger, B., Moisture level has greatest impact on boiler design. (In: Power from wood—a special report) Power Mag. 124:54–57.
22. Lindley, J. A., L. F. Backer, and D. W. Johnson. 1994. A Review of Agricultural Crop Harvest and Collection Technology: *Proc. Bioenergy '94*, pp. 257–264. Golden, CO: Western Regional Biomass Energy Program, Western Area Power Administration.
23. Council for Agricultural Science and Technology. 1995. Crop Production—Waste Management and Utilization, *Waste Management and Utilization in Food Production and Processing*, Task Force Report No. 124.
24. Hill, P. R. 1996. The Economic Value of Crop Residue. *Nat. Conserv. Tillage Dig.* 3:9–10.
25. Nikolaisen, L., ed., C. Nielsen, M. G. Larsen, P. Keller, and L. K. Henriksen. 1992. *Straw for Energy Production: Technology-Environment-Economy*. Copenhagen: The Centre of Biomass Technology, Danish Energy Agency.
26. Coates, W. 1995. Harvesting cotton stalks for use as a biomass feedstock *Proc. Second Biomass Conference of the Americas*, pp. 130–139. Golden, CO: National Renewable Energy Laboratory.
27. Mahin, D. B. 1986. Cane energy systems. *Bioenergy Systems Report*. Arlington, VA: US Agency for International Development.
28. Kitani, O. and C. W. Hall, eds. 1989. Physical properties of biomass. *Biomass Handbook*, pp. 880–882. New York: Gordon and Breach.
29. *Chemical Engineering Progress*. December 1993. Food byproducts "bearing fruit" As Feedstocks for Industry, p. 1. New York: American Institute of Chemical Engineers.
30. Meade, G. P. and J. C. P. Chen. 1977. Bagasse and its uses. *Cane Sugar Handbook*, pp. 95–108. New York: Wiley.
31. Eniasivam, S. 1996. India-bagasse based cogeneration opportunities and obstacles. *World Cogeneration* 8(3):16–17.
32. Margaret A. Bagasse, 1964, *Kirk-Othmer Encyclopedia of Chemical Technology*, 2nd Completely Revised Edition, Vol. 3 pp. 434–438. New York: Wiley.
33. Mahin, D. B. 1986. Power from rice husks. *Bioenergy Systems Report*. Arlington, VA: US Agency for International Development. USA, April.
34. Bailey, R. W. 1988. *Particulate Waste Product Combustion System Used in Malaysian Drying System*, product literature. Stuttgart, AR: PRM Energy Systems.

35. Harp, S. L. 1993. Pilot Scale Cotton Gin Trash Energy Recovery. *Proc. First Biomass Conference of the Americas*, pp. 712–717. Golden, CO: National Renewable Energy Laboratory.
36. Abbott, R. and R. Quick. 1989. Grain wastes for Fuel. *Bio-Joule*, 11:4.
37. Miles, T. R., T. R. Miles, Jr., L. L. Baxter, R. W. Bryers, B. M. Jenkins, and L. L. Oden. 1995. *Alkali Deposits Found in Power Plants: A Preliminary Investigation of Their Extent and Nature, Summary Report*. Golden, CO: National Renewable Energy Laboratory.
38. Manure Production and Characteristics. 1990. ASAE D384.1; Standards 1990: *Standards, Engineering Practices, and Data*; 37th ed. St. Joseph, MI: ASAE.
39. *Regional Assessment of Nonforestry-Related Biomass Resources, Summary Volume*. 1990. Prepared by JAYCOR, Vienna, Virginia, for the Southeastern Regional Biomass Energy Program, Tennessee Valley Authority, Muscle Shoals, AL.
40. Authors. 1989. Cow manure burned in place of oil, *Waste Age* 20:16.
41. Blair, J. F. 1987. Manure-fueled power plant to be built at Hereford, Texas. *Feedstuffs* 59:7.
42. Kreith, F. 1992. Solid waste management in the U.S. and 1989-1991 State Legislation. *Energy*, 17:427–476.
43. Holmes, J. R. 1984. *Managing Solid Wastes in Developing Countries*, p. 15. New York: J. Wiley.
44. Bushnell, D. J., C. Haluzok, and A. Dadkhah-Nikoo. 1989. *Biomass Fuel Characterization: Testing and Evaluating the Combustion Characteristics of Selected Biomass Fuels.* Portland, OR: Bonneville Power Administration.
45. *Bioenergy in the Southeast: Status, Opportunities and Challenges; Recommendations of the Southeast Bioenergy Roundtable 1996*. Muscle Shoals, AL: Southeastern Regional Biomass Energy Program, Tennessee Valley Authority.
46. Badger, P. C. 1995. Marketing wood waste for fuel. *BioCycle*, 36:71.
47. Atkins, R. S., C. T. Donovan, and J. M. Peterson. 1992, *Wood Products in the Waste Stream: Characterization and Combustion Emissions*, Report 92–8. Albarey, NY: New York State Energy Research and Development Authority.
48. Fransham, P. 1995. Recycling of treated wood. *Proc. Second Biomass Conference of the Americas*, pp. 1069–1078, Golden, CO: National Renewable Energy Laboratory.
49. Wright, L. L. 1994. Production technology status of woody and herbaceous crops. *Biomass Bioenergy* 6:191–209.
50. White, E. H., L. P. Abrahamson, and D. J. Robison. 1995. Commercialization of willow bioenergy—A dedicated feedstock supply system. *Proc. Second Biomass Conference of the Americas*, pp. 1534–1546, Golden, CO: National Renewable Energy Laboratory.
51. Meridian Corporation. 1986. *Short-Rotation Intensive Culture of Woody Crops for Energy: Principles and Practices for the Great Lakes Region*. Chicago, IL: Great Lakes Regional Biomass Energy Program, Council of Great Lakes Governors.
52. Licht, L. A., J. L. Schnoor, D. R. Nair, and M. F. Madison. 1992. *Ecoltree Buffers for Controlling Nonpoint Sediment and Nitrate*. Paper 922626. St. Joseph, MI: ASAE.

53. Herdman, R. C. (Director). 1993. *Potential Environmental Impacts of Bioenergy Crop Production—Background Paper*, OTA-BP-E-118. Washington, DC: US Government Printing Office.
54. Arnold, R. B. 1995. Investment in fast-growing trees offers future wood procurement advantages. *Pulp & Paper Mag.* 69:136.
55. Curtin, D. T. and P. E. Barnett. 1986. *Development of Forest Harvesting Technology: Application in Short Rotation Intensive Culture (SRIC) Woody Biomass*. Tech. Note B58. Muscle Shoals, AL: Tennessee Valley Authority, Office of Agricultural and Chemical Development.
56. Sutor, P. 1995. A new system of harvesting biomass is born: *Proc. Second Biomass Conference of the Americas*, pp. 1228–1235, Golden, CO: National Renewable Energy Laboratory.
57. Woods, J. E. and C. H. Wilson. 1997. West Wind Technology, Athens, TN, personal communication.

## 3.4 Green Manures and Forage Legumes

*F. R. Magdoff*

### 3.4.1 Background

Green manure crops, now also commonly referred to as cover crops, are those crops grown specifically to improve the soil and are generally not harvested for sale. Although the terms green manures and cover crops are now used interchangeably, they originally referred respectively to crops grown to add organic matter (and/or N) to a soil and those that were grown primarily to maintain soil cover for erosion control. In addition, the term catch crops was commonly used to refer to crops grown for the purpose of taking up available nutrients following the growth of the main economic crop. Because soil-improving crops have many different effects on soils and subsequent crops and they are usually grown with multiple purposes in mind, the original distinctions between the various terms have become less significant. Sod crops grown as part of a rotation may also have many of the same effects as those of green manure crops. Forage legumes are therefore included in the discussion below.

Green manure crops have been grown since antiquity, with indications of their use in ancient Chinese as well as Roman agricultural systems [1]. The introduction of forage legumes, especially the clovers, into the medieval European agriculture permitted continuous cropping to replace alternate-year fallow or 2 years of crops alternating with a fallow year [2]. The enhanced productivity of the land that resulted from this new system created the agricultural surplus thought to have been an essential element in the early stages of the Industrial Revolution [2].

Rotations with forage legumes and the use of green manures was an important part of agriculture in the advanced capitalist countries until the post-World War II period. Two separate factors occurred in the last half of the 20th century that have had a huge impact on the use of such crops. Synthetic N fertilizers became widely available at relatively low prices, decreasing the importance of $N_2$-fixing crops for supplying N to cereal crops. In addition, the change from mixed livestock–crop farming systems to those on which animal feed crops are raised on farms geographically remote from animal production facilities means that there is little reason to grow forage crops on grain farms.

A number of factors have led to the renewed interest in green manures. Soil quality on farms growing grain under continuous cultivation has deteriorated as manure was eliminated as a fertility source, soil-building forage legumes and grasses were eliminated from rotations, and organic matter was depleted. Accelerated erosion on many soils has only made the situation worse. Many farmers are aware of the negative effects on their soil's structure and nutrient-supplying ability. There have also been concerns about the adverse impact that conventional agriculture has had on water quality, with $NO_3$–N contamination of groundwaters as well as problems with both N and P enrichment of surface waters. There has been an interest in trying to develop systems that have more biological diversity both aboveground and belowground than conventional agricultural methods. The new orientation of many farmers to reduce purchases from off the farm and to rely more on on-farm resources has created added interest in using $N_2$-fixing forage legumes as green manures or rotation crops [3].

### 3.4.2 Effects on Soil and Crops

There are many potential effects of forage legumes in rotations and green manures used with annual or perennial crops. The magnitude of the effects may be so large that when combined with conservation tillage may significantly help reclaim severely eroded soils [4].

*Biomass Production and Influences on Soil Organic Matter Content*

Considerable amounts of biomass can be produced by a cover crop, providing sufficient time is available and the weather is appropriate. In a comparison of three green manures in North Carolina (USA), Wagger [5] found that early killing of winter rye, crimson clover, and hairy vetch resulted in approximately 6.3, 3.8, and 3.3 Mg/ha, respectively. However, delaying the killing of the cover crop for a few weeks resulted in biomass increases of from 1.5 to 2.7 Mg/ha. The biomass produced helps increase soil organic matter, and when left on the surface can act as a mulch to help water infiltration and retention and weed control.

Although there may be considerable biomass production, there are some reports in the literature that there is little to no increase in soil organic matter following green manures [6]. This is partly due to the readily decomposable nature of green manures. However, after a number of years of a forage legume or green manure crops, total soil organic matter increases occur [7].

*Nitrogen Contribution to Following Crops*

Forage legumes as well as legume cover crops have received prominent attention in recent years because of their potential to contribute significant amounts of biologically fixed N to the subsequent summer crop [8,9]. The relatively rapid release of N from legume cover crops allows them to provide N in a timely manner for the following crop [5]. Average estimates of the equivalent amount of fertilizer N replaced by hairy vetch (*Vicia Villosa Roth*) and crimson clover (*Trifolium incarnatum L.*) in these studies were 90 to 100 kg/ha for corn (*Zea mays L.*) and 72 kg/ha for grain sorghum [*Sorghum bicolor* (*L.*) *Moench*], respectively. Following 3 years of alfalfa, first-year corn can obtain enough N to produce ~9-tons dry-weight grain/ha and have a total harvested N uptake of ~194 kg N/ha [10]. Three years of alfalfa was estimated to provide first-, second-, and third-year corn with the equivalent of 136-, 36-, and 16-kg N fertilizer/ha respectively, applied to continuous corn. Other estimates for the N contribution to first-year corn following multiyear alfalfa are 70–84 kg/ha [11] to over 150 kg/ha [12]. The N contributions of other forage legumes such as red clover may be significantly less than those for alfalfa [10]. In a rice–wheat system, sesbania was found to contribute half the N (58 kg/ha) or more of the N needs of rice [13].

*Using Cover Crops to Deplete Soil $NO_3$–N*

Following a harvest of heavily fertilized annual crops there is frequently a significant amount of $NO_3$–N still present in the soil profile [14]. There is some indication that the use of winter rye cover crops may result in lower $NO_3$–N concentrations in ground water [15]. While experiments with wheat, rye, or barley indicate that from 10 to ~70 kg of N may be removed from the profile [16], in northern areas much of the growth and removal of $NO_3$–N from the profile will occur in the spring near the end of the period of most active leaching losses.

*Effects on Soil Physical Properties*

The influence of green manure crops on the physical characteristics of soils has been previously reviewed [17]. The binding together of the sand, silt, and clay particles into stable aggregates helps to maintain good tilth. Polysaccharides produced during the breakdown of organic residues plus fungal hyphae promote the development of these stable soil aggregates. The increase in the mean weight diameter of water-stable aggregates under alfalfa was found to proceed faster than increases in total soil organic matter [7], indicating that the changes in particular soil organic matter fractions may be more important than the changes in total amounts of organic matter. The increase in the positive effects of alfalfa on soil aggregation occurred mainly during the first 3 years of the stand [7].

Many soils under cultivation tend to develop a thin surface layer or crust [18]. Raindrop impact on bare soil plays a central role in promoting clay dispersion, breakdown of surface aggregates, and filling of macropores near the surface. Under these conditions, water permeability decreases drastically, and the crust is sometimes of such strength that the emergence of seedlings is restricted. Green manure crops can help to reduce the strength of the restricting layer and increase infiltration of water into the soil [19]. A living green manure crop or its residues intercepts raindrops and decreases the impact on surface aggregates. In addition, the action of living roots and root exudates and mycorrhizal fungi associated with green manures helps to maintain soil structural integrity near the soil surface.

Using green manures and forage legumes tends to increase soil organic matter compared with annual grain cropping or clean cultivation in vineyards and orchards. A soil that is high in organic matter will have better aggregation and tend to be less dense and allow better root penetration and development than in depleted organic matter situations. In addition, the soil will have higher infiltration rates because of a more stable surface structure that is able to resist the dispersive force of raindrop impact. The activities of larger soil organisms, such as earthworms and ants, also help to improve water infiltration. With more water infiltrating instead of running off the field, the soil will be less prone to erosion.

Sandy soils with higher levels of organic matter will have more small pores to store plant-available water and be less prone to drought. On the other hand, more clayey soils will have better internal drainage when large amounts of organic matter are present than when in an organic-matter-depleted condition.

Green manures or permanent sods between rows of tree crops also give better traction for equipment, especially in wet weather. This probably results mainly from better soil drainage and greater stability of the soil structure when roots and mycorrhizae are plentiful.

*Effects on Soil Chemical Properties*

The cation exchange capacity as well as pH buffering of many soils is strongly influenced by the content of soil organic matter [20]. Thus the accumulation of soil organic matter that occurs under frequent green manuring or including forage legumes in a crop rotation helps the soil to retain cations as well as providing buffering against rapid changes in pH. Inorganic nutrients are also taken up by green manures and converted to organic forms, which must be mineralized by soil organisms before utilization by the succeeding crops. Greater accumulation of nutrients in legume green manures also means that larger quantities of P, K, Ca, and Mg may be available following legume cover crops than grasses [21].

*Effects on Soil Biology*

Green manure crops are normally killed or incorporated into soils long before maturity. The addition of fresh, physiologically young residues causes a burst of biological activity in the soil.

The presence of a growing crop on the soil during the part of the year when the soil would normally be bare maintains higher biological activity, presumably mainly because of the stimulation of rhizosphere organisms. Microbial biomass and soil respiration was increased by use of a green manure instead of a fallow year between crops of wheat [22]. In another experiment, surface soil where crimson clover was grown during the off season had significantly larger populations of bacteria and a number of microbial enzymes than did soil from a well-fertilized control treatment [23]. Including red clover and hairy vetch in a corn–soybean cash grain system was found to increase microbial biomass as well as organic N and available N [24].

Some of the positive effects of crop rotation may be due to the colonization of beneficial organisms around crop roots, providing enhanced protection from potentially harmful organisms [25].

Growing a legume cover crop appears to increase the vesicular-arbuscular mycorrhizae (VAM) spore abundance [26]. Legumes may contribute to VAM diversity and abundance because they are highly responsive to VAM [27].

*Effects on Weeds and Insects*

A good stand of green manure is able to suppress weed growth by competition for light and nutrients. Residues of some cover crops such as winter rye are able to suppress germination of weed seeds through an allelopathic effect. When used in a no-till cropping system, the residues remaining on surface can be an effective weed control. With cover crops that decompose rapidly, such as hairy vetch, although weed suppression during the early season can be excellent, weed growth in the later part of the growing season may be a problem [28].

The possible influences of green manures on predators of insect pests have received attention in recent years. Both pest and beneficial arthropods may be stimulated by green manures [29]. By selection and management of green manures it is possible to enhance pest management. For example, legumes such as crimson clover and hairy vetch attract aphids that then attract predators and parasitoids. The aphids that are attracted by the legume green manures do not feed on pecan trees. But the predators of these aphids are also predators of other aphid species that do damage pecans [30, 31]. Thus using legume green manures to help develop the population of aphid predators and parasitoids means that these organisms will switch over to feed on the aphids in the trees after the green manure senesces or is killed.

*Rotation Effect*

Compared with monoculture cropping, there is usually a positive yield effect on the main crop of interest that results from growing another crop as part of a rotation. There are many potential positive influences of green manures aside from the N contribution of legumes. These include breaking pest and weed cycles, better soil physical structure (influencing water infiltration and storage as well as root growth), and enhanced colonization by beneficial micro-organisms. The influence of alfalfa on succeeding wheat crops can be substantial, with one report of finding a positive effect on growth of 12 subsequent crops of wheat [32].

Because of the numerous effects of green manures on soil biological, physical, and chemical properties, it is frequently not easy to find out the reasons for beneficial influences on crop yield under field conditions.

### 3.4.3 Types of Green Manure and Forage Legume Crops

There are many different species that have been evaluated or used as green manure crops. Legumes and grasses, including cereals, have been the most extensively used green manures, although other species are sometimes useful. The selection of a particular species depends on the main economic crop(s) being grown as well as the soil, climate, and the desired objectives. There are two major distinctions among the common green manures: annuals versus biennials or perennials and legumes versus nonlegumes.

As discussed above, one of the main reasons for growing legume cover crops is because of their contribution of fixed-N to the succeeding crops. Summer annuals, crops usually grown only during the summer, include soybeans, peas, and beans. Winter annuals, legumes which are normally planted in the fall and counted on to overwinter, include berseem clover, crimson clover, hairy vetch, and subterranean clover. Some, like crimson clover, can overwinter only in regions with mild frost. However, hairy vetch is able to withstand fairly severe winter weather. Biennial and perennial legumes include red clover, white clover, sweet clover, and alfalfa. Crops that are usually used as winter annuals are sometimes grown as summer annuals instead of leaving the soil fallow in cold, short-season regions. Also, summer annuals that are easily damaged by frost, such as cowpeas, can be grown as a winter annual in the deep South of the US and in the subtropics.

Nonleguminous crops used as green manures include the cereal grasses, such as rye, wheat, oats, and barley, as well as other grass family species, such as rye grass. Other crops sometimes grown as green manures include buckwheat, rape, and turnips.

### 3.4.4 Using Green Manure Crops in Different Cropping Systems

There are many ways of integrating green manures into cropping systems. A few examples of how they work are discussed below.

*Humid Temperate Region Annual Economic Crop*

Green manure crops can frequently fit into summer annual cropping systems in humid temperate regions. They can work especially well if the climate following harvest is sufficiently mild to allow for postharvest green manure establishment. However, some green manures can be successfully established by seeding into row crops, such as corn, before the canopy has completely closed. The green manure germinates but is then suppressed because of a lack of light until the economic crop is harvested. Although yields may sometimes be reduced when interseeding into a row crop, reduction of N fertilizer rates as well as reduced run-off and erosion may make this system attractive for many growers.

In an annual cropping system, the farmer usually suppresses the green manure by either plowing it under or by mowing or using herbicides for no-till systems. There has been limited success in growing crops with a living mulch between crop rows (e.g., [33]). In general, the green manure needs to be suppressed by either physical means such as mowing or rototilling or by use of herbicides.

The date of killing (incorporation) of the green manure is an important issue. It is desirable to let the green manure grow as long as possible to get maximum biomass and production or $N_2$-fixation. However, when green manures grow too long in the spring one or more of the following problems may occur:

1. it may become too mature, resulting in N immobilization (for grass species);
2. delayed planting of economic crop past the optimal date;
3. too great a depletion of soil water;
4. it may go to seed and become a weed problem in the future.

As noted above, some species of both legumes and nonlegumes are not very winter hardy. In cold climates, their use offers the possibility of relying on winter killing to eliminate the need for killing the crop in the spring. Although this may be a reliable system in certain regions, winter killing means that there will be no chance for spring-time biomass production or $N_2$-fixation. Thus this system is probably of greatest use in connection with the growth of a very early economic crop.

*Semiarid Region Annual Economic Crop*

In many cases the use of a green manure means either eliminating what would have been a fallow year in a rotation or changing crops to include forages in the rotation. In very low rainfall areas, such as those where a winter-wheat-fallow system is used to allow storage of moisture during the fallow year, there are still possibilities for incorporation of green manures or forage legumes into the rotation. Alfalfa may be grown as a rotational crop under these conditions. Also a number of green manures have been evaluated or are being used under these systems. However, with water storage the purpose of the fallow year, allowing the growth of a very water-demanding crop such as sweet clover to utilize too much water may result in grain yield reductions [34].

Case studies of nine dry-land North American farms indicate creative ways to incorporate green manures and forage legumes [35]. These farms are either utilizing grain legumes for cash crops or green manures or are also growing forage legumes as part of their rotation. Spring crops (wheat, barley, peas) are sometimes underseeded with clover and the clover worked into the soil or mowed the following year. Sometimes annual legumes such as lentil or flatpea are seeded into the previous year's stubble and allowed to winter kill and remain over winter to help trap snow in the field.

*Perennial Economic Crop*

Orchards and vineyards are frequently planted with the aim of maintaining the soil surface free of weeds or other cover. Because of the many potential benefits of green manures, especially positive effects on control of insect pests by beneficial organisms, their integration into orchard and vineyard systems are being evaluated. Where rainfall during the growing season is plentiful, perennials such as white clover may provide good ground cover as well as additional N for the main crop. However, even in this situation, care is usually taken to provide as little competition to trees as possible until they are well established. Where growing-season water is limited (as in California's Mediterranean-type climate), winter annuals will be more appropriate. In this situation, crimson clover, subterranean clover, and annual medics have been found useful, and, under the proper conditions, they may reseed themselves [36].

### 3.4.5 Energy Implications of Use of Green Manures and Forage Legumes

Because of the numerous potential positive effects of green manures and forage legumes it is difficult to assess their energy implications fully. In addition to the N contribution to the following crops from legumes, other effects include better soil structure with enhanced water infiltration, enhanced weed control and decreased water loss in no-till systems, better soil biological and chemical conditions, some control of insect pests, etc. For each cropping system there are also differences in the number of passes over the field for soil preparation, planting, mowing, harvest, etc. However, when energy savings from the N contribution from green manure or forage legumes in a cash grain system are considered, the effects may be considerable. Nitrogen has been estimated to account for approximately 1/3 of the total energy used in corn production [37]. Thus it is not surprising that the use of green manures to supply most of the N for corn results in considerable energy savings [38]. Under no-till conditions, when the legume green manure may also be able to substitute for black plastic mulch or herbicide [39], the energy savings can be even greater.

## References

1. Pieters, A. J. 1927. *Green Manuring Principles and Practices.* New York: Wiley.
2. Bairoch, P. 1976. Agriculture and the Industrial Revolution 1700–1914. *The Fontana Economic History of Europe*, vol. 3, ed. Cipolla, C. M., pp. 452–506. Sussex: Harvester.
3. Power, J. F. 1987. Legumes: their potential role in agricultural production. *Am. J. Alt. Agric.* 2:69–73.
4. Langdale, G. W., L. T. West, R. R. Bruce, W. P. Miller, and A. W. Thomas. 1992. Restoration of eroded soil with conservation tillage. *Catena* 5:81–90.
5. Wagger, M. G. 1987. Timing effects of cover crop dessication on decomposition rates and subsequent nitrogen uptake by corn. *The Role of Legumes in Conservation Tillage Systems*, ed. Power, J. F., pp. 35–37. Ankeny: Soil Conservation Society of America.
6. Allison, F. E. 1973. *Soil Organic Matter and Its Role in Crop Production.* Amsterdam: Elsevier.
7. Angers, D. A. 1992. Changes in soil aggregation and organic carbon under corn and alfalfa. *Soil Sci. Soc. Am. J.* 56:1244–1249.
8. Ebelhar, S. A., W. W. Frye, and R.L. Blevins. 1984. Nitrogen from legume cover crops for no-till corn. *Agron. J.* 76:51–55.
9. Hargrove, W. L. 1986. Winter legumes as a nitrogen source for no-till grain sorghum. *Agron. J.* 78:70–74.
10. Fox, R. H. and W. P. Piekielek. 1988. Fertilizer N equivalence of alfalfa, birdsfoot trefoil, and red clover for succeeding corn crops. *J. Prod. Agric.* 1:313–317.
11. Baldock, J. O., R. L. Higgs, W. H. Paulson, J. A. Jackobs, and W. D. Shrader. 1981. Legume and mineral N effects on crop yields in several crop sequences in the upper Mississippi Valley. *Agron. J.* 73:885–890.
12. Voss, R. D. and W. D. Shrader. 1984. Rotation effects and legume sources of nitrogen for corn. *Organic Farming: Current Technology and Its Role in a Sustainable Agriculture*, eds. Bezdichek, D.F. *et al.*, pp. 61–68. Special Pub. No. 46. Madison, WI: American Society of Agronomy.
13. Hussain, T., G.Jilani, J. F. Parr, and R. Ahmad. 1995. Transition from conventional to alternative agriculture in Pakistan: The role of green manures in substituting for inorganic N fertilizers in a rice-wheat farming system. *Am. J. Alt. Agric.* 10:133–137.
14. Magdoff, F. R. 1991. Understanding the Magdoff pre-sidedress nitrate soil test for corn. *J. Prod. Agric.* 4:297–305.
15. Meisinger, J. J., W. H. Hargrove, R. L. Mikkelsen, J. R. Williams, and V. W. Benson. 1991. Effects of cover crops on groundwater quality. *Cover Crops for Clean Water*, ed. Hargrove, W. L., pp. 57–68. Ankeny: Soil and Water Conservation Society.
16. Wagger, M. G. and D. B. Mengel. 1988. The role of nonleguminous cover crops in the efficient use of water and nutrients. *Cropping Strategies for Efficient Use of Water and Nutrients*, ed. Hargrove, W. L., pp. 129–154. Special Pub. No. 51. Madison, WI: American Society of Agronomy.

17. MacRae, R. J. and G. R. Mehuys. 1985. The effect of green manuring on the physical properties of temperate-area soils. *Adv. Soil Sci.* 3:71–94.
18. Sumner, M. E. and B. A. Stewart, eds. 1992. Soil crusting: physical and chemical processes. *Advances in Soil Science*. Boca Raton: Lewis.
19. Folorunso, O. A., D. E. Rolston, T. Pritchard, and D. T. Louie. 1992. Cover crops lower soil surface strength, may improve soil permeability. *Calif. Agric.* 46(6): 26–27.
20. Magdoff, F. R. and R. J. Bartlett. 1985. Soil pH buffering revisited. *Soil Sci. Soc. Am. J.* 49:145–148.
21. Groffman, P. M., P. Hendrix, Chun-Ru, and D. A. Crossley Jr. 1987. Nutrient cycling processes in a southeastern agroecosystem with winter legumes. *The Role of Legumes in Conservation Tillage Systems*, ed. Power, J. F., pp. 7–8. Ankeny: Soil Conservation Society of America.
22. Campbell, C. A., V. O. Biederbeck, R. P. Zentner, and G. P. Lafond. 1991. Effect of crop rotations and cultural practices on soil organic matter, microbial biomass and respiration in a thin black chernozem. *Can. J. Soil Sci.* 71:363–367.
23. Kirchner, M. J., A. G. Wollum II, and L. D. King. 1993. Soil microbial populations and activities in reduced chemical input agroecosystems. *Soil Sci. Soc. Am. J.* 57:1289–1295.
24. Doran, J. W., D. G. Fraser, M. N. Culik, and W. C. Liebhardt. 1987. Influence of alternative and conventional agricultural management on soil microbial processes and nitrogen availability. *Am. J. Alt. Agric.* 2:99–106.27.
25. Jawson, M. J., A. J. Franzluebbers, D. K. Galusha, and R. M. Aiken. 1994. Soil fumigation with monoculture and rotations: response of corn and mycorrhizae. *Agron. J.* 85:1174–1180.
26. Galvez, L., D. D. Douds, Jr., P. Wagoner, R. L. Longnecker, L. E. Drinkwater, and R. R. Janke. 1995. An overwintering cover crop increases inoculum of VAM fungi in agricultural soil. *Am. J. Alt. Agic.* 10:152–156.
27. Rabatin, S. C. and B. Stinner. 1989. The significance of vesicular-arbuscular mycorrhizal fungal-soil macroinvertebrate interactions in agroecosystems. Agric. *Ecosyst. Environ.* 27:195–204.
28. Worsham, A. D. 1991. Role of cover crops in weed management and water quality. *Cover Crops for Clean Water*, ed. Hargrove, W. L., pp. 141–145. Ankeny: Soil and Water Conservation Society.
29. Bugg, R. L. 1991. Cover crops and control of arthropod pests of agriculture. *Cover Crops for Clean Water.*, ed. Hargrove, W. L., pp. 157–163. Ankeny: Soil and Water Conservation Society.
30. Bugg, R. L., S. C. Phatak, and J. D. Dutcher. 1990. Insects associated with cool-season cover crops in southern Georgia: implications for pest control in truck-farm and pecan agroecosystems. *Biol. Agric. Hortic.* 7:17–45.
31. Bugg, R. L., M. Sarrantonio, J.D. Dutcher, and S. C. Phatak. 1991. Understory cover crops in pecan orchards: possible management systems. *Am. J. Alt. Agric.* 6:50–62.
32. Hoyt, R. B. 1990. Residual effects of alfalfa and bromegrass cropping on yields of wheat grown for 15 subsequent years. *Can. J. Soil Sci.* 70:109–113.

33. Grubinger, V. P. and P. L. Minotti. 1990. Managing white clover living mulch for sweet corn production with partial rototilling. *Am. J. Alt. Agric.* 5:4–12.
34. Biederbeck, V. O., H. A. Bjorge, S. A. Brant, J. L. Henry, G. E. Hultgreen, J. A. Kielly, and A. E. Slinkard. 1995. Farm facts: soil improvement with legumes, Including legumes in crop rotations. *Canada–Saskatchewan Agreement on Soil Conservation.*
35. Matheson, N., B. Rusmore, J. R. Sims, M. Spengler, and E. L. Michalson. 1991. Cereal-legume cropping systems: nine farm case studies in the dryland Northern Plains, Canadian Prairies, and Intermountain Northwest. Helena: Alternative Energy Resources Organization.
36. Miller, P. R., W. L. Graves, W. A. Williams, and B. A. Madson. 1989. *Covercrops for California Agriculture.* Division of Agriculture and Natural Resources, University of California. Leaflet 21471.
37. Pimentel, D., L. E. Hurd, A. C. Bellotti, M. J. Forster, I. N. Oka, O. D. Sholes, and R. J. Whitman. 1973. Food production and the energy crisis. Science 182:443–449.
38. Ess, D. R., D. H. Vaughan, J. M. Luna, and P. G. Sullivan. 1994. Energy and economic savings from the use of legume cover crops in Virginia corn production. *Am. J. Alt. Agric.* 9:178–185.
39. Abdul-Baki, A. A., J. R. Teasdale. R. Korcak, D. J. Chitwood, and R. N. Huettel. 1996. Fresh-market tomato production in a low-input alternative system using cover-crop mulch. *Hortic. Sci.* 31:65–69.

## 3.5 Biomass Feedstocks

### 3.5.1 Biocrude Oil

*D. S. Scott and R. L. Legge*

***Definitions***

Biocrude oil (or bio-oil for short) can be defined as the liquid fraction of the product obtained from the pyrolytic decomposition of biomass. Historically, this liquid was known as pyroligneous liquor and was an early source of several organic chemicals, principally methanol and acetic acid. The pyrolytic methods that gave this type of liquid are rarely used now and have been supplanted in the past two decades by fast pyrolysis technologies. In the following section the term bio-oil is used only for the liquid product from such fast pyrolysis processes. The other products of pyrolysis, namely char, or charcoal, and noncondensible gases, are also discussed where appropriate.

***Raw Materials***

Any lignocellulosic material can be used as a feedstock for fast pyrolysis processes, although some are easier to handle than others in a continuous system. Most early work was done on wood wastes, either wood or bark, but subsequently various agricultural wastes were used. Mention might be made of sugar cane bagasse, sorghum bagasse, wheat straw, corn stover, sunflower seed hulls, almond shells, rice hulls, coconut shell, palm tree waste, etc., as typical materials for which experimental test results have been published. In general, an easily ground material is preferred, as fast pyrolysis methods require relatively small particle sizes.

It follows that biomass which is highly fibrous, oily or sticky, or has other features that make grinding and handling difficult, is less desirable as raw material for the production of bio-oils. Most biomass is low in sulfur, so sulfur content is generally not a problem for bio-oil use as a fuel. However, in some potential biomass feedstocks, nitrogen content can be high, and the effect of this factor on the bio-oil fuel quality must be considered.

***Methods of Production***

Fast pyrolysis methods for the conversion of biomass to organic liquids in high yields became a technical reality only near 1980. About that time, research on pyrolysis in sand-fluidized beds showed that higher yields of organic liquids could be obtained than were previously possible if the residence time of volatiles was kept below 1 s and if a reactor capable of high heating rates for small biomass particles was used [1]. The necessary reaction conditions can be achieved in fluidized beds, cyclonic reactors, or transport reactors. Typically, a fluidized bed of sand at 500°C and atmospheric pressure with a volatiles' residence time of ~0.5 s will give a liquid yield from clean sawdust of 70%–75% of the weight of the dry feed. Of this liquid yield, ~10% is water of reaction. If the feed is not dry, then the moisture content of the feed will also be present. The bio-oil so produced is a homogeneous liquid of high density (sp. gr., 1.2) and low pH (2.5–3.5), and the total water content is below ~30% with a heating value similar to that of wood with the same moisture content [2, 3].

These fast pyrolysis methods are in marked contrast to early methods that are classed as slow pyrolysis. In fact, such slow pyrolysis methods have been practiced throughout human history for the manufacture of charcoal. In these processes, charcoal is the main product, and liquid yields and thermal efficiencies are low. Fast pyrolysis produces a uniquely high liquid yield and also a liquid of different composition and properties than is obtained with historical pyrolysis methods.

A typical flow chart for a fluidized-bed fast pyrolysis system is shown in Fig. 3.37. Reaction conditions and yields for such a system are given in Table 3.60 for a variety of lignocellulosic biomass materials that are usually considered as waste or of low value and that are suitable feedstocks. It is apparent that the pyrolytic liquid yield (organics plus water of reaction) is roughly proportional to the cellulose and hemicellulose content (ash-free basis).

***Analyses of Typical Bio-Oils***

Table 3.61 gives the composition of the pyrolytic bio-oil with respect to major components for the same materials as those in Table 3.60 [4–6].

One of the unusual features of fast pyrolysis is the high concentration of some unique chemicals in the bio-oil, notably hydroxyacetaldehyde (glycolaldehyde) and levoglucosan (anhydroglucose). It has been shown that both of these products arise primarily from the thermal decomposition of the cellulosic fraction [7, 8]. The pyrolytic lignin, defined as the precipitate formed when water is added to the bio-oil, is aromatic in character and rich in phenolics and is assumed to arise mostly from decomposition of the lignin portion of the biomass. It is clear that the bio-oil, in addition to being a potentially useful liquid fuel, could also be a feedstock for the recovery of several unique, high-value chemicals.

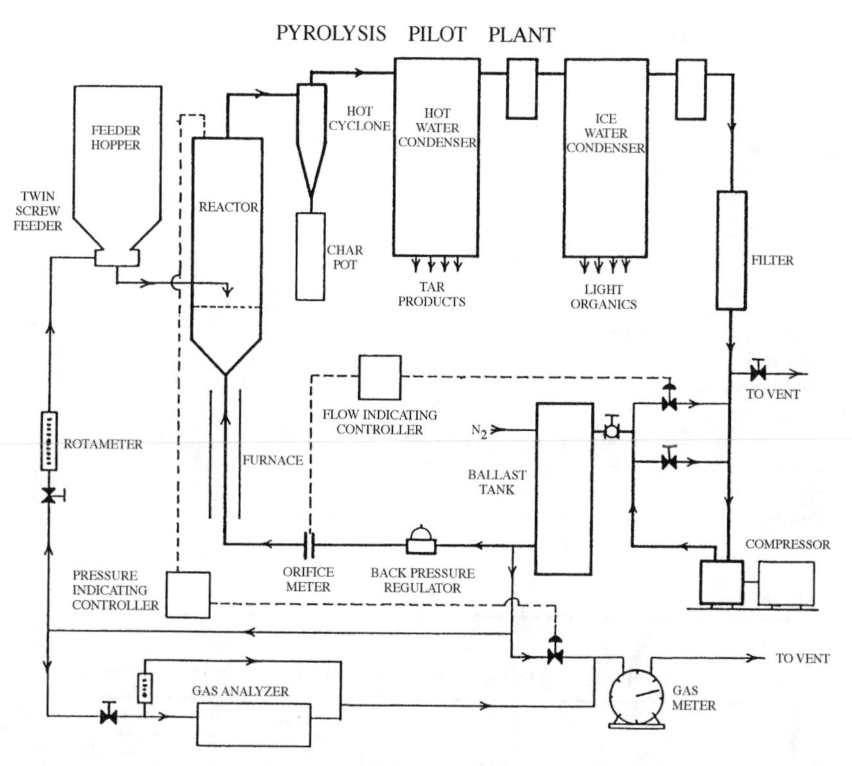


**Figure 3.37. Flow chart for a typical fluidized-bed fast pyrolysis process (Waterloo FPP).**

A further aspect of fast pyrolysis was also developed much more fully in the past decade. If the indigenous alkaline cations naturally present in biomass, mostly potassium and calcium, are removed before pyrolysis, either by acid prehydrolysis, weak acid wash, or by an ion-exchange process, then the resulting bio-oil will have a completely different composition [5]. The mechanistic changes resulting from the absence of the catalytic effect of alkaline cations has been discussed by Radlein *et al.* [9]. A simplified representation of the possible processes that can occur during fast pyrolysis are shown in Fig. 3.38 for cellulose.

The fragmentation mode of reaction occurs when alkaline cations are present, and it yields a bio-oil high in low molecular weight carbonyl and carboxylate functionalities. In the absence of alkaline catalysis, depolymerization of the natural polymers is predominant and monomeric fragments are found in the highest yields. In particular, the natural polymers of sugars give the monomeric form predominantly, such as anhydroglucose (levoglucosan) from cellulose, and anhydropentoses, e.g., mannosan, from hemicelluloses. Some anhydrodisaccharides and higher are also formed, e.g., cellobiosan, as well as glucose. A comparison of the liquid yields and compositions from a pretreated poplar wood and a normal wood is given in Table 3.62. It is apparent that there is a marked change in the mechanism of the decomposition caused by the acid prehydrolysis that removed the inorganic cations and most of the hemicellulose. Over 80% of the cellulose

**Table 3.60. Yields from fast pyrolysis of different biomass feeds**

| Parameter | Poplar Sawdust | Spruce Sawdust | Sorghum Bagasse | Wheat Chaff | Sunflower Hulls |
|---|---|---|---|---|---|
| Moisture (wt. %) | 4.6 | 7.0 | 10.1 | 6.9 | 11.1 |
| Ash (wt. % moisture-free basis) | 0.46 | 0.50 | 9.2 | 22.5 | 4.0 |
| Maximum particle size (mm) | 1.0 | 1.0 | 0.5 | 1.0 | 1.0 |
| Temperature (°C) | 504 | 500 | 510 | 515 | 500 |
| Yields (wt. % moisture- and ash-free) | | | | | |
| Gas | 11.0 | 7.8 | 11.7 | 15.9 | 19.5 |
| Char | 11.8 | 12.2 | 13.4 | 17.6 | 23.2 |
| Water (reaction product) | 10.7 | 11.6 | 10.6 | 15.7 | 9.8 |
| Organic Liquid | 66.2 | 66.5 | 58.8 | 51.0 | 46.3 |
| Total Recovery, % | 99.7 | 98.1 | 94.5 | 100.2 | 98.8 |
| Liquid Properties | | | | | |
| Total liquid yield (wt. % as-fed) | 77.6 | 79.2 | | | |
| Water content (wt. %) | 18.6 | 21.9 | | | |
| pH | 2.4 | 2.1 | | | |
| Density ($g/cm^3$) | 1.23 | 1.22 | | | |
| Elemental analysis (wt. % moisture-free) | | | | | |
| Carbon | 53.6 | 54.0 | | | |
| Hydrogen | 7.0 | 6.8 | | | |

**Table 3.61. Composition of fast pyrolysis liquids from different biomass feedstocks[a]**

| | Feedstock | | | | |
|---|---|---|---|---|---|
| Yields (wt. % Dry Biomass) | Poplar Sawdust | Spruce Sawdust | Sorghum Bagasse | Wheat Chaff | Sunflower Hulls |
| Hydroxyacetaldehyde | 10.0 | 7.7 | 7.2 | 6.5 | 0.8 |
| Levoglucosan | 3.0 | 4.0 | 1.7 | 2.0 | 0.3 |
| Formic acid/formaldehyde | 3.1 | 7.2 | 2.7 | 1.3 | ND |
| Acetic acid | 5.4 | 3.9 | 2.3 | 6.1 | 2.1 |
| Hydroxy propanone (acetol) | 1.4 | 1.2 | 1.5 | 3.2 | 1.2 |
| Cellobiosan | 1.3 | 2.5 | 0.4 | 0.4 | 0.1 |
| Pyrolytic lignin[b] | 16.2 | 20.6 | 14.9 | 15.1 | 38.4 |

[a] From [6].
[b] Pyrolytic lignin is defined as the water-insoluble fraction that forms when water content exceeds 50%. It consists primarily of phenolic material formed from the decomposition of lignin.

is converted to anhydrosugars, which can be hydrolyzed easily to fermentable hexoses, compared with the untreated biomass, which yielded only an ~20% conversion.

In addition to the changes in product spectra that can be accomplished by various pretreatments, the product yields and yield ratios for a given feedstock can also be changed by control of the time–temperature history of the biomass particle [9]. As discussed below, the fast pyrolysis process can be tailored to produce a range of chemical products. The recovery and the purification of some of the more valuable components from the complex bio-oil liquids represent a current area of research.

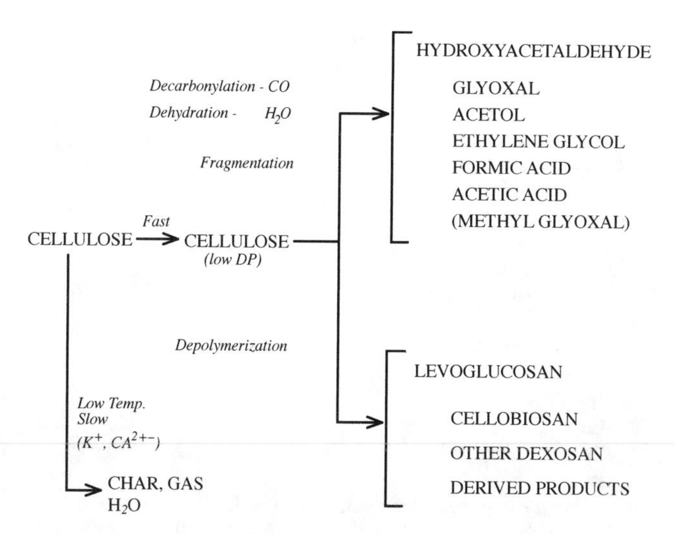


**Figure 3.38. Proposed simplified reaction paths for the thermal decomposition of cellulose.**

### *Applications of Bio-Oil for Energy*

The use of bio-oil from high-yield pyrolysis processes as a primary fuel for the production of energy is not yet fully commercialized. The two main areas of potential application, as a boiler fuel or as a fuel for internal-combustion engines, especially stationary diesel units, is now under intensive development, particularly in Europe through EU agencies. Two demonstration-scale plants are known, one in Spain, which has been operated by a utility company for over 3 years from 1993–1996 (an atmospheric-pressure fluidized-bed unit) [11], and one in Italy to be operated by Italian State Energy Co., which was reported to have started up sometime in 1997. Few details are available for this unit. Extensive testing of the bio-oil as a heavy diesel fuel has been carried out in the UK and in Finland, and initial boiler combustion tests have been done in Canada, Italy, and Finland. Accounts of some of the early developmental work can be found in a compilation of reports from a 1994 meeting (*Proceedings of the Conference on Biomass Pyrolysis Oil Properties and Combustion*, available from U.S. Dept. of Commerce, National Technical Information Service, Springfield, VA).

In summary, tests have succeeded to the point at which the fast pyrolysis method is now ready for full commercial implementation, given the particular set of economic and local conditions that will make it profitable on a large scale.

### *Problems of Small-Scale Production and Use of Bio-Oil*

The capital cost for a fast pyrolysis plant is approximately equally divided among the drying and grinding step, the pyrolysis step, and the product collection and storage section. If a biomass waste material is available in the form of small particles, e.g., sawdust, chaff, or straw, needs little drying, and has a low or no value, then small-scale operation may be feasible. In many cases, the supply of suitable biomass feed is limited

**Table 3.62. Comparison of the pyrolysis products from raw and pretreated poplar wood [5]**

| | Poplar Wood | |
|---|---|---|
| Parameter | Untreated | Pretreated |
| Temperature (°C) | 497 | 501 |
| Vapor residence time (s) | 0.46 | 0.45 |
| Particle size (μm) | −590 | −590 |
| Moisture (%) | 3.3 | 16.5 |
| Cellulose (% moisture free) | 49.1 | 62.8 |
| DP | — | — |
| Ash (% moisture-free) | 0.46 | 0.04 |
| Yields (% moisture-free wood) | | |
| Organic liquid | 65.8 | 79.6 |
| Water | 12.2 | 0.9 |
| Char | 7.7 | 6.7 |
| Gas | 10.8 | 6.5 |
| Yields of tar components (% moisture-free feed) | | |
| Oligosaccharides | 0.7 | 1.19 |
| Cellobiosan | 1.3 | 5.68 |
| Glucose | 0.4 | 1.89 |
| Fructose (?) | 1.31 | 3.89 |
| Glyoxal | 2.18 | 0.11 |
| 1,6-anhydroglucofuranose | 2.43 | 4.50 |
| Levoglucosan | 3.04 | 30.42 |
| Hydroxyacetaldehyde | 10.03 | 0.37 |
| Formic acid | 3.09 | 1.42 |
| Acetic acid | 5.43 | 0.17 |
| Ethylene glycol | 1.05 | |
| Acetol | 1.40 | 0.06 |
| Methylglyoxal | 0.65 | 0.38 |
| Formaldehyde | 1.16 | 0.8 |
| Pyrolytic lignin | 16.2 | 19.0 |
| Totals | 51.5 | 69.9 |
| Percentage of moisture-free pyrolysis oil | 78.3 | 87.8 |
| Sugars, glucose-equivalent yield (% cellulose) | 20.4 | 83.4 |

in a locality, and it is estimated that it is not worthwhile to transport a potential feedstock more than ~50 km. In this event, a relatively small-scale operation may be all that is possible, e.g., 50 dry tonnes/day or less. In smaller scale plants, it is anticipated that the cost of the feed and feed preparation, and the cost of labor, would be the major operating costs. Various estimates give figures for the probable cost of production of bio-oil in the range of $50/tonne to $150/tonne, depending on local conditions, and the scale of operations. Small plant production costs would likely fall in the range of $100–$150/tonne of bio-oil. It is obvious that one of the major advantages of producing a bio-oil from waste biomass for energy purposes is that the biomass feedstock supply can be decoupled from its energy-producing application. Thus a number of small-scale pyrolysis plants, designed for a local supply of biomass waste, could ship their bio-oil production to a central power station and thus overcome the problems of seasonal feedstock supply and storage.

***Upgrading of Bio-Oil Fuel Characteristics***

Bio-oil as produced from a fast pyrolysis process has certain disadvantages for use as a fuel, particularly for diesel applications. The water content must be controlled, preferably between ~15% and 25%. The water is not just a diluent—it acts as a cosolvent and gives a homogeneous liquid product of relatively low viscosity. If the water content is too low, the viscosity will be very high, and too high a water content may cause phase separation and reduce the calorific value of the bio-oil. Also, the bio-oil is corrosive because of the low pH and is unstable in storage or at higher temperatures. The addition of 5%–10% ethanol has been found both to aid in control of the viscosity and to give greater stability over time.

Another solution to improve the fuel qualities of the bio-oil is to alter it chemically by derivatization or by hydrogenation, to reduce the considerable reactivity of the carbonyl and phenolic components. Both approaches have been demonstrated to be technically achievable by several researchers, and to yield satisfactory liquids [12–14]. However, adequate supplies of the upgraded fuels have not yet been prepared to allow commercial scale testing or to allow full evaluation of the process economics, although such programs are underway (1997).

***Bio-Oils as Chemical Feedstocks***

As already pointed out, it is possible to obtain widely different bio-oil compositions by pretreating the biomass feedstock in various ways before pyrolysis. In addition to these processes, sand or other solid heat carriers used in a fast pyrolysis unit can be replaced by a solid catalyst, or by a reactive atmosphere, or both. This technology is particularly easy to apply in a fluidized-bed process. The catalyst can then be selected for low-temperature gasification to make synthesis gas [15–17]; if a hydrogen atmosphere is used with nickel catalyst, a high carbon conversion to methane can be accomplished [18], or if hydrotreating catalysts are used in a hydrogen atmosphere, good yields of light hydrocarbons can result [10]. All the above processes take place at atmospheric pressure and at temperatures from 400 to 700°C. Other controlled hydrogenations and oxidations of the bio-oil are also possible to yield a variety of potentially valuable organic chemicals.

One of the important preliminary steps for many routes to recovering chemicals from bio-oil is a simple water separation that will precipitate highly phenolic material and extract water-soluble low molecular weight organics such as hydroxyacetaldehyde. A typical phase diagram for such an extraction is shown in Fig. 3.39 for a poplar wood bio-oil. Clearly any water content over 50% in this case will result in phase separation. The two phases can then be processed separately for chemical recovery.

An attempt is made to summarize many of the potential routes for production of chemicals from bio-oils in Fig. 3.40. At the present time, the commercial production of anhydrosugars has been undertaken, particularly levoglucosan by Resource Transforms International Ltd. (Canada) and of phenolics from the lignin fraction by the National Renewable Energy Laboratory (USA). However, it is clear from Fig. 3.40 that there are many potential applications for bio-oil as a chemical feedstock that await only the development of economically attractive recovery processes.

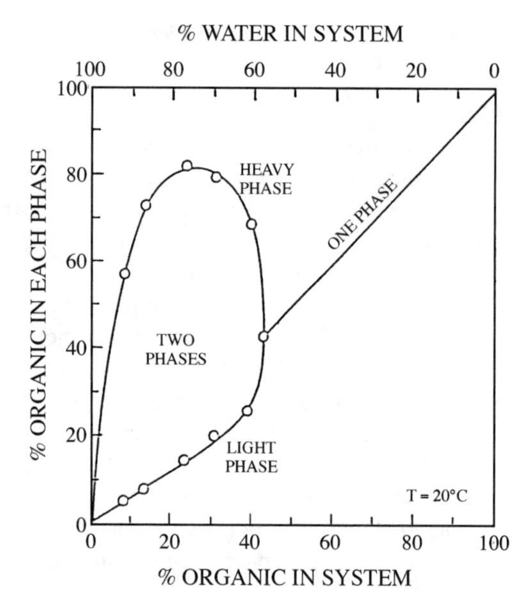


**Figure 3.39. Phase diagram for wood bio-oil from poplar with water at 25°C.**

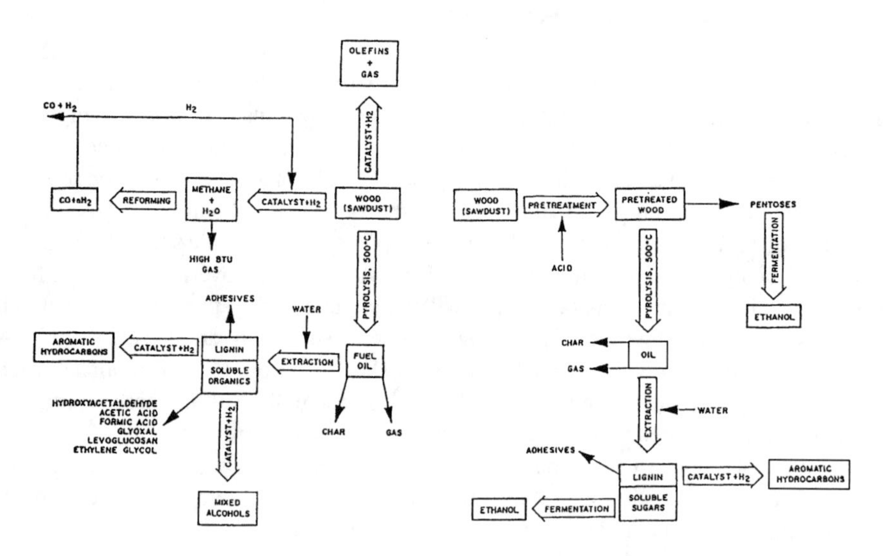


**Figure 3.40. Chart showing potential for chemical recovery from fast pyrolysis bio-oils.**

### 3.5.2 Bioplastics

*D. S. Scott and R. L. Legge*

Over the past couple of decades there has been increased interest in the production of biodegradable polymeric materials or plastics that are appropriate for pharmaceutical, medical, electrical, or agrochemical applications. This has fueled research in the development of polymeric materials that can be derived from microbial sources. Microbial polymers of possible interest in this area include microbial cellulose, polyhydroxyalkanoates, and thermoplastic starches.

#### *Microbial Cellulose*

Much of the interest in microbial cellulose is related to the ability to modify characteristics of the cellulose during its production and that pretreatment processes are not required for removing lignin or other biogenic components traditionally found in plant sources. Aspects of microbial cellulose as they relate to the genus *Acetobacter* can be found in the comprehensive treatise dealing with the genus [19]. A topical review dealing with the development of microbial cellulose as a specialty chemical is also available [20]. Microbial cellulose production is not restricted to the genus *Acetobacter*, as there are reports of an anaerobic coccoid bacterium, *Sarcina ventriculi*, producing intercellular fibrous material that is likely cellulose [21]. Microbial cellulose has served as an excellent model for understanding cellulose biosynthesis and function [22].

There are few reports in the public domain regarding conditions for optimal cellulose production by *Acetobacter*. Under static conditions, this organism produces an interwoven network of cellulose fibers that form a cohesive pellicle. It has been suggested that noncellulose-producing mutants show a high frequency of reversion to cellulose producers when subcultured through static culture [23]. Appropriate considerations for reactor choice may be key in the scale-up of cellulose production by *Acetobacter* [24]. Masaoka *et al.* [25] have compared cellulose production of both *Acetobacter* and *Agrobacterium* species with various carbon sources. The maximum cellulose production rate reported in this study was 36 g/day $m^2$. The most effective carbon sources were glucose, fructose, and glycerol. Comparison of cellulose productivity from *D*-arabitol and *D*-glucose under the same culture conditions showed a productivity of $\sim$6 times that for arabitol versus glucose. *D*-glucose serves as a precursor to gluconic acid production, which may inhibit bacterial growth and cellulose production.

A number of examples of the ability to modify microbial cellulose during synthesis can be found in the literature. For example, fluorescent brightening agents, such as Calcofluor, have been found to perturb cellulose production during biosynthesis, producing sheetlike or band material [27, 28]. Other agents that have been shown to alter microfibril assembly include Congo red and carboxymethyl cellulose [29, 30].

Modification of microbial cellulose postsynthesis has been demonstrated by a number of groups. Geyer *et al.* [31] have shown that it is possible to dehydrate microbial cellulose by using a stepwise exchange with nonpolar solvents and that this cellulose is conducive to a number of derivitization procedures including carboxymethylation, silylation, and acetylation. Characteristics of the resulting derived cellulose depend on the swelling character and type of pretreatment. Others have show that simple washing and caustic treatment of the cellulose pellicle can produce a bacterial cellulose membrane suitable for use as a dialysis membrane for aqueous applications [32]. This membrane has mechanical strength superior to that of a dialysis grade regenerated cellulose membrane, allowing higher permeation rates due to thinner membranes with comparable strength.

Yoshino *et al.* [33, 34] have demonstrated the suitability of bacterial cellulose for the production of graphite films. Bacterial cellulose films were found to be converted into highly conductive films with conductivities exceeding 6000 S/cm by heat treatment at temperatures above 2400°C. Changes in the electrical conductivity were observed by intercalation of $FeCl_3$ and K. The properties of the resulting films were highly dependent on the heat treatment temperature.

***Polyhydroxyalkanoates***

There has been considerable interest in the production of a group of linear polyesters termed polydroxyalkanoates (PHA's). These materials are of interest because they serve as candidates for a biodegradable plastic material [35]. For a recent review of the occurrence, metabolism, metabolic role, and industrial uses see [36]. One important and widely studied example is poly(3-hydroxybutyrate) (PHB) [37]. PHB has physical and chemical properties that are similar to polypropylene and polyethylene, allowing it to be formed into films, fibers, sheets, or molded into various shapes [38]. A variety of organisms have been investigated that can produce PHB, including *Alcaligenes eutrophus*, recombinant *Escherichia coli*, and *Methylobacterium extorquens*. ICI patented a process for PHB production by *Methylobacterium organophilum*, which can accumulate 30% or more of its dry weight as PHB [39]. ICI produces PHB and uses *Alcaligenes eutrophus* with a glucose–salt medium in the fed-batch mode [40]. More recent studies have shown that under potassium limitation yields of 52%–56% on a dry weight basis can be achieved [37].

One of the advantages of microbial plastic is that molecular weights of the polymer are influenced by the nature of the organism and growth conditions. In *Alcaligenes eutrophus* the molecular mass of PHB appears to be a consequence of turnover [41]. In batch culture, degradation of PHB occurs after the cessation of PHB accumulation, resulting in an accumulation of low molecular mass PHB. Nitrogen appears to play an important part in the regulation of PHB production and degradation. Shimizu *et al.* [42] have shown that the addition of a nitrogen source once adequate PHB accumulation has occurred results in simultaneous PHB production and degradation. The degradation rate of low molecular weight PHB occurred at a greater rate than high molecular weight PHB, resulting in a shift to PHB of a higher molecular weight.

Current factors that influence the economics of PHA are a result of the high costs associated with production. This includes raw material cost(s), complexity of the technology and separation, and purification costs [43]. Considerable effort has focused on the use of methanol for the production of PHB. Bourque *et al.* [44] have developed a process that uses methanol as the sole carbon and energy source, with a minimal medium to produce a very high molecular mass PHB. Under optimal conditions, including multivariable and adaptive control strategies, Bourque *et al.* were able to achieve PHB production of the order of 40%–53% dry weight. There has been interest in the use of waste products as an alternate carbon and energy source. For example, Kim and Chang [45] have obtained yields of 58% dry weight with *Alcaligenes eutrophus* on tapioca hydrolysate with controlled glucose feeding by using exit gas data. Other approaches to reducing production cost have been to shift to *Alcaligenes latus* in which PHB is growth associated [46]. Another advantage of *A. latus* is that it can use sucrose directly as its carbon source. Yamane *et al.* [46] have achieved yields of 50% on a dry cell weight basis and PHB productivity is in the range of 4 g PHB/L h with a high-cell-density pH-stat fed-batch approach.

One significant disadvantage to the widespread application of PHA are some undesirable mechanical properties such as brittleness. A potential solution to reducing brittleness is the generation of polymer blends. Blends of PHB with cellulose acetate butyrate, generated by compression molding, produced either single-phase amorphous mixtures or partially crystalline mixtures [47, 48]. These mixtures showed considerable recalcitrance to biodegradation over a broad composition range. Gatenholm and Mathiason, [49] have investigated the effect of PHB processing in the presence of cellulose. They propose that processing of PHB with cellulose can be considered as reactive with the formation of functional groups resulting in chain scission and volatile production. A better understanding of this process may lead to processes for tailored polymer blends with optimal mechanical properties and biodegradability.

An alternative strategy to traditional fermentation approaches for the production of biodegradable polyesters has recently been proposed by van der Leij and Witholt [50]. This proposal advances the option of harnessing plants as a host for large-scale production of new and renewable bulk raw materials. Some work along these lines has already been successful [51]. The engineering challenge will be large-scale, economical recovery of PHB from plant biomass.

Bacterial polymers offer the advantage of modification by using physiological alterations to the organism. This feature has been documented by a number of laboratories [52]. Although much of the focus on PHA modification has focused on polymers chosen for their biodegradability, more recent interest has focused on modification to produce optical and smart chiral materials. Kim *et al.* [53] have demonstrated the feasibility of producing chiral polymer structures for nonlinear optical applications by using two stage fermentation approaches.

## References

1. Scott, D. S. and J. Piskorz. 1982. The flash pyrolysis of aspen poplar wood. *Can. J. Chem. Eng.* 60:666–674.
2. Scott, D. S., J. Piskorz, and D. Radlein. 1985. Liquid products from the continuous flash pyrolysis of biomass. *Ind. Eng. Chem. Proc. Des. Devel.* 24:581–588.
3. Scott, D. S. and J. Piskorz. 1985. Production of liquids from biomass by continuous flash pyrolysis. *BioEnergy 84*, eds., Egneus H. and A. Ellegard, vol. 3, pp. 15–33. London: Elsevier.
4. Piskorz, J., D. S. Scott, and D. Radlein. 1989. The composition of oils obtained by the fast pyrolysis of different woods. *ACS Symp. Series* No. 376, eds. Soltes, E. J. and T. A. Milne (eds), pp. 167–178. Washington, DC: American Chemical Society.

5. Piskorz, J., D. Radlein, D. S. Scott, and S. Czernik. 1989. Pretreatment of wood and cellulose for production of sugars by fast pyrolysis. *J. Anal. Appl. Pyrolysis* 16: 127–142.
6. Piskorz, J., P. Majerski, D. Radlein, and D. S. Scott. 1990. Fast pyrolysis of some agricultural and industrial materials. *Proc. 8th Canadian Bioenergy Seminar*, Location: CANMET, Natural Resources Canada.
7. Piskorz, J., D. S. Scott, and D. Radlein. 1986. On the mechanism of the rapid pyrolysis of cellulose. *J. Anal. Appl. Pyrolysis* 9:121–137.
8. Radlein, D., A. Grinshpun, J. Piskorz, and D. S. Scott. 1987. On the presence of anhydro-oligosaccharides in sirups from the fast pyrolysis of cellulose. *J. Anal. Appl. Pyrolysis* 12:39–49.
9. Radlein, D., J. Piskorz, and D. S. Scott. 1991. Fast pyrolysis of natural polysaccharides as a potential industrial process. *J. Anal. Appl. Pyrolysis* 20:41–61.
10. Radlein, D., S. Mason, J. Piskorz, and D. S. Scott. 1991. Hydrocarbons from the catalytic pyrolysis of biomass. *Energy Fuels* 5:760–764.
11. Cuevas, A., C. Reinoso, and D. S. Scott, 1994. The production and handling of WFPP bio-oil and its implications for combustion. Proc. *Biomass Pyrolysis Oil Properties and Combustion Conference*, NREL-CP-430–7215, 151–156.
12. Elliott, D, E. Baker, J. Piskorz, D. S. Scott, and Y. Solantausta. 1988. Production of liquid hydrocarbon fuels from peat. *Energy Fuels* 8:234–235.
13. Piskorz, J., P. Majerski, D. Radlein, and D. S. Scott. 1989. Conversion of lignins to hydrocarbon fuels. *Energy Fuels* 3:723–726.
14. Radlein, D., J. Piskorz, P. Majerski, and D. S. Scott. 1996. Method of upgrading biomass pyrolysis liquids for use as fuels and as a source of chemicals by reaction with alcohols. UK Provisional Patent to Resource Transforms Inter. Ltd., Waterloo, Ontario, Canada.
15. Arauzo, J., D. Radlein, J. Piskorz, and D. S. Scott. 1994. A new catalyst for the gasification of biomass. *Energy Fuels* 8:1192–1196.
16. Arauzo, J., D. Radlein, J. Piskorz, and D. S. Scott. 1997. Catalytic pyrogasification of biomass-evaluation of modified nickel catalysts. *Ind. Eng. Chem. Res.* 36:67–75.
17. Mudge, L., E. Baker, M. Brown, and W. Wilcox. 1987. Bench scale studies on gasification of biomass in the presence of catalysts. Final report, Contract DE-AC0676RLO-1830, Pacific Northwest Laboratory, Battelle Memorial Institute.
18. Garg, M., J. Piskorz, D. S. Scott, and D. Radlein. 1988. The hydrogasification of wood. *Ind. Eng. Chem. Res.* 27:256–264.
19. Asai, T., 1968. *Acetic Acid Bacteria.* Baltimore: University Park Press.
20. Legge, R. L. 1990. Microbial cellulose as a specialty chemical. *Biotechnol Adv.* 8:303–319.
21. Canale-Parola, E. and R. S. Wolfe. 1964. Synthesis of cellulose by *Sarcina ventriculi. Biochim. Biophys. Acta.* 82:403–405.
22. Ross, P., Mayer, R., and Benziman, M. 1991. Cellulose biosynthesis and function in bacteria. Microbiol. Rev., 55:35–58.
23. Cook, K. E. and J. R. Colvin. 1980. Evidence for a beneficial influence of cellulose production on growth of *Acetobacter xylinum* in liquid medium. *Current Microbiol.* 3:203–205.

24. Moo-Young, M. and Y. Chisti. 1988. Considerations for designing bioreactors for shear sensitive culture. *Biotechnology* 6:1291–1296.
25. Masaoka, S., T. Ohe, and N. Sakota. 1993. Production of cellulose from glucose by *Acetobacter xylinum. J. Ferment. Bioeng.* 75:18–22.
26. Oikwa, T., T. Morino, and M. Ameyama. 1995. Production of cellulose from D-arabitol by *Acetobacter xylinum* KU-1. *Biosci. Biotechnol Biochem.* 59:1564–1565.
27. Benziman, M., C. H. Haigler, R. M. Brown, J. A. R. White, and K. M. Cooper. 1980. Cellulose biogenesis: polymerization and crystallization are coupled processes in *Acetobacter xylinum. Proc. Natl. Acad. Sci. USA.* 77:6678–6682.
28. Haigler, C. H. and M. Benziman. 1982. Biogenesis of cellulose I microfibrils occurs by cell-directed self-assembly in *Acetobacter xylinum. Cellulose and Other Natural Polymer Systems*, ed. R. Malcolm. New York: Plenum Press.
29. Brown, Jr., R. M., C. H. Haigler, J. Suttie, A. L. White, E. Roberts, C. Smith, T. Itoh, and K. Cooper. 1983. The biosynthesis and degradation of cellulose. *J. Appl. Polymer Sci.* 37:33–78.
30. Haigler, C. H. and H. Chanzy. 1988. Electron diffraction analysis of altered cellulose synthesized by *Acetobacter xylinum* in the presence of fluorescent brightening agents and direct dyes. *J. Ultrastruct. Mol. Struct.* 98:299–311.
31. Geyer, U., Th. Heinze, A. Stein, D. Klemm, S. Marsch, D. Schumann, and H.-P. Schmauder. 1994. Formation, derivatization and applications of bacterial cellulose. *Int. J. Biol. Macromol.* 16:343–347.
32. Shibazaki, H., Kuga, S., Onabe, F., and Usuda, M. 1993. Bacterial cellulose membrane as separation medium. J. Appl. Poly. Sci. 50:965–969.
33. Yoshino, K., Matsuoka, R., Nogami, K., Yamanaka, S., Watanabe, K., Takahashi, M., and Honma, M. 1990. Graphite film prepared by pyrolysis of bacterial cellulose. J. Appl. Phys. 68:1720–1725.
34. Yoshino, K., Matsuokao, R., Nogami, K., Araki, H., Yamanaka, S., Watanabe, K., Takahashi, M., and Honma, M. 1991. Electrical property of pyrolyzed bacterial cellulose and its intercalation effect. Synth. Met. 41–43:1593–1596.
35. Brandl, H., R. Bachofen, J. Mayer, and E. Wintermantel. 1995. Degradation and applications of polyhydroxyalkanoates. *Can J. Microbiol.* 41(Suppl. 1):143–153.
36. Anderson, A. J. and E. W. Dawes. 1990. Occurrence, metabolism, metabolic role, and industrial uses of bacterial polyhydroxyalkanoates. *Microbiol. Rev.* 54:450–472.
37. Kim, S. W., P. Kim. S. L. Hyun, and J. H. Kim, 1996. High production of poly-ß-hydroxybutyrate (PHB) from *Methylobacterium organophilum* under potassium limitation. *Biotechnol Lett.* 18:25–30.
38. Byrom, D. 1987. Polymer synthesis by microorganisms: technology and economics. TIBTECH 5:246–250.
39. Powell, K. A., B. A. Collinson, and K. R. Richardson. 1980. European Patent, Application 80 300 432.4.
40. Mulchandani, A., J. H. T. Luong, and C. Groom. 1989. Substrate inhibition kinetics for microbial growth and synthesis of poly-ß-hydroxybutyric acid by *Alcaligenes eutrophus* ATCC 17697. *Appl. Microbiol. Biotechnol.* 30:11–17.
41. Taidi, B., Mansfield, D. A., and Anderson, A. J. 1995. Turnover of poly(3-hydroxybutyrate) (PHB) and its influence on the molecular mass of the polymer

accumulated by *Alcaligenes eutrophus* during batch culture. FEMS Microbiol. Lett. 129:201–206.
42. Kinetic study of poly-D(-)-3-hydroxybutyric acid (PHD) production and its molecular weight distribution control in a fed-batch culture of *Alcaligenes eutrophus*. J. Ferm. Bioeng. 76:465–469.
43. Yamane, T. 1992. Cultivation engineering of microbial bioplastics production. FEMS Microbiol. Rev. 103:257–264.
44. Bourque, D., Y. Pomerleau, and D. Groleau. 1995. High-cell-density production of poly-ß-hydroxybutyrate (PHB) from methanol by *Methylobacterium extorquens*: production of high-molecular-mass PHB. *Appl. Microbiol. Biotechnol.* 44:367–376.
45. Kim, B. S. and H. N. Chang. 1995. Control of glucose feeding using exit gas data and its application to the production of PHB from tapioca hydrolysate by *Alcaligenes eutrophus. Biotechnol Technol* 9:311–314.
46. Yamane, Y., Fukunaga, M., and Lee, Y. W. 1996. Increased PHB productivity by high-cell-density fed-batch culture of *Alcaligenes latus*, a growth-associated PHB producer. Biotech. Bioeng. 50:197–202.
47. Scandola, M., Ceccorulli, G., and Pizzoli, M. 1992. Miscibility of bacterial poly(3-hydroxybutyrate) with cellulose esters. Macromolecules 25:6441–6446.
48. Tomasi, G. and Scandola, M. 1995. Blends of bacterial poly(3-hydroxybutyrate) with cellulose acetate butyrate in activated sludge. Pure Appl. Chem. A32:671–681.
49. Gatenholm, P. and A. Mathiasson. 1994. Biodegradable natural composites. II. Synergistic effects of processing cellulose with PHB. *J. Appl. Poly. Sci.* 51:1231–1237.
50. van der Leij, F. R. and Witholt, B. 1995. Strategies for the sustainable production of new biodegradable polyesters in plants: a review. Can J. Microbiol. 41(Suppl. 1): 222–238.
51. Nawrath, C., Y. Poirier, and C. Somerville. 1994. Targeting of the polyhydroxybutyrate biosynthetic pathway to the plastids of *Arabidopsis thliana* results in high levels of polymer accumulation. *Proc. Natl. Acad. Sci. USA*. 91:12760–12764.
52. Doi, Y. 1990. *Microbial Polyesters.* New York: VCH.
53. Kim, O., R. A. Gross, D. R. Rutherford. 1995. Bioengineering of poly(ß-hydroxyalkanoates) for advanced material applications: incorporation of cyano and nitrophenoxy side chain substituents. *Can. J. Microbiol* 42(Suppl. 1):32–43.

### 3.5.3 Chemical Ingredients from Biomass

*V. Vidrich and M. Michelozzi*

Vegetal biomass is not only a source of energy and a unique resource for various chemicals but also the only renewable carbon source in space and in time.

During this era of petrochemicals, almost all chemicals are produced from petroleum and natural gas, and only modest quantities are still obtained from wood, and most of them as by-products of the pulp and paper industry. In a world of diminishing resources it is expected that, in the future, vegetal biomass will have a major role in contributing chemicals to the raw material needs of industry and advances in biomass conversion technology will offer a greater range of higher-quality products to successive generations.

Biomass can be distinguished as cell wall components, which can account for as much as 95% of plant material, and other constituents including extractives, bark, non-structural carbohydrates, and foliage. Cellulose, hemicellulose, and lignin, which are structural materials of cell walls, are used for the majority of chemicals. Cellulose is the principal component of plant cell walls and is the most abundant organic matter on Earth. Hemicelluloses are other structural polysaccharides consisting of pentoses and hexoses. Extractives are the soluble compounds (in organic solvent or in water) that represent from 5% to 15% of the biomass. They include oils, terpenes, fatty acids, unsaponifiables, polyhydric alcohols, nitrogen compounds, hydrocarbons, sterols, tannins, flavonoids, etc. Bark is characterized by a higher concentration in extractives and a lower concentration in cell wall components than wood. Foliage is an important source for essential oil, glucosides, chlorophylls, carotenoids, proteins, vitamins, resin acids, and other chemicals that have applications in pharmacy, herbalism, veterinary medicine, perfumery, cosmetics, flavoring industries, as adhesives, etc.

Since there are so many ingredients, it seems appropriate in this chapter to review only the compounds which are the most significant as chemicals from a commercial point of view.

Chemicals obtainable from forestry biomass are summarized in Fig. 3.41.

***Chemicals from Pyrolysis***

Pyrolysis is understood to be the incomplete breakdown of plant material, with the generation of charcoal, condensable (pyroligneous), and noncondensable (combustible) gases and tar. The various processes are described in another part of this volume, but suffice it to say that, through these processes, from pyroligneous, acetic acid, methanol and acetone are obtained, as well as lesser quantities of aldehydes, ketones and acids. From tar, in addition to its classic cyclic compounds, it is possible to obtain levoglucosans from crystaline cellulose, or from resinous wood, pine oils. Essential oils are then separated from tar by steam distillation and their composition is made up of monoterpenes, and sesquiterpenes.

The Italian National Industry for Electric Energy, in cooperation with the Umbria Regional Agency for Development and Innovation in Agriculture and with financial support from the European Union, is currently experimenting with a flash-pyrolysis system at Bastardo di Giano near Perugia in Umbria (Italy) where, in addition to charcoal and tar, hydrocarbons are obtained at $\sim$600°C with pyroligneous by working in a fluidized bed. These hydrocarbons can be utilized as combustion material for energy. This system is not limited only to woody material, but can also utilize any plant material including that from agriculture (e.g., refining wastes, straw, chaff, bran, pruning wastes, etc.).

*Cellulose*

When cellulose undergoes pyrolysis reactions of oxidation, dehydration, depolymerization and decarboxylation take place. Generally the afforded products are of low molecular weight, such as $CO$, $CO_2$, $H_2O$, $CH_3COOH$, $CH_3COCH_3$ and levoglucosans.

The degradation that takes place below 300°C involves above all the amorphous areas of the cellulose, affording condensable and noncondensable gaseous products. Above 300°C, an attack of areas with high packing density takes place with depolymerization

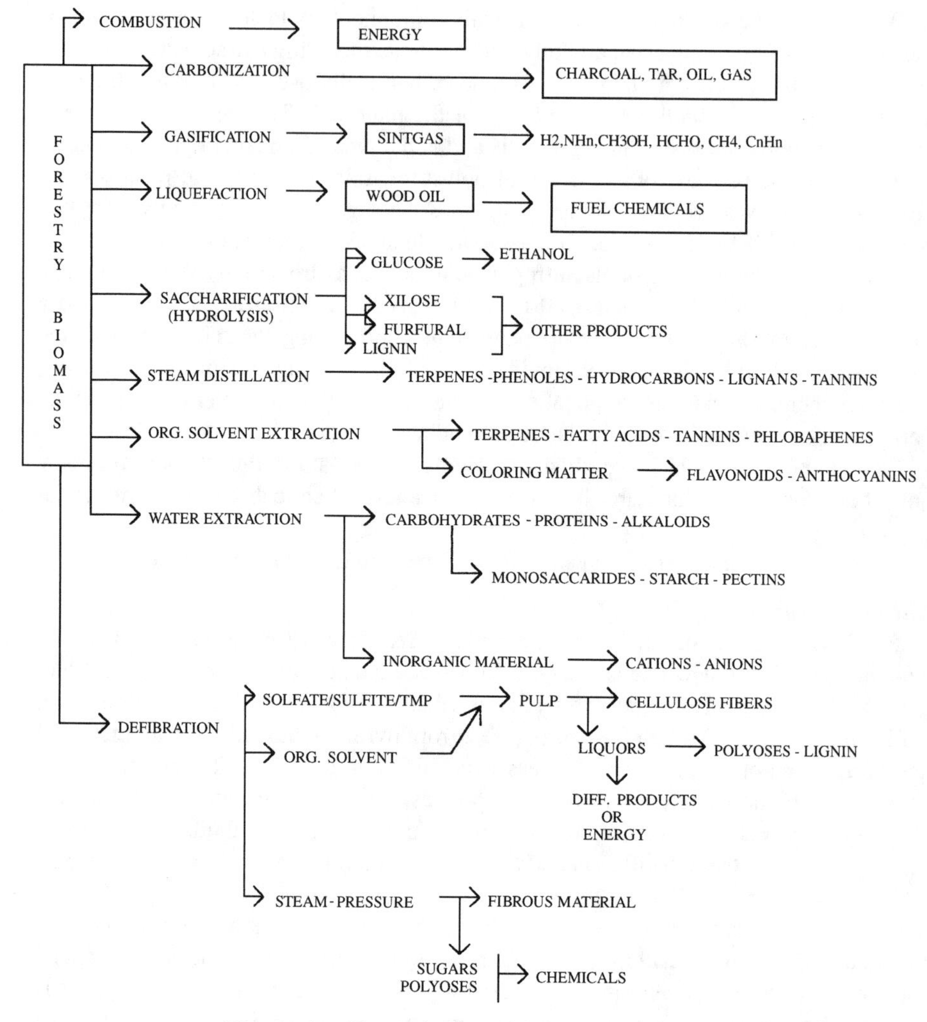


**Figure 3.41. Chemicals obtainable from forestry biomass.**

and oxidation of the glucose and with the formation of levoglucosans, which are then found in the tar. Levoglucosans are of considerable industrial interest [1–3].

*Hemicelluloses*

When woody substances undergo pyrolysis, the hemicelluloses depolymerize and decompose at temperatures of 200–270°C. First, there is depolymerization into soluble fragments to the point of the single constituent monomers; then there is thermal decomposition with the formation of condensable and noncondensable volatile composites. Derivatives of furan and acetic acid come from the pentoses present in hardwood. Each pentose (xylose and arabinose) decomposes with temperature into two molecules of acetic acid and one of formaldehyde. Furan is found in the tar, from which it can be distilled. Its esosans do not give rise to levoglucosans since they are amorphous.

Hemicelluloses play a part in the charcoal and tar yield, in addition to the above-mentioned products. Among the noncondensable gases there are $CO_2$, CO, and hydrocarbons, depending on the temperature of pyrolysis. Most chemicals are produced at a temperature of about ~170°C.

*Lignin*

From studies published by various authors [4–6] it is known that lignin is responsible for 50% of charcoal and 10%–15% of acetic acid, acetone and methanol, as well as non-condensable gases like $CO_2$, CO, $CH_4$, and others. Other important products are phenolic compounds: *o*- and *p*-cresol, 2,4 xylenol, guaiacol, 4-ehylgaiaeoe, 4-propyl guaiacol. etc.

From experiments [6] subjecting black liquor from paper mills to pyrolysis at 400–500°C, approximately 10% of pyroligneous containing guaiacol was obtained, with 4-methylguaiacol and 4-ethylguaiacol at a yield of up to 6% of lignin from the black liquor.

When lignin is subjected to flash pyrolysis, there is a high yield of ethylene. Currently, as mentioned above, flash pyrolysis is being tested in Italy to produce combustible material for energy.

***Gasification***

The process of gasification of plant by-products can be of considerable importance, especially in developing countries, for energy as well as for the production of methanol. In fact, without going into the particulars of the various possible processes (static or moving-bed gasogenis), it can be said that the final products of greatest interest are CO and $H_2$. These can in turn be exploited to drive internal-combustion engines or gas turbines connected to electric generators. Moreover, within the gasogenis mantle tubes carrying water, for conversion into steam, may be located and the steam can then be utilized in various ways.

CO and $H_2$ can undergo the catalytic process of liquification with the production of methanol, a useful product for the production of other chemicals and as an additive to gasoline.

The use of gasogenis can be carried out in fixed power units. The gas produced has a caloric potential of 1000–1500 Kcal/$m^3$ for air gas, while for blue gas the range is 2200–4500 Kcal/$m^3$ with a yield of 60%-70% [7].

***Hydrogenolysis***

This process could in the not-too-distant future be a source for a considerable number of vital chemical products obtainable from mill residues. Depending on the refining conditions, mixtures of -oses, acids or aromatic compounds could be obtained, with the difficulty in industrial application being the separation to purity of the single components.

***Hydrolysis***

The processes of acid hydrolysis of wood, with both concentrated and dilute acids, give rise to glucose and, through enzymatic reaction, ethanol and/or levulinic acid. Generally, when cellulose undergoes hydrolysis a prehydrolysis step is undertaken to distance the hemicellulose and extracts that would give undesirable products to the final result. Levulinic acid and furfural can be obtained from the esosans and the pentosans, respectively.

*Hydrolysis of Cellulose*

Hydrolysis can be carried out with either concentrated or dilute acids, in the latter case at high pressure. The empirical reaction that deals with the formation of glucose from cellulose requires an activation energy of 42.9 Kcal/mol, while that of glucose degradation is 32.8 Kcal/mol. In addition to glucose, by hydrolysis it is also possible to obtain approximately 15% hydroximethylfurfural and 40% levulinic acid.

The kinetics for the reaction to obtain glucose is favored, with dilute acids, by temperature while the selectivity is influenced by the concentration of the agent acid. Yields of 52% glucose at 232°C for 16 s with 1.2% $H_2SO_4$ are possible. According to the US Department of Agriculture's Forest Products Laboratory, at 175°C with 2% $H_2SO_4$ glucose is increased by 23%. Hardwood contains 30% fermentable sugars, while that of conifers 70% [8–11]. Carrying out the reaction at high temperature and in a short time gives maximum yields. With concentrated acids, hydrolysis affects principally the crystalline areas of cellulose, giving an industrial yield of 85%.

*Hydrolisis of Hemicellulose*

Currently the most important product obtained from hardwood, which has a pentosan content of 13%–30%, is furfural. In Italy it is produced from *Castanea sativa L.* wood with a yield $= p/3$, where $p$ is the percentage of pentosans present in the raw material to be refined.

Regarding the possible chemicals that can be obtained from spent liquor, the production of xylose and xylitol is undoubtedly interesting. Leikin, Jaffe, Steiner, and Lindlar have experimented with the purification of xylose and xylitol with exchange resin for the separation of the acids [12]. Numerous patents exist for the preparation of xylitol for medical and animal feed use. Lindlar and Steiner, for example, pass the hydrolized compounds in cation-exchange resin before crystallization of the xylitol. There is also a patent that is based on catalytic hydrogenation containing hexitols and penitols that are treated in the presence of nickel-activated charcoal at a temperature of 100–350°C and a pressure of between 100 and 350 atm. With this last system, the following glycols are obtained: xyletol, glycerol, mannitol, sorbitol, erytrol, ethylene glycol, and other compounds.

The potential of using spent liquor from resinous wood to obtain derivatives of mannoses and arabinogalactans should also be mentioned. Possibilities are being studied for their production for use as soap builders, stabilizers, ligands, and products for animal feed.

By-products of the pulping industry can be potential sources of -oses, organic acids, and methanol deriving from lignin and pectine hemicellulose. Currently these by-products are being used as combustible material for energetic needs within the refining process.

***Glucose and Its Derivatives***

Currently ethanol is produced by synthesis from ethylene, but in the near future, with an increase in world population and as a result a greater need for raw material, acid hydrolysis of celluloid materials may find new applications. Glucose obtained from acid hydrolysis can be converted, by fermentation, into ethanol. Ethanol is an important base product for the chemical industry and it can be, among other things, converted into butadiene, which is used for the production of synthetic rubber [13].

Through hydrogenation of glucose at a high temperature, 1,2-propanediol, together with a small quantity of glycerol and glyethylenglycol, can be obtained. Another product from this process is sorbitol, an intermediate in the production of vitamin C (ascorbic acid) [14].

***Chemicals from Other Conversion Processes***

A potential source of hemicellulose derivatives can be spent liquor from Kraft processes and bisulfite or soda sulfate from the pulp industry. Alternatively, it is used as a combustion material for energy to drive the principal productive process [15].

Clearly the separation and the purification of the various hemicellulose components of spent liquor require considerable cost and energy but, as briefly outlined, they will continue to become more and more economical and feasible.

Lignin as raw material for the production of chemicals is tied above all to the recovery of black liquor in the pulp industry [16]. The lignin obtained as a by-product of the refining of Kraft pulp has a wide range of uses: as a soap builder, surface-active agent and stabilizer for soil, wax, oily emulsions in water, soap, latex, fire-retardant foam, and rubber; as a dispersant for insecticides, herbicides, clay, pigments, ceramics, and pesticides; as a sequestrant for metals in industrial waters; as an additive to cement, in industrial purifiers, in the manufacturing of rubber, and in vinyl plastics; as an adhesive and binding agent in plastic laminates, in printing ink, in feed pellets for animals; as coreactant for phenols, urea formaldehyde, urethanes, and epoxides; in ion-exchange resins, as an oxygen acceptor, as a protector of colloids in steam boilers, and as coagulant for proteins. Lignin from black liquor can be transformed by hydrogenolysis into phenol and benzene.

***Chemicals from Extractives***

*Tannins*

A descriptive treatment of natural tanning agents would require much space, but in this context their chemical nature and industrial importance can be briefly mentioned.

Tannins, or tanning agents of plant origin, are divided into two main groups: hydrolysable tannins that include gallic and ellagic acids and phlobaphenes, which have their skeleton shaped by flavone. Tannins are found in almost all plant species in their various parts: bark, wood, leaves, fruit, and galls formed as a result of oviposition by certain insects. Also lichens contain tannins and are used at high altitudes to tan skins. From an industrial perspective, tannin content that exceeds 10%/dw is considered feasible for production [16].

In Italy tannins are produced from the wood of the chestnut (hydrosoluble) and from the leaves of sumac (*Rhus coriaria* L. and *Rhus cotinus* L.). In addition, tannins are produced from imported materials such as the wood of quebracho (*Schinopsis* spp. Engl.), divi divi, and numerous tropical acacias.

Tannins, in addition to tanning skins, are used as floats for drilling of the soil, in the preparation of germanium for semiconductors, for medical use as astringents (although their use in excess, particularly some flavonoids, seems to lead to cancer of the esophagus) [17] and in the manufacturing of low-cost adhesives for wood with formaldehyde.

*Flavonoids*

These polyphenolic compounds are present principally in the bark [17]. Quercetin has been widely studied and is obtainable from the dihydroquercetin in the bark of numerous conifer and hardwood species.

Flavones, flavonols, isoflavones, catechin and 3,4-flavandiol are known in medicine for their properties regarding the increase of resistance of capillary vessels. In fact they act as inhibitors to blood coagulation. In Italy flavones are used medically as vasodilators for coronaries or as alternatives to nitroglycerin, since they are not toxic [18]. Moreover, some flavonoids have shown to possess antitumor properties while others are considered cancerous; further study is under way [19, 20].

*Terpenoids*

More than 30,000 compounds are classified as terpenoids (terpenes or isoprenoids), the largest group of chemicals from vegetal biomass.

Terpenes are derived from the $C_5$ unit isopentenyl pyrofosfate by means of mevalonic acid, which is converted from acetyl-coenzyme A [21]. The assembly of different numbers of $C_5$ units results in $C_{10}$ monoterpenes, $C_{15}$ sesquiterpenes, $C_{20}$ diterpenes, $C_{30}$ triterpenes, $C_{40}$ tetraterpenes, and polyterpenes containing more than 40 carbons (Fig. 3.42).

Terpenoids are widely distributed in the plant and the animal kingdoms; they occur in both gymnosperms and angiosperms and they are particularly abundant in conifers. Terpenoids are present in all organs of the plant and they occur in various tissues such as cortex, xylem, and foliar tissue, and are generally stored in specialized secretory structures.

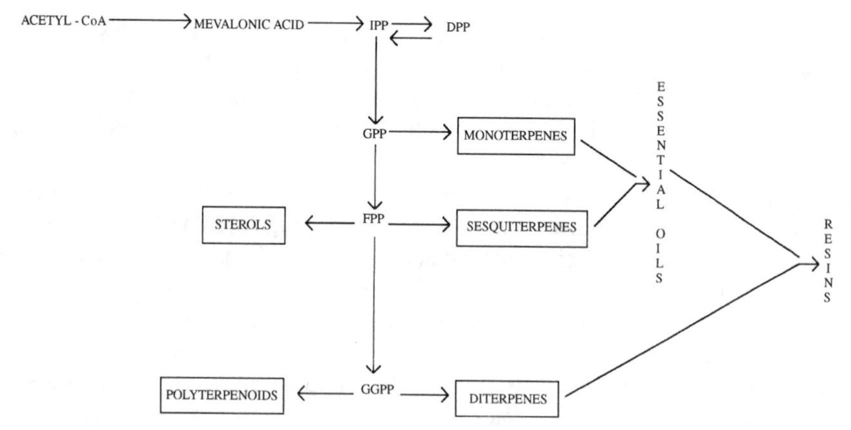


**Figure 3.42. Principal steps of the terpenoids biosynthetic pathway.**

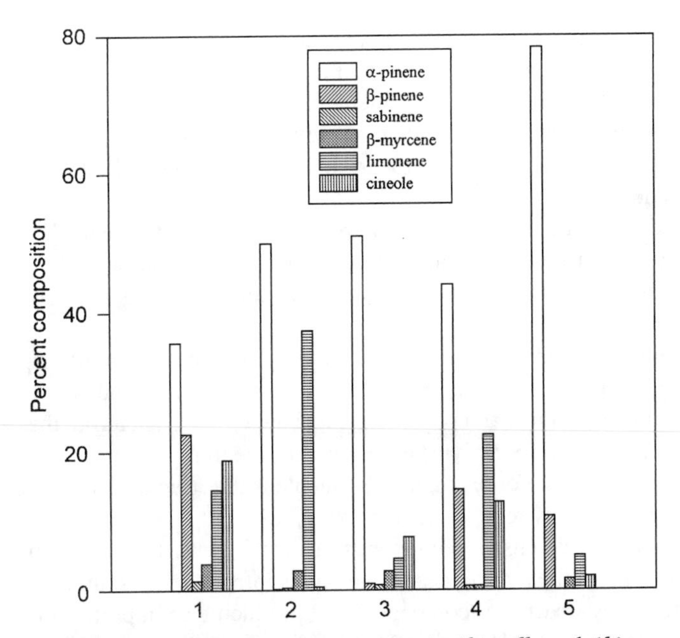


**Figure 3.43. Monoterpene composition in *Abies alba* and *Abies nebrodensis* populations: 1, Lavarone; 2, Abetone; 3, Abeti Soprani and Collemeluccio; 4, Aspromonte, Gariglione, and Serra San Bruno; 5, Madonie (*A. nebrodensis*).**

The relative proportions of monoterpenes (percentages) are strongly inherited, and variation in terpene profiles can be important for commercial application [22]. An example of the discriminating power of terpenes is illustrated in the study of genetic variability in *Abies nebrodensis* Lojac. and *Abies alba* Mill. [23].

In Fig. 3.43 differences in monoterpene profiles between *A. nebrodensis* and Italian provenances of *A. alba* are clearly evident.

Essential oils consist of monoterpenes and sesquiterpenes and were among the first foliage chemicals utilized. Since ancient times great attention has been directed to the possible utilization of these plant substances as flavoring compounds and in making perfumes. Approximately half of the essential oil-producing genera are found in Mediterranean-type ecosystems and a number of aromatic plants occur in the tropics.

Volatile terpenes (monoterpenes and sesquiterpenes) and nonvolatile terpenoids (diterpenoids and triterpenoids) constitute resins. Resins play an important role in protecting plant tissues from herbivores since both constitutive and induced terpenes act as chemical defenses [24,25]. Resin has been used since antiquity in food preparation and was added to wine jars in early societies for its preservative properties and also as flavor constituent [26]. Resin-producing plants occur predominantly among conifers and tropical angiosperms.

Volatile oils serve as solvents and as a source of products for the paint and the pharmaceutical industries. Resins are processed to rosin and turpentine, which are important sources of chemical raw material. Turpentine and rosin components and their chemically modified forms encompass a wide variety of uses as solvents, flavors and fragrances, in mineral flotation, pharmacy, and in synthetic rubber manufacture.

Terpenes are involved in primary metabolism: some are growth regulators as abscisic and gibberellic acids and others are found in chlorophyll (phytol). Terpenoids include carotenoids and sterols.

Monoterpenes, sesquiterpenes, and diterpenes are classified as secondary compounds and play important roles in the ecosystem. Many terpenes show toxic, deterrent, inhibitory, or attractive effects toward other organisms, affecting a broad spectrum of ecological interactions.

Seed germination and plant growth are strongly inhibited by some terpenoids. Terpenes show allelophatic activities, and several of these compounds are highly phytotoxic by causing anatomical and physiological changes [27]. Therefore these secondary metabolites with herbicidal activity have a great potential as sources of new environmentally safe herbicides. For example, 1,8-cineole is the structural base of cinmethylin, an herbicide that severely reduces plant growth [28].

Terpenoids act as defensive compounds against insects, fungi, and microbes. The insect-repellent properties make these compounds interesting as sources of insecticides [29]. Monoterpenes are used as components for producing an important class of insecticides: the pyrethroids [30]. However, the considerable importance of these compounds as natural pest control agents is related to the effect on the behavior of insects by a nontoxic mode of action. Terpenoidal pheromones are used in attractant traps to reduce the population of insects [31, 32].

A number of terpenes display antibacterial, antifungal and antiviral activities. These antimicrobial and other pharmacological properties, are used in pharmacy for the production of preparations with a broad spectrum of activities. For example, they are used as anti-inflammatories, decongestants for the respiratory tract, sedatives, carminatives and cardiotonic agents. Interestingly, taxol, a diterpene, has recently been shown to be toxic to human tumor cells [33]. Artemisin is a sesquiterpene extracted from *Artimisia annua* L., which shows antimalarial properties [34]. Some terpenoids display inhibition of rumen micro-rganisms responsible for the digestion of plant material. These molecules are also related to feeding preferences affecting the taste and the odor of food. Therefore terpenoids become important in food preparation since they are used as fragrances and preservatives, e.g., cineole, which is used as flavoring and antioxidant.

It may be concluded that the increase in demands to reduce any possible hazards to the environment will require more extensive use of terpenoids and other metabolites as natural preservatives and fragrances in food preparation, a source of new environmentally safe pest control agents and herbicides.

In short, it can be said that among the important extract components, in addition to tannins, flavonoids and terpenoids, are organic acids, dyes, and alkaloids which can have important commercial applications. As mentioned in the introduction, it is impossibile to review all the ingredients, however for further reference regarding products obtainable from the forest biomass, see the paper written by Vidrich [35].

***Utilization of Foliage and Small Branches for Fodder and Chemicals***

Foliage and branches are potential sources of fodder supplements and chemicals with numerous commercial applications. Foliage from forest-grown plants contains vitamins, mineral elements, protein, soluble nonstructural carbohydrates, etc., with great significance as fodder additives for poultry and domestic animals.

Foliage utilization in animal and poultry feeds started in Russia more than 60 years ago, and its value as fodder additive is due to the nutrients it contains [36].

Tree foliage is an important source of carotenoides, which are the plant pigments responsible for yellow red colors. They are found in chloroplasts and also in non-photosynthetic tissues such as flowers, fruits, roots and seeds. Beta $\beta$-carotene and other carotenoids are of considerable importance since they are converted to vitamin A. Carotenoids have industrial importance since they are used as non toxic colorant in food preparation and in cosmetics. The antioxidant properties of carotenoids are used in pharmacological preparations against several diseases [37].

Foliage from forest plants is also a source of vitamins C and E, while vitamin B group, K, and provitamin D occur in small concentrations.

Proteins occur in considerable amounts in forest foliage and although conifer needles show a lower protein content (6%–12%) than leaves from deciduous species (12%–20%), they can be harvested any time of the year in order to have fresh fodder year round.

Chlorophylls are another group of chemicals of interest as fodder and in pharmaceutical industries. These pigments seem to increase the growth of animals and they have several pharmacological activities. Chlorophyll and chlorophyll derivates are of considerable commercial importance as a bioactive component in creams, shampoos, toothpastes, etc., and in medical treatments for several diseases including ulcers, tuberculosis, and other disorders [38].

Foliage also contains other extractives such as glucosides, alkaloids and phenols, some of which are important or potentially important in pharmaceutical industries [39]. Contents of these ingredients vary among species and are affected by environmental conditions. A complete review of chemicals that occur in the leaves of forest plants is quite impossible since all the classes of vegetal organic compounds are present in foliage and many of these metabolites have or potentially have interesting applications in industry.

For reasons of economic efficiency, use is required of modern whole-tree technologies that make it possible, for example, to produce essential oils and to use the waste spent foliage as a fodder additive. Other uses for the waste spent foliage are in adhesives as an extender because of the occurrence of compounds that show adhesive properties [39]. Figure 3.44 shows various options of processing leaves and small branches.

In conclusion, it can be said that the importance of forest and other biomasses for the production of chemicals will certainly increase in the future. Improvements in technology and the increase in the cost of petroleum-based products, because of the availability of this finite resource, will make chemicals from forestry biomass able to compete with petrochemicals. Additionally, knowledge of the consequences to the environment and public health due to the use of synthetic compounds in agriculture and the food industry increases interest in the search of natural compounds that can probably be considered

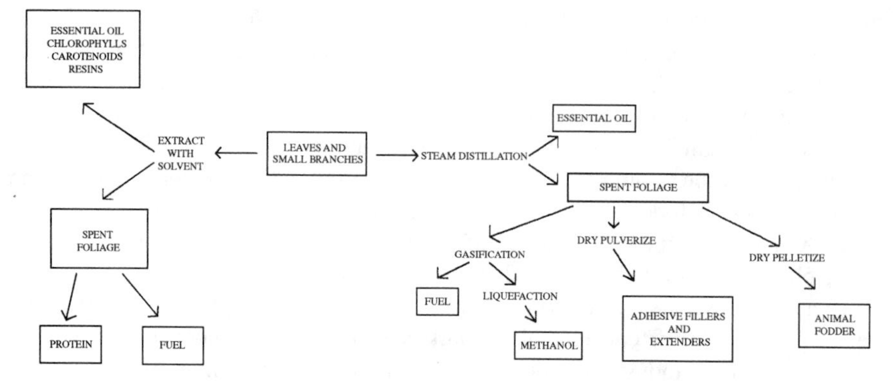


**Figure 3.44. Various options to process leaves and small branches.**

safe and have low environmental toxicity. In light of what has been said, strategies for use of biomass from short rotation forestry should be considered.

## References


1. Vidrich, V. 1988. *Il Legno ed i Suoi Impieghi Chimici*, pp. 111–112. Bologna: Edagricole.
2. Shafizadeh, F. 1968. Pyrolisis and combustion of cellulosic materials. *Adv. Carbohydr. Chem.* 29:419.
3. Golova, O. P. and R. G. Krylova. 1957. Thermal decomposition of cellulose and its structure. *Dokl. Akad. Nauk SSSR* 115:419.
4. Heuser, E. and C. Skioelderbraud. 1919. Destructive distillation. of lignin. *Z. Angew. Chem.* 321:41.
5. Fischer, F. and H. Schrader. 1920. The dry distillation. of lignin and cellulose, *Gesammelte Abh. Kennt. Kohle* 5:106.
6. Raffael, E., W. Rauch, and S. O. Beyer. 1974. Lignin containing phenolic- formaldehyde resins as adhesives for gluing veneers. Part I. *Holz Roh Werkst* 32(6):225.
7. Vidrich, V. 1988. *Il Legno ed i Suoi Impieghi Chimici*, pp. 125–131. Bologna: Edagricole.
8. Harris, J. F. 1975. Acid hydrolisis and degradation reactions for utilizing plant carbohydrates. *Appl. Polym. Symp.* 28:131.
9. Lebedev, N. V. and A. A. Bannikova. 1960. Hydrolysis of cellulose with concentrated HC1 at different temperatures. *Zh. Tr. Gos. Nauchn. Issled. Inst. Gidrol. Sulfit. Spirt Prom.* 8:47.
10. Vyrodova, L. P. and V. I. Sharkov. 1964. Effect of the concentrated sulfuric acid ratio and the presence of sugars on the solubility of cellul., *Zh. Tr. Gos. Nauchn. Issled. Inst. Gidrol.* Sulfit. *Spirt Prom.* 12:40.
11. Sakai, Y. 1965. Combination of sulfuric acid with cellulite during hydrolysis with small amount of concentrated sulfuric acid. Bull. Chem. Soc. Jpn. 38:863.
12. Steiner, K. and H. Lindlar. 1971. Title of patent. US Patent 3,586,537.

13. Vidrich, V. 1988. *Il Legno ed i Suoi Impieghi Chimici*, p. 133. Bologna: Edagricole.
14. Clark, I. T. 1958. Hydrogenolysis of sorbitol. *Ind. Eng. Chem.* 50:1125.
15. Vidrich, V. 1987. Gli aspetti merceologici dei prodotti del forteto e di altri boschi cedui. (Utilizzazione chimico-forestale del legno dei cedui). Accademia Economico-Agraria Dei Georgofili, vol. XXXIII, Serie settima: 1–21.
16. Vidrich, V. 1988. *Il Legno ed i Suoi Impieghi Chimici*, p. 24, Bologna: Edagricole.
17. Anon, C., 1978. OSHA issues tentative carcinogen list, *Chem. Eng. News.* 56(**X**):20.
18. Venkataraman, K. 1975. Flavones. the flavonoids, eds. Harborne, J. B. *et al.* Chap. 6. New York: Academic.
19. Hufford, C. D. and W. L. Lasswell. 1976. Uvaretin and isouvaretin, two novel cytotoxic C-benzylflavanones from *Uvaria chamae. J. Org. Chem.* 41:1297.
20. Wattenberg, L. W. and J. C. Leong. 1970. Inhibition of the carcinogenic action of benzolapyrene by flavones, *Cancer Res.* 30.
21. Chappell, J. 1995. The biochemistry and molecular biology of isoprenoid metabolism. *plant physiol.* 107:1–6.
22. Hanover, J. W. 1992. Applications of terpene analysis in forest genetics. *New Forests* 6:159–178.
23. Vendramin, G. G., M. Michelozzi, L. Lelli, and R. Tognetti. 1995. Genetic variation in *Abies nebrodensis*: a case study for a highly endangered species. *Forest Genet.* 2:171–175.
24. Langenheim, J. H. 1994. Higher plant terpenoids: A phytocentric overview of their ecological roles. *J. Chem. Ecol.* 20:1223–1280.
25. Threlfall, D. R. and I. M. Whitehead. 1991. Terpenoid phytoalexins: aspects of biosynthesis, catabolism, and regulation. *Ecological Chemistry and Biochemistry of Plant Terpenoids*, eds. Harborne, J. B. and F. A. Tomes-Barberan, pp. 159–208, Oxford: Clarendon.
26. Bower, B. 1996. Wine making's roots age in stained jar. Sci. News 149:359.
27. Fischer, N. H.1991. Plant terpenoids as allelopathic agents. *Ecological Chemistry and Biochemistry of plant Terpenoids*, eds. Harborne, J. B. and F. A. Tomes-Barberan, pp. 377–399. Oxford: Clarendon.
28. Duke, S. O. and J. Lydon. 1987. Herbicides from natural compounds. Weed Technol. 1:122–128.
29. Pickett, J. A. 1991. Lower terpenoids as natural insect control agents. *Ecological Chemistry and Biochemistry of Plant Terpenoids*, eds. Harborne, J. B. and F. A. Tomes-Barberan, pp. 297–313. Oxford: Clarendon.
30. Elliott, M., N. F. Janes, and C. Potter, 1978. The future of pyrethroids in insect control. Annu. Rev. Entomol. 23:443–469.
31. Bakke, A. and J. H. Gorbitz. 1986. Composition for attraction of pine shoot beetles. International patent, Application Number PCT/N086/00063.
32. Lofqvist, J., G. Birgersson, J. Byers, and G. Bergstrom. 1987. A method and composition for observation and control of Pityogenes chalcographus. European patent, Application number 87850032.1.
33. Kingston, D. G. I., G. Samaranayaka, and C.A. Ivey. 1990. The chemistry of taxol, a clinically useful anti-cancer agent. *J.Nat.Prod.* 53:1–12.

34. Trigg, P. I. 1989. Qinghaosu (artemisin) as an antimalarial drug. *Econom. Med. Plant. Res.* 3:20–56.
35. Vidrich, V. 1989. Chemicals from forestry biomass under temperate climates. Biomass *Handbook*, eds. Kitani, O. and C. W. Hall, pp. 665–672. New York: Gordon & Breach.
36. Keays, J. L. and G. M. Barton. 1975. Recent advances in foliage utilization. *Can. For. Serv. West. For. Prod. Lab.* Rep. VP-X-137.
37. Krinsky, N. I., M. M. Mathews-Roth, and R. F. Taylor, eds. 1990. *Carotenoids: Chemistry and Biology.* New York: Plenum Press.
38. Ievin, I. K., M. O. Daugavietis, O. R. Polis, and V. J. Deruma. 1981. Tree verdure as a source of organic raw material. *Am. Soc. Agric Eng. Pap.* 81–1593. St. Joseph, MI: ASAE.
39. Barton, G. M. 1981. Foliage. *Organic Chemicals from Biomass*, ed. Goldstein, J.

# Index